Methods in Molecular Biology

For further volumes:
http://www.springer.com/series/7651

For over 35 years, biological scientists have come to rely on the research protocols and methodologies in the critically acclaimed *Methods in Molecular Biology* series. The series was the first to introduce the step-by-step protocols approach that has become the standard in all biomedical protocol publishing. Each protocol is provided in readily-reproducible step-by-step fashion, opening with an introductory overview, a list of the materials and reagents needed to complete the experiment, and followed by a detailed procedure that is supported with a helpful notes section offering tips and tricks of the trade as well as troubleshooting advice. These hallmark features were introduced by series editor Dr. John Walker and constitute the key ingredient in each and every volume of the *Methods in Molecular Biology* series. Tested and trusted, comprehensive and reliable, all protocols from the series are indexed in PubMed.

Super-Resolution Microscopy

Methods and Protocols

Edited by

Frédéric A. Meunier

Queensland Brain Institute, University of Queensland, St Lucia QLD, QLD, Australia

Humana Press

Editor
Frédéric A. Meunier
Queensland Brain Institute
University of Queensland
St Lucia QLD, QLD, Australia

ISSN 1064-3745 ISSN 1940-6029 (electronic)
Methods in Molecular Biology
ISBN 978-1-0716-5267-1 ISBN 978-1-0716-5268-8 (eBook)
https://doi.org/10.1007/978-1-0716-5268-8

This Humana imprint is published by the registered company Springer Science+Business Media, LLC, part of Springer Nature.
The registered company address is: 1 New York Plaza, New York, NY 10004, U.S.A.

Preface

As I started my PhD in Neurobiology about three decades ago, I could only imagine what the world of molecules would look like. Back then, we could not visualize individual molecules at the resolution required to directly observe them as they performed their vital functions. Thirty years later, we no longer need to rely on our imagination to see the choreography of life unfold at the scale of individual molecules. Seeing the scaffolding of critically important proteins in live cells, the flux of vesicles, and the clustering of proteins is now possible. We can resolve intricate molecular structures and behaviors with incredible resolution, no longer having to infer it from blurry images. Super-resolution microscopy has turned our inferred dreams into reality. However, like all dreams, they take time and effort to implement. While a few instruments can now achieve almost subnanometer resolution, it is time to take a step back and help the growing number of laboratories that are starting, or have already embarked upon, this empowering journey.

Far-field optical microscopy combined with the development of immunocytochemistry and the genetic tagging of proteins with green fluorescent protein has been one of the most impactful tools in the history of science. It opened a window into the microscopic world, allowing us to access and understand the structures that animate our cells. However, for a long time, that window remained blurred by the fundamental diffraction limit of light. This was the physical barrier that, for over a century, dictated what we could and could not see. Individual structures that were closer together than a couple of hundred nanometers merged into a single fuzzy blob. And we have seen far too many of these blobs that have prevented us from understanding the most intricate principles of life.

Then, the revolution began. Super-resolution microscopy, also called optical nanoscopy, emerged at the turn of the twenty-first century and quite literally rewrote the rulebook. Techniques like STimulated Emission Depletion (STED), PhotoActivated Localization Microscopy (PALM), STochastic Optical Reconstruction Microscopy (STORM), Structured Illumination Microscopy (SIM), and Expansion Microscopy (ExM) each took different routes, but converged on a single goal: circumventing or simply breaking the diffraction limit of light. Through the clever use of light, chemistry, and computation, structures at the scale of tens of nanometers could be resolved, peering deep into that which was previously invisible. More than this, we could now see with our own eyes the nanoscale world in which our molecules of interest lived.

This scientific leap for me felt as if gazing for the first time through a telescope and seeing not just a point of light, but other planets and stars. It has opened a new frontier, revealing the fine meshwork of the cytoskeleton, the patchwork of protein nanoclusters, and the choreography of small organelles such as synaptic vesicles with a resolution that was once unimaginable.

But as with all great revolutions, many challenges remain. Although the instruments themselves are powerful, they can be expensive and super-resolution techniques can also be challenging to implement. More difficult even is the data that these techniques generate: terabytes of blinking fluorophores and shifting molecules that require sophisticated tools and deep expertise to analyze and interpret. To truly democratize super-resolution microscopy, we must make it accessible, usable, and interpretable to all biologists.

Democratising Super-Resolution Microscopy and Analysis is our small contribution to that mission. It is part technical manual, part conceptual analysis, and part invitation to explore. Each chapter is written by experts deeply engaged in the daily practice of pushing the limits of resolution and extracting biological information at the single molecule level.

Readers will gain insights into the physics and photophysics that underpin techniques like Single-Molecule Localization Microscopy (SMLM), STED, and SIM. These sections will, I hope, equip newcomers and experienced users alike with the tools to run and analyze data with clear instruction and explanation on why these methods work. You will find detailed protocols for SMLM techniques including PALM and STORM, guidance on live-cell compatible fluorophores, and descriptions of cutting-edge variations like single particle tracking PALM (sptPALM) for tracking single molecules in living cells and small organisms. Most importantly, we cover the analysis of single molecule tracking data to derive essential spatiotemporal nanoclustering metrics. We describe a way to perform multiplexing immunostaining, nanoscale visualization of brain architecture and tracking small organelles such as synaptic vesicles. We included new ways to image endogenous proteins at the nanoscale, including a chapter devoted to ExM, where samples are physically enlarged to reveal details that even the best optics cannot resolve. We venture into how SIM and τ-STED bring speed and volumetric information to living cells, and how super-resolution spinning disk microscopy (SoRa) enables high-throughput imaging.

Throughout, the emphasis is not just on the technology but on how to gather biological quantitative data. Super-resolution microscopy is brought to bear on real, pressing questions: how nanoclusters at the plasma membrane organize signaling, how protein condensates form in the nucleus, and how synaptic vesicles move and transmit bits of information from neuron to neuron. There are examples from neurons, muscle, epithelium, and brain tissue, testaments to how super-resolution microscopy is reshaping all fields of biology.

Importantly, we devote considerable attention to the often overlooked but absolutely vital world of data analysis. Super-resolution datasets are rich but noisy and extracting reliable information from them requires expertise in computational image processing, localization fitting, trajectory mapping and cluster analysis.

We also showcase innovations in multiplexing. You will learn about NanoPlex, a clever strategy for iterative imaging of endogenous proteins, using cleavable nanobodies that allows dozens of targets to be visualized from the same sample, all while preserving fine nanoscale structure. This opens the door to single-cell spatial proteomics, a long-sought goal that is now within reach.

And finally, we look ahead. How do we ensure that super-resolution microscopy becomes a routine tool in biology labs everywhere? What training, infrastructure, and analysis packages are required?

In the end, super-resolution microscopy is more than a set of techniques, it is a way of directly seeing biology in action. It reminds us that life stems from macromolecules operating at the scale of nanometers in the chaos generated by thermal energy. This book is a first step into getting novel insights into this nanoworld by providing a practical guide to exploring it.

Welcome to *Democratising Super-Resolution Microscopy and Analysis.* We hope it helps you unlock new ways of seeing, and therefore understanding, the living world.

St Lucia QLD, QLD, Australia ***Frédéric A. Meunier***

Contents

Contributors

RUMELO AMOR • *Queensland Brain Institute, The University of Queensland, Brisbane, QLD, Australia*

ADEKUNLE T. BADEMOSI • *Clem Jones Centre for Ageing Dementia Research (CJCADR), Queensland Brain Institute, The University of Queensland, Brisbane, QLD, Australia*

THIBAULT BRUGIERE • *Interdisciplinary Institute for Neuroscience, CNRS UMR 5297 and University of Bordeaux, Bordeaux, France*

MO CHEN • *Clem Jones Centre for Ageing Dementia Research (CJCADR), Queensland Brain Institute, The University of Queensland, Brisbane, QLD, Australia; Institute for Biomedicine and Glycomics, Griffith University, Nathan, QLD, Australia*

PARISA DASHTESTANI • *Institute of Neuro- and Sensory Physiology, University Medical Center Göttingen, Göttingen, Germany; International Max Planck Research School for Molecular Biology, Göttingen, Germany*

YULIA DEMBITSKAYA • *Interdisciplinary Institute for Neuroscience, CNRS UMR 5297 and University of Bordeaux, Bordeaux, France; Institut für Anatomie und Zellbiologie, Universitätsmedizin Göttingen, Göttingen, Germany*

PATRICK EIRING • *Department of Biotechnology & Biophysics, Biocenter, University of Würzburg, Würzburg, Germany*

PRANAY GAIKWAD • *Interdisciplinary Institute for Neuroscience, CNRS UMR 5297 and University of Bordeaux, Bordeaux, France; Institut für Anatomie und Zellbiologie, Universitätsmedizin Göttingen, Göttingen, Germany*

JORDAN GIRARD • *Interdisciplinary Institute for Neuroscience, CNRS UMR 5297 and University of Bordeaux, Bordeaux, France*

RACHEL S. GORMAL • *Clem Jones Centre for Ageing Dementia Research (CJCADR), Queensland Brain Institute, The University of Queensland, Brisbane, QLD, Australia*

DANIELA HACKER • *AG Optobiology, Institute of Biology, Humboldt-Universität zu Berlin, Berlin, Germany*

MARVIN JUNGBLUT • *Rudolf Virchow Center, Research Center for Integrative and Translational Bioimaging University of Würzburg, Würzburg, Germany*

KYE KUDO • *Clem Jones Centre for Ageing Dementia Research (CJCADR), Queensland Brain Institute, The University of Queensland, Brisbane, QLD, Australia*

SHANLEY F. LONGFIELD • *Clem Jones Centre for Ageing Dementia Research (CJCADR), Queensland Brain Institute, The University of Queensland, Brisbane, QLD, Australia*

ANUSHA MALAPAKA • *Clem Jones Centre for Ageing Dementia Research (CJCADR), Queensland Brain Institute, The University of Queensland, Brisbane, QLD, Australia*

ALEX J. MCCANN • *Clem Jones Centre for Ageing Dementia Research (CJCADR), Queensland Brain Institute, The University of Queensland, Brisbane, QLD, Australia*

FRÉDÉRIC A. MEUNIER • *Clem Jones Centre for Ageing Dementia Research (CJCADR), Queensland Brain Institute, The University of Queensland, Brisbane, QLD, Australia; School of Biomedical Sciences, The University of Queensland, Brisbane, QLD, Australia*

MARINA MIKHAYLOVA • *AG Optobiology, Institute of Biology, Humboldt-Universität zu Berlin, Berlin, Germany*

NIKOLAOS MOUGIOS • *Institute of Neuro- and Sensory Physiology, University Medical Center Göttingen, Göttingen, Germany; Center for Biostructural Imaging of Neurodegeneration (BIN), University of Göttingen Medical Center, Göttingen, Germany*
VASILEIOS MOYSIDIS • *AG Optobiology, Institute of Biology, Humboldt-Universität zu Berlin, Berlin, Germany*
U. VALENTIN NÄGERL • *Interdisciplinary Institute for Neuroscience, CNRS UMR 5297 and University of Bordeaux, Bordeaux, France; Institut für Anatomie und Zellbiologie, Universitätsmedizin Göttingen, Göttingen, Germany*
KOSUKE OKUDA • *Interdisciplinary Institute for Neuroscience, CNRS UMR 5297 and University of Bordeaux, Bordeaux, France; Institut für Anatomie und Zellbiologie, Universitätsmedizin Göttingen, Göttingen, Germany*
FELIPE OPAZO • *Institute of Neuro- and Sensory Physiology, University Medical Center Göttingen, Göttingen, Germany; Center for Biostructural Imaging of Neurodegeneration (BIN), University of Göttingen Medical Center, Göttingen, Germany; NanoTag Biotechnologies GmbH, Göttingen, Germany*
SABER H. SABER • *Clem Jones Centre for Ageing Dementia Research (CJCADR), Queensland Brain Institute, The University of Queensland, Brisbane, QLD, Australia; Molecular Cell Biology Lab, Zoology and Entomology Department, Faculty of Science, Assiut University, Assiut, Egypt*
MARKUS SAUER • *Department of Biotechnology & Biophysics, Biocenter, University of Würzburg, Würzburg, Germany; Rudolf Virchow Center, Research Center for Integrative and Translational Bioimaging University of Würzburg, Würzburg, Germany*
MARK A. SCHEMBRI • *Institute for Molecular Bioscience, The University of Queensland, Brisbane, QLD, Australia; School of Chemistry and Molecular Biosciences, The University of Queensland, Brisbane, QLD, Australia*
CHRISTOPHER SMALL • *Clem Jones Centre for Ageing Dementia Research (CJCADR), Queensland Brain Institute, The University of Queensland, Brisbane, QLD, Australia*
SEVANNAH A. STEEVES • *Clem Jones Centre for Ageing Dementia Research (CJCADR), Queensland Brain Institute, The University of Queensland, Brisbane, QLD, Australia*
PARNAYAN SYED • *Clem Jones Centre for Ageing Dementia Research, Queensland Brain Institute, The University of Queensland, Brisbane, QLD, Australia*
DANUSH TABAN • *Department of Biotechnology & Biophysics, Biocenter, University of Würzburg, Würzburg, Germany*
ANDREW THOMPSON • *Queensland Brain Institute, The University of Queensland, Brisbane, QLD, Australia*
PATRIK VERSTREKEN • *VIB-KU Leuven Centre for Brain and Disease Research, Leuven, Belgium; KU Leuven, Department of Neurosciences, Leuven Brain Institute, Mission Lucidity, Leuven, Belgium*
TRISTAN P. WALLIS • *Clem Jones Centre for Ageing Dementia Research (CJCADR), Queensland Brain Institute, The University of Queensland, Brisbane, QLD, Australia*

Chapter 1

Structured Illumination Microscopy

Andrew Thompson and Rumelo Amor

Abstract

Structured illumination microscopy (SIM) is a super-resolution technique that uses patterned excitation light to achieve an optical resolution that is twice the diffraction limit. It does not require special buffers or dyes. In this chapter, we discuss the implementation of structured illumination in fluorescence microscopy, show how to perform SIM acquisition and processing, and discuss typical SIM acquisition and processing artefacts and how to address them.

Key words Structured illumination microscopy, Fluorescence microscopy, Super-resolution microscopy, Volumetric imaging

1 Introduction

Structured illumination was implemented in fluorescence microscopy by Heintzmann and Cremer in 1999 [1] and Gustafsson [2], Frohn et al. [3], and Gustafsson et al. [4] in 2000 to extend the lateral resolution beyond the Abbe diffraction limit [5]. Conceptually, structured illumination microscopy (SIM) is elegantly illustrated by the moiré effect. If two optically unresolvable patterns of very high frequency were superposed on each other, there arises a beat pattern from the superposition: the moiré pattern (Fig. 1a). This moiré pattern is of a lower frequency than the two patterns and is therefore resolvable. In addition, the frequency of the moiré pattern is defined by the difference in the frequencies of the two original patterns, so if one of the patterns is that of the specimen structure, and the other, light of a known, pre-determined structure, then the unknown structure of the specimen can be determined and made resolvable.

The lateral resolution of an optical microscope is defined as the minimum separation distance between two distinct points that are resolvable from each other, and is dependent on the numerical

Frédéric A. Meunier (ed.), *Super-Resolution Microscopy: Methods and Protocols*, Methods in Molecular Biology, vol. 3034,
https://doi.org/10.1007/978-1-0716-5268-8_1,

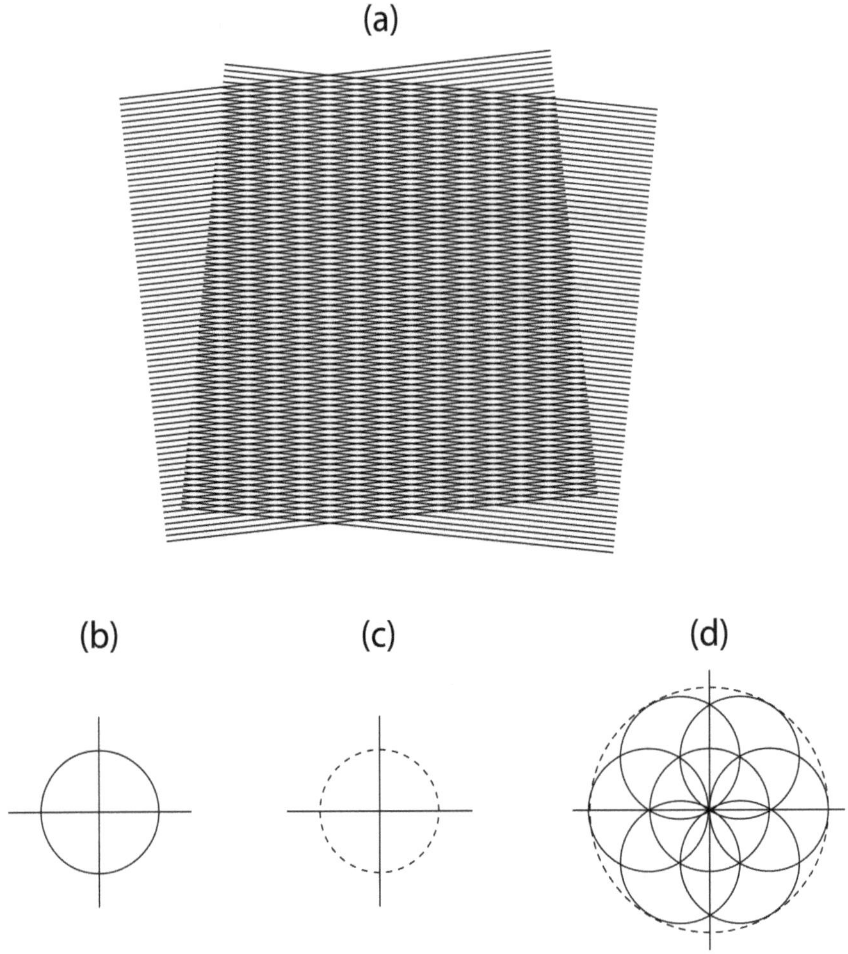

Fig. 1 Extending the resolution beyond the diffraction limit via structured illumination microscopy. (**a**) Moiré fringes (dark vertical lines) arising from the superposition of two patterns have a lower frequency than the original patterns (light horizontal lines) and are more resolvable. (**b**) Circular observable region of reciprocal space on a diffraction-limited microscope. (**c**) The three Fourier components of a sinusoidally striped illumination pattern define a circle with the same limits as the observable region. (**d**) A sequence of images acquired with the specimen illuminated with structured light with different orientations and phases produces moiré fringes representing information that have moved in reciprocal space, in turn producing an observable region that is twice the size of that in a diffraction-limited microscope, doubling the resolution

aperture of the microscope objective and the wavelength of emitted light [6]:

$$r_{xy} = \frac{1.22\lambda}{2\mathrm{NA}},$$

where r_{xy} is the lateral resolution, λ is the emission wavelength, and NA is the numerical aperture. On a microscope that uses an oil-immersion objective of numerical aperture equal to 1.4 and green light with a wavelength of around 550 nm, the lateral resolution is about 240 nm. This is the resolution limit of a classical wide-field microscope and is represented by a circular observable region in reciprocal space (Fig. 1b). In SIM, the specimen is illuminated with structured light, for example, a sinusoidally striped illumination pattern, which has three Fourier components. The possible positions of the structured light's two side components in reciprocal space also define a circle with the same limits as the observable region (Fig. 1c). However, if a sequence of images of the specimen is acquired using structured light of different orientations and phases, moiré patterns would arise, corresponding to information that has changed position in reciprocal space, and the circles defined by the three Fourier components effectively allow information to be acquired from an area that is double the size of the native observable region, doubling the resolution (Fig. 1d). The typical lateral resolution in SIM is around 120 nm.

In 2008, Gustafsson et al. [7] and Schermelle et al. [8] extended the resolution improvement in SIM to three dimensions by using three-beam interference, creating structured light in both the lateral and axial directions. Three-dimensional structured illumination microscopy (3D-SIM) doubled the lateral and axial resolution in wide-field fluorescence microscopy. The technique also removes out-of-focus blur, achieving optical sectioning. The axial resolution of an optical microscope is about 2.5 times poorer than its lateral resolution, which makes it 600 nm for the conditions described above. The axial resolution in 3D-SIM is around 300 nm, a two-fold improvement.

2 Materials

2.1 The Structured Illumination Microscope

The basic elements of a structured illumination microscope comprise a laser for fluorescence excitation, a collimator, a polariser, a phase grating, a focusing lens, the objective lens, a dichroic mirror, a tube lens, and a camera for image acquisition (Fig. 2). A linearly polarized laser illuminates a phase grating, with the grating made to transmit only the diffraction orders +1 and −1, thereby generating a sinusoidally striped illumination pattern of high contrast at the specimen plane (*see* **Note 1**). The grating is mounted on a precision stage that allows rotation and translation, providing control of the grating's orientation and position, and as a result, the illumination pattern's orientation and phase. The resulting fluorescence signal emitted by the specimen, with the inherent moiré pattern generated by its illumination with structured light, is directed by a dichroic mirror to the camera sensor via the tube lens. The camera acquires several images as the grating is rotated and shifted to change the orientation and phase of the illumination.

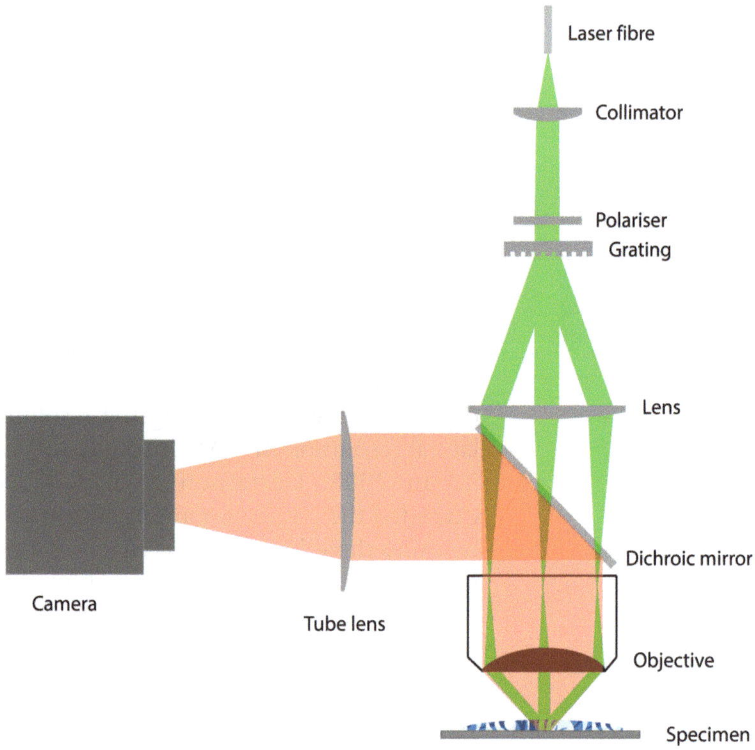

Fig. 2 Structured illumination microscope setup. The output from the excitation laser is collimated onto a linear phase grating and focused by a lens onto the back aperture of the microscope objective. The objective. The beams are collimated again after the objective and intersect at the focal plane, illuminating the specimen. Fluorescence emission is collected by the objective and directed by the dichromatic mirror and tube lens onto the camera sensor

2.2 Beads Slide for Calibration

1. Cover glasses, thickness no. 1.5H, high-performance, 18 mm × 18 mm, thickness 0.170 ± 0.005 mm.
2. Glass microscope slides, white, 1 mm thickness, 75 mm × 25 mm.
3. Carboxylate-modified microspheres, 100 nm diameter, blue fluorescence, 350 nm excitation, 440 nm emission.
4. Carboxylate-modified microspheres, 100 nm diameter, green fluorescence, 505 nm excitation, 515 nm emission.
5. Carboxylate-modified microspheres, 100 nm diameter, red fluorescence, 580 nm excitation, 605 nm emission.
6. Carboxylate-modified microspheres, 200 nm diameter, far red fluorescence, 660 nm excitation, 680 nm emission.
7. Ethanol, absolute, EMSURE ACS, ISO, Reag. Ph Eur, ≥ 99.5% v/v.

8. Glycerol, purism. p.a., ACS reagent, anhydrous, ≥ 99.5% v/v.
9. Immersion oil, fluorescence-free, $n = 1.518$, for 23 °C.
10. Immersion oil, fluorescence-free, $n = 1.518$, for 30 °C (*see* **Note 2**).

2.3 Beads Slide for Chromatic Aberration Correction

1. Cover glasses, thickness no. 1.5H, high-performance, 18 mm x 18 mm, thickness 0.170 ± 0.005 mm.
2. Glass microscope slides, white, 1 mm thickness, 75 mm × 25 mm.
3. Microspheres, multi-spectral, 200-nm diameter, blue (360 nm excitation, 430 nm emission), green (505 nm excitation, 515 nm emission), red (560 nm excitation, 580 nm emission), and far red (660 nm excitation, 680 nm emission) fluorescence.
4. Ethanol, absolute, EMSURE ACS, ISO, Reag. Ph Eur, ≥ 99.5% v/v.
5. Glycerol, puriss. p.a., ACS reagent, anhydrous, ≥ 99.5% v/v.
6. Immersion oil, fluorescence-free, $n = 1.518$, for 23 °C.
7. Immersion oil, fluorescence-free, $n = 1.518$, for 30 °C (*see* **Note 2**).

2.4 Standard Cell Slide

1. GATTA-4C multi-color huFIB displaying 4 colors, nucleus: DAPI, 358 nm excitation, mitochondria: TOM20 with Alexa Fluor 488, 488 nm excitation, microtubules: α-Tubulin with Alexa Fluor 555, 555 nm excitation, actin: Alexa Fluor 647, 647 nm excitation (GATTAquant GmbH, article number: 4262).
2. Immersion oil, fluorescence-free, $n = 1.518$, for 23 °C.
3. Immersion oil, fluorescence-free, $n = 1.518$, for 30 °C (*see* **Note 2**).

3 Methods

3.1 Preparation of Beads Slide for Calibration

1. Dilute 1 μL of 100-nm green fluorescent microspheres and 1 μL of 100-nm red fluorescent microspheres in 98 μL of distilled water to make 100 μL of pre-stock solution.
2. Take 0.5 μL of the pre-stock solution, 10 μL of 100-nm blue fluorescent microspheres, and 10 μL of 200-nm far red fluorescent microspheres and these to 29.5 μL of distilled water to make 50 μL of stock solution.
3. Sonicate the stock solution for 10 min. Take 1 μL of the stock solution and add it to 99 μL of absolute ethanol.
4. Take a 5-μL drop of this solution and air-dry it on a cover glass.

5. Apply 10 μL of glycerol on the cover glass. Gently lower the cover glass onto a microscope slide using forceps, making one edge touch first to avoid air bubbles.
6. Wait for the glycerol to spread between the cover glass and the slide. Remove excess glycerol from the edges using small, folded strips of lens tissue.
7. Seal with non-auto-fluorescing nail varnish.

3.2 Preparation of Beads Slide for Chromatic Aberration Correction

1. Dilute 5 μL of 200-nm multi-spectral microspheres in 95 uL of distilled water to make 100 μL of stock solution. Sonicate for 10 min.
2. Take 2 μL of the stock solution and add it to 10 μL of absolute ethanol.
3. Take a 5-μL drop of this solution and air-dry it on a cover glass.
4. Apply 10 μL of glycerol on the cover glass. Gently lower the cover glass onto a microscope slide, avoiding the formation of air bubbles.
5. Let the glycerol spread evenly between the cover glass and the slide. Remove any excess using lens tissue.
6. Seal with nail varnish.

3.3 Acquisition of an Experimental Point Spread Function (PSF)

Because custom-built SIM setups require significant in-house expertise to build and maintain, most SIM experiments are performed on commercial systems, of which the most widely installed worldwide are the Zeiss Elyra 7 SIM, Nikon N-SIM, and GE Healthcare OMX Blaze SIM. The protocol described here is based on the use of the Zeiss Elyra PS.1 SIM controlled by Zen Black 2012 software [9], but because all three systems have similar functions and share a similar approach to SIM acquisition and processing, the protocol should be applicable to any system.

1. Apply immersion on the 63×/1.4 NA oil-immersion objective and mount the bead slide for calibration. Acquire a z-stack of the beads in the required spectral channels in SIM mode, using a step size of 110 nm and 41 steps.
2. Navigate to Processing > Structured Illumination > Experimental PSF. Select the beads z-stack as input.
3. To perform a radial symmetrization of the PSF, that is, to calculate a round PSF from an oval PSF, tick the "Symmetrical" box.
4. To smooth the PSF to zero at the edge, tick the "Damp Edge" box.
5. To interpolate the PSF result to the same pixel size as the original image or to two, three, or four times smaller than the original pixel size, choose the desired value in the "PSF Sampling" menu.

6. Set the mask size used to select the beads in "Feature Size."
7. Set the threshold of the intensity over the background in "Noise Level."
8. Define how different the beads' intensity and size can be from the median values as calculated for every localized bead in "Beads Quality."
9. Assign the best focal plane within the stack in "Focal Plane."
10. Inspect the intermediate results in the "Preview" tab until they are satisfactory.
11. Press "Apply" to compute an average PSF from all the beads that have been selected. A separate PSF will be generated for each spectral channel.
12. Save the experimental PSF to be used later in SIM processing if a theoretical PSF is not desired.

3.4 SIM Acquisition and Processing

1. To prepare fixed cell specimens for SIM experiments, grow the cells of interest on cover glasses, fix them, label the structures of interest with fluorescence, and mount them using standard protocols. To perform live-cell SIM experiments, culture the cells in an incubator at 37 °C and 5% CO_2 until they are ~80% confluent, seed them onto cover glasses, and label the structures of interest with fluorescence via transfection and/or incubation using standard protocols. Perform the imaging experiment on a SIM system with incubation and temperature and CO_2 control. In the protocol described here, we have used the GATTA-4C multi-colour standard cell slide of human fibroblasts.
2. Apply immersion on the 63×/1.4 NA oil-immersion objective and mount the specimen. Select an appropriate region and acquire a z-stack in the required spectral channels in SIM mode, choosing the optimal grating for each wavelength (23-μm grating for 405 nm, 28 μm for 488 nm, and 34 μm for 561 and 640 nm) and setting the number of grating rotations. For multi-channel SIM acquisition, acquire the longest wavelength first to minimize photobleaching.
3. Navigate to Processing > Structured Illumination > Structured Illumination. Select the cell z-stack as input.
4. Select Automatic or Manual in "Processing." Use automatic processing on image data with high modulation contrast. Use manual processing if there is low modulation contrast. Manual processing allows broad control over the reconstruction parameters.
5. Select how the individual frames within the z-stack are processed in "Mode." In 2D mode, each frame is reconstructed using information from that frame only. In 3D mode,

information from adjacent z-planes in terms of structure and out-of-focus contributions is used in addition to information from the frame being reconstructed, producing better lateral and axial resolution and image contrast by removing out-of-focus signal more effectively. 3D mode requires more computational power.

6. In manual processing mode, enable or disable the "Max. Isotrop" filter. If enabled, the software computes the super-resolution image with the same resolution in every rotation direction (isotropic), and the Fourier transform will be a circle. If disabled, the software will use the high frequencies at the rotation angles, and the Fourier transform will appear as a flower as it extends wider at the rotation angles.
7. Untick "Auto Noise Filter" to access the slider that controls the noise filter for Wiener deconvolution. The more negative numbers on the left of the slider allow higher frequencies, yielding noisier and more punctate images.
8. Set the value on the "SR Frequency Weighting" filter, which behaves like a cone that multiplies the Fourier transform of the reconstructed image and affects the higher frequencies more prominently. The weighting filter changes the slope of the cone: a weighting filter of 0 does no filtering, a filter that is less than 1 makes the cone convex, a filter equal to 1 makes a line, and a filter greater than 1 makes the cone concave.
9. Choose between "Baseline Shifted" or "Baseline Cut." In baseline-shifted mode, negative values that may arise as noise from the Wiener filter used are preserved and an offset is added to the image, so the minimum intensity is zero. Baseline-cut mode clips the negative values to zero.
10. The "Sectioning" filter values determine the strength of filtering applied on the 0-frequencies of the zeroth, first, and second orders. The higher the filter value, the stricter the filtering applied, and more lower frequencies are removed.
11. The "OTF display" window shows the shape of the optical transfer function in the lateral direction. All filter values affect the OTF used for the reconstruction.
12. Choose whether to display the SR-SIM image only, or with the DCV image (deconvolution), or with the WF image (wide-field), or with both DCV and WF in "Output."
13. Select between a theoretical or experimental PSF in "PSF." If selecting Experimental PSF, point the software to a pre-saved PSF file, for example, a PSF experimentally determined as described in Subheading 3.3.

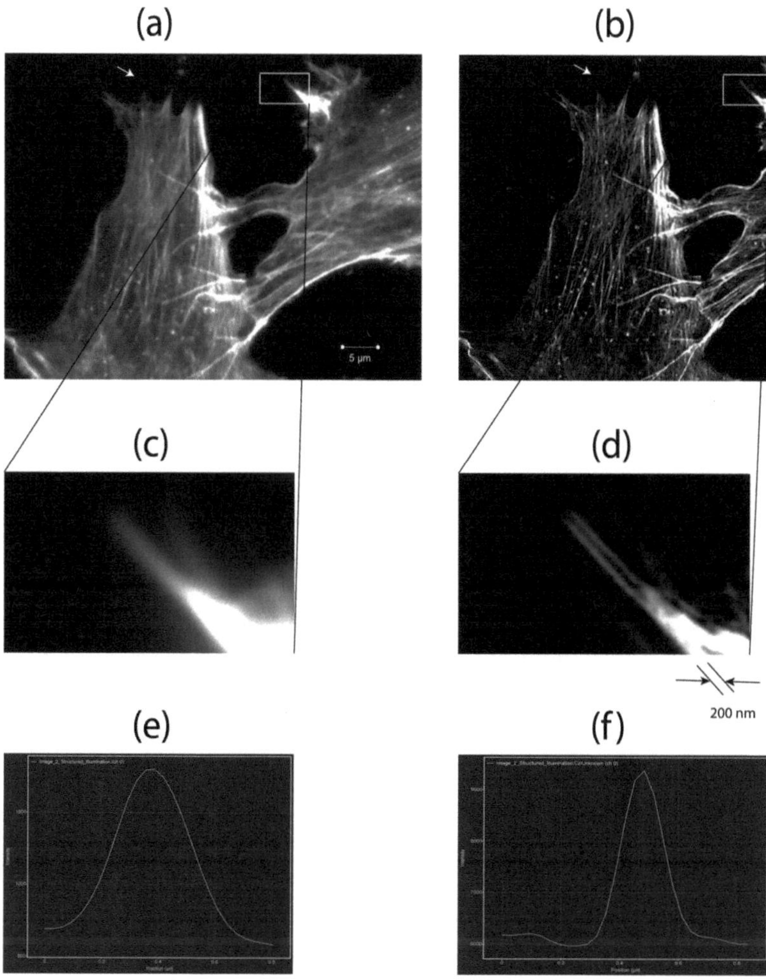

Fig. 3 Images of the actin cytoskeleton in human fibroblast cells, acquired using wide-field (**a, c**) and structured illumination microscopy (**b, d**). Magnified views of the boxed regions in (**a**) and (**b**) show that fibers separated by less than the diffraction limit appear fused into one in wide-field (**c**) but are well-resolved in the SIM image (**d**). Intensity profiles across the finest fibers in (**a**) and (**b**) (white arrows) show the full-widths at half-maximum to be approximately 300 nm in wide-field (**e**) and 160 nm in SIM (**f**)

14. Press "Apply" to compute the super-resolution SIM image. Raw wide-field and super-resolution SIM images are shown in Fig. 3, demonstrating the resolution improvement in SIM.

3.5 Chromatic Aberration Correction

Light waves of different wavelengths behave differently as they propagate through optics, and therefore chromatic aberration is inherently present in multi-colour optical images including SIM data as different wavelengths of light come to a focus at different locations laterally and axially as they exit individual optical elements. A brief outline of chromatic aberration correction in SIM is presented here (*see* **Note 3**).

1. Apply immersion on the 63×/1.4 NA oil-immersion objective and mount the bead slide for chromatic aberration correction. Acquire a multi-channel z-stack of the beads in the requisite spectral channels in SIM mode, using a step size of 110 nm and 41 steps.
2. Navigate to Processing > Structured Illumination > Structured Illumination and select the beads stack as input.
3. Set the SIM processing parameters and click "Apply" to compute the super-resolution image.
4. Navigate to Processing > Channel Alignment and select the SIM-processed beads data as input (*see* **Note 4**).
5. Tick the "Fit" box. This aligns all the channels of the image data with respect to the channel that is listed first, which is treated as the reference.
6. For "Alignment Mode," select between "Lateral" and "Affine." Lateral alignment corrects for linear lateral and axial displacements. Affine alignment corrects for linear lateral and axial displacements, rotation in the image plane in x and y, and stretching in the image plane in x and y (*see* **Note 5**).
7. Click "Apply" to calculate the offsets, which are displayed in a coefficient table. Confirm that the correction is optimal by comparing the raw (Fig. 4a) and chromatic aberration-corrected image data (Fig. 4b).
8. Click "Save" to save the coefficient table in a.bin file to be used later to correct for chromatic aberration in SIM data.
9. Apply fresh immersion on the 63×/1.4 NA oil-immersion objective and mount the multi-colour cell slide. Select an appropriate region and acquire a z-stack in the required spectral channels in SIM mode, again choosing the optimal grating for each wavelength and the number of rotations.
10. Perform SIM processing as described above. There will be chromatic shifts between the channels, in the lateral and axial directions, and these will be particularly bad as the wavelength difference becomes greater.
11. Navigate to Processing > Channel Alignment and select the SIM-processed multi-spectral cell data as input.
12. Leaving the "Fit" box unticked, click "Load" to open the.bin file that contains the coefficient table for correcting the shifts.
13. In the assignment table that is underneath the coefficient table, make sure that the Laser ID associated with each Channel (CH) agrees with the Laser ID on the coefficient table.
14. Click "Apply" to perform the correction. Confirm that the applied shifts are optimal by comparing the raw data (Fig. 4c) with the corrected image data (Fig. 4d).

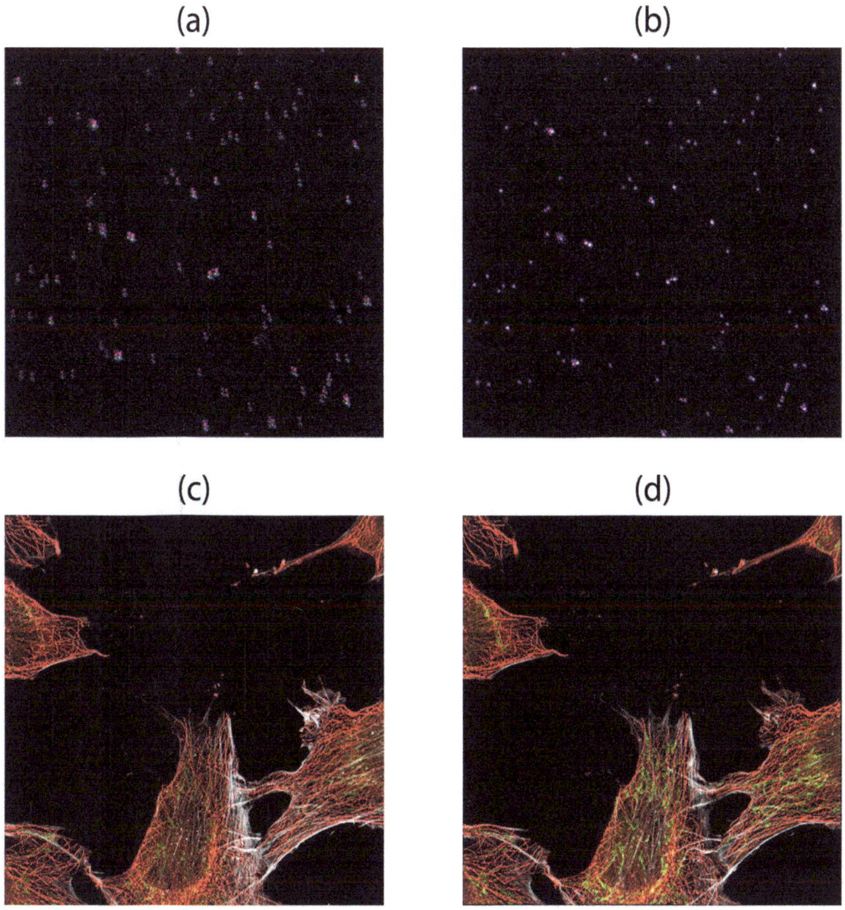

Fig. 4 Multi-spectral 200-nm beads before (**a**) and after (**b**) chromatic aberration correction. Human fibroblast cells labelled for mitochondria (TOM20 with Alexa Fluor 488, green), microtubules (a-tubulin with Alexa Fluor 555, red), and actin (Alexa Fluor 647, greyscale) before (**c**) and after (**d**) chromatic aberration correction. Note the presence of both lateral and axial displacements

3.6 Artefacts in SIM Acquisition and Processing

There are several artefacts that may arise during SIM acquisition and processing. Here we enumerate the most common artefacts and discuss how to address them.

1. Hammerstroke artefacts (Fig. 5a) arise from high-frequency noise contributions and may be caused by low modulation contrast in the raw SIM data due to low contrast in the structure of interest itself or due to out-of-focus contributions, weak fluorescence labelling, low signal dynamic range, and a too-low Wiener filter (noise filter) in SIM processing [10]. Approaches to removing hammerstroke artefacts include selecting specimen structures that have higher contrast, reducing the range of the z-stack acquisition to reduce out-of-focus blur (*see* **Note 6**), increasing signal strength by using brighter dyes, higher

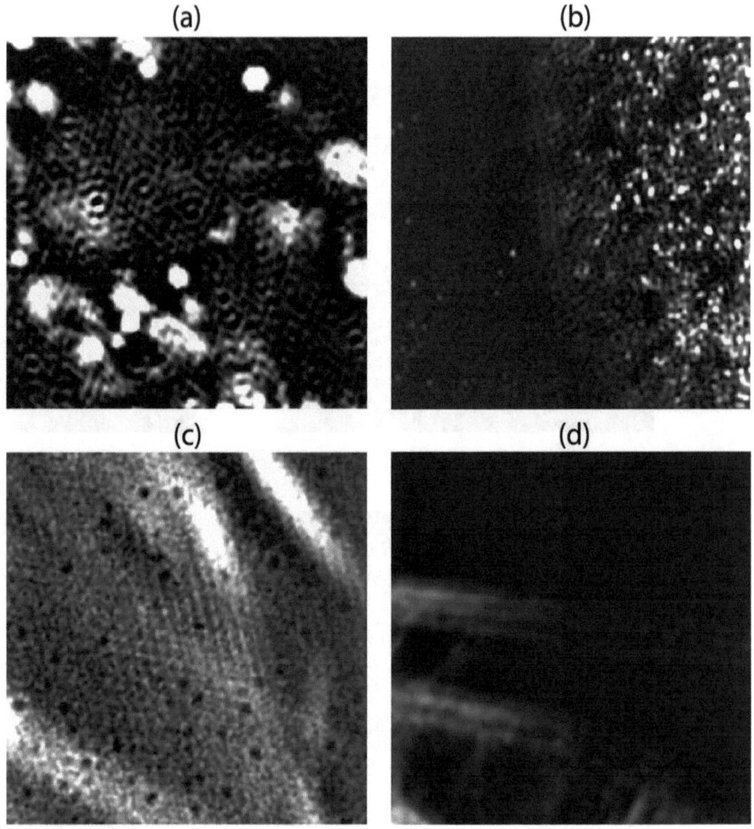

Fig. 5 Artefacts in structured illumination microscopy. (**a**) hammerstroke; (**b**) hatching; (**c**) honeycomb; and (**d**) ghosting

laser powers, or longer exposure times, and increasing the Wiener filter in SIM processing.

2. Hatching artefacts (Fig. 5b) are stripes in one or more of the angles and may be caused by an incorrect angle setting and a too-high Wiener filter in SIM processing. Test the result with Max. Isotrop enabled and reduce the Wiener filter value.
3. Honeycomb artefacts (Fig. 5c) are repeating hexagonal patterns in the reconstructed image and may be caused by missing frequencies in the axial direction. Possible strategies to remove honeycomb artefacts include using three-beam interference in 3D-SIM to address the lack of frequency contributions axially and reducing out-of-focus signal contributions by reducing the range of the z-stack acquisition to a stack that is tighter around the specimen's top and bottom planes.
4. Ghosting artefacts (Fig. 5d) are duplications of the structures of interest where there should be only one and arises from refractive index mismatch. Change the immersion medium to one that has a refractive index that best matches that of the specimen (*see* **Note 2**).

4 Notes

1. Structured illumination microscopy has been implemented using a programmable spatial light modulator in place of the phase grating for rapid and precise control of the excitation pattern. See, for example, Forster and co-workers [11].
2. The temperature type of immersion oil required would depend on the temperature of the specimen on the microscope stage. For example, if the temperature is 27 °C, the 30 –°C immersion oil would give more symmetrical PSFs than the 23 –°C type, and less ghosting.
3. The chromatic correction for SIM not only compensates for chromatic aberrations directly from the objective, but also from switching positions on the motorized filter turret. If one is using single-band emission filters for each channel, there will be a shift between channels due to the change in filter turret positions, which introduces changes in the angles as the fluorescence signal is directed to the camera sensor. A way of minimizing the shift due to filter-turret switching is to use dual-band emission filters, especially if visualizing only two channels.
4. The channel alignment procedure should be performed for each combination of frame size, objective, and filter cube.
5. It is recommended that "Affine" be used by default, as this method takes stretching and rotation effects into account when aligning image data, making it a more accurate and robust solution.
6. The z-stack for SIM acquisition should start and stop at the first and last planes with modulation, not from "black to black" as in confocal acquisition.

Acknowledgments

The authors are grateful to the Australian Government through the Australian Research Council for generously supporting the purchase of the QBI Advanced Microscopy Facility's Zeiss Elyra PS.1 SIM/PALM/STORM through the Linkage Infrastructure, Equipment and Facilities (LIEF) grant LE 130100078 led by Professor Frédéric A. Meunier. We thank Mr. Gavin Symonds, Dr. Debora Olivier, and Dr. Klaus Weisshart for helpful discussions on the technical editing of the manuscript.

Conflict of Interest The authors declare no conflict of interest.

References

1. Heintzmann R, Cremer CG (1999) Laterally modulated excitation microscopy: improvement of resolution by using a diffraction grating. In: Optical biopsies and microscopic techniques III, Proc SPIE, vol 3568, pp 185–196. https://doi.org/10.1117/12.336833
2. Gustafsson MGL (2000) Surpassing the lateral resolution limit by a factor of two using structured illumination microscopy. J Microsc 198(2):82–87. https://doi.org/10.1046/j.1365-2818.2000.00710.x
3. Frohn JT, Knapp HF, Stemmer A (2000) True optical resolution beyond the Rayleigh limit achieved by standing wave illumination. Proc Natl Acad Sci USA 97(13):7232–7236. https://doi.org/10.1073/pnas.130181797
4. Gustafsson MGL, Agard DA, Sedat JW (2000) Doubling the lateral resolution of wide-field fluorescence microscopy using structured illumination. In: Three-dimensional and multidimensional microscopy: image acquisition processing VII, Proc SPIE, vol 3919, pp 141–150. https://doi.org/10.1117/12.384189
5. Abbe E (1873) Beiträge zur Theorie des Mikroskops und der mikroskopischen Wahrnehmung. Arch Mikrosk Anat 9(1):413–468. https://doi.org/10.1007/BF02956173
6. Inoué S (2006) Foundations of confocal scanned imaging in light microscopy. In: Pawley J (ed) Handbook of biological confocal microscopy, 3rd edn. Springer, Boston, pp 1–19. https://doi.org/10.1007/978-0-387-45524-2_1
7. Gustafsson MGL, Shao L, Carlton PM et al (2008) Three-dimensional resolution doubling in wide-field fluorescence microscopy by structured illumination. Biophys J 94(12):4957–4970. https://doi.org/10.1529/biophysj.107.120345
8. Schermelleh L, Carlton PM, Haase S et al (2008) Subdiffraction multicolor imaging of the nuclear periphery with 3D structured illumination microscopy. Science 320(5881):1332–1336. https://doi.org/10.1126/science.115694
9. Carl Zeiss Microscopy GmbH (2013) ELYRA Operating Manual: Zen 2012 (black edition)
10. Demmerle J, Innocent C, North AJ et al (2017) Strategic and practical guidelines for successful structured illumination microscopy. Nat Protoc 12(5):988–1010. https://doi.org/10.1038/nprot.2017.019
11. Förster R, Lu-Walther HW, Jost A et al (2014) Simple structured illumination microscope setup with high acquisition speed by using a spatial light modulator. Opt Express 22(17):20663–20677. https://doi.org/10.1364/OE.22.020663

Chapter 2

Multicolor Super-resolution Fluorescence Imaging of Cells by Expansion Microscopy

Danush Taban, Marvin Jungblut, Patrick Eiring, and Markus Sauer

Abstract

This chapter presents a step-by-step protocol to resolve the organization of the nuclear pore complex (NPC) in HeLa-NUP107-GFP cells by eight-fold expansion microscopy (ExM). To enhance the efficiency of the procedure, we introduce an accelerated expansion protocol designed to allow rapid iteration and improve experimental throughput. This protocol involves both pre-expansion staining with wheat germ agglutinin (WGA) and post-digestion staining with nanobodies against GFP to visualize the NPC. The subsequent use of confocal microscopy enables multicolor super-resolution imaging with approximately 30 nm spatial resolution.

Key words Expansion microscopy, Super-resolution microscopy, Multicolor fluorescence imaging, Nuclear pore complex

1 Introduction

1.1 Expansion Microscopy

Light microscopy has been among the most used techniques in the toolbox of biological and medical research for decades. However, many complex cellular compartments and molecular machinery exist at length scales well below the diffraction limit of light. Super-resolution microscopy has since aided in investigating cellular organelles, deciphering the distribution of receptors or plasma membrane components, and tracking the dynamics of individual molecules. Despite its usefulness, most of these methods require expensive equipment and a high level of expertise, especially for multicolor sub-diffraction limit imaging [1].

Expansion microscopy (ExM) was introduced by the group of Edward Boyden in 2015 to address these limitations and has since been adapted for a wide range of bioimaging applications [2]. ExM has democratized super-resolution microscopy for a broader audience of researchers. Expanded samples can be imaged using conventional wide-field or confocal microscope setups, which are

Frédéric A. Meunier (ed.), *Super-Resolution Microscopy: Methods and Protocols*, Methods in Molecular Biology, vol. 3034,
https://doi.org/10.1007/978-1-0716-5268-8_2,

accessible to most laboratories, and the chemical components for ExM are commercially available. In recent years, the method has also seen extensive developments. For instance, while the original expansion protocol was limited to a four-fold expansion factor, different groups have developed methods to increase the expansion factor either by modulating the composition of the hydrogel or by performing iterative rounds of expansion [3–8].

Regardless of the individual protocols, the concept of ExM remains the same. Biomolecules within a sample are covalently linked into a polymer matrix through chemical modification, imprinting their location into the matrix. The mesh, made of charged polymers, forms a swellable hydrogel that increases in volume when submerged in deionized water. As the gel expands, the location of these biomolecules is retained in the matrix, causing the sample structure to enlarge proportionally with the gel.

A homogenization step, either through protein denaturation or digestion, is performed before the sample expands to ensure uniformity. This process dissolves inter- and intramolecular bonds, allowing ideally isotropic expansion of the sample.

While most ExM applications involve staining the structure of interest prior to gelation, several recent studies have implemented protocols that allow staining after gelation and homogenization. This offers several advantages compared to pre-staining. First, the linkage error between the dye and the protein of interest is reduced, as the label is not anchored into the gel and thus does not expand along with it. Second, recent studies have suggested that more binding epitopes can be made available if immunostaining is performed after expansion [9]. These protocols perform denaturation by SDS and heat to homogenize the sample, rather than an enzymatic cleavage of the peptide bonds [9–11].

The availability of the components needed for expansion and the ease of implementing standard immunostaining protocols in the expansion workflow enable even laboratories not specialized in microscopy to generate multicolor super-resolution data with a spatial resolution of up to 30 nm [11–13]. This protocol demonstrates the capability of multicolor super-resolution imaging by combining eight-fold expansion with Airyscan confocal microscopy to resolve the nuclear pore complex and its ultrastructural context. This is achieved through a combination of staining techniques: WGA staining to visualize the nuclear pore and DAPI staining to highlight the distribution of DNA in the proximity of the pore.

1.2 Nuclear Pore Complex as a Reference Structure for Super-resolution Microscopy

Nuclear pore complexes (NPCs) are large protein assemblies, approximately 120 MDa in size, that form pores within the nuclear envelope [14]. They serve as the sole mediator for transport between the cytoplasm and the nucleus. Each NPC is composed of 30–32 nucleoporins (NUPs) that assemble to form a cylindrical scaffold with an eight-fold symmetry with respect to the long axis of the pore and a two-fold symmetry with respect to the plane of the nuclear lamina. Overall, the dimensions of an NPC are about 125 nm in diameter and ~ 45 nm in height [15].

Due to their biological significance and status as one of the largest protein complexes in vertebrates, NPCs have long been a major topic of research in the field of structural biology. Advances in techniques like crystallography and cryogenic electron microscopy (cryo-EM) have significantly contributed to our understanding of the organization and assembly of NPCs at near-atomic resolution [14, 16].

Beyond structural biology, NPCs are also ideal targets for benchmarking super-resolution fluorescent microscopy techniques [8, 17–19]. Not only can resolving its three-dimensional organization present a significant challenge to some super-resolution techniques, but its well-defined eight-fold pore symmetry makes it a valuable model for evaluating the labeling efficiency.

1.3 Sample Expansion Using Ten-Fold Robust Expansion Protocol (TREx)

One of the previously mentioned iterations to the original expansion protocol was described by Damstra et al. in 2022 [4]. In this study, the monomer composition was optimized to produce a gel that can expand up to ten-fold while remaining mechanically stable. They initially described two distinct gel compositions with different concentrations of N,N′-methylenebisacylamide: 50 ppm for tissues and cultured cells, and 90 ppm for cultured cells. The former produces ten-fold expanded gels, the latter gels with a maximal expansion of approximately eight-fold. Due to the higher crosslinker density in the 90 ppm N,N′-methylenebisacrylamide formulation, we chose this protocol for our experiments. While this results in a slightly reduced expansion factor, the hydrogels are more robust, and the denser crosslinked polymer mesh suggests more anchor points are available to integrate the functionalized cellular proteins into the gel matrix. Airyscan confocal microscopy alone yields a moderate resolution improvement of ~1.4×, translating to an effective lateral resolution of 150 nm. Thus, combining Airyscan confocal imaging with eight-fold TREx expansion, we can anticipate a maximal lateral resolution of 20–30 nm.

1.4 Visualization of the Nuclear Pore Complex (NPC) by Expansion Microscopy

We utilize a HeLa cell line with a stable expression of NUP107, genetically tagged with GFP [20]. Although GFP fluorescence is significantly reduced after expansion, we are still able to visualize the NUP107-GFP fusion protein using anti-GFP nanobodies conjugated to ATTO643. Interestingly, we found that this staining can be performed both before expansion, using modified nanobodies equipped with an expansion cassette to crosslink the nanobody into the gel matrix, or after a short digestion step. Surprisingly, despite the digestion of proteins during the homogenization, which would typically be expected to destroy binding epitopes, we observe that a short but intense digest at 50 °C—the optimal temperature for Proteinase K activity—preserves enough binding epitopes to

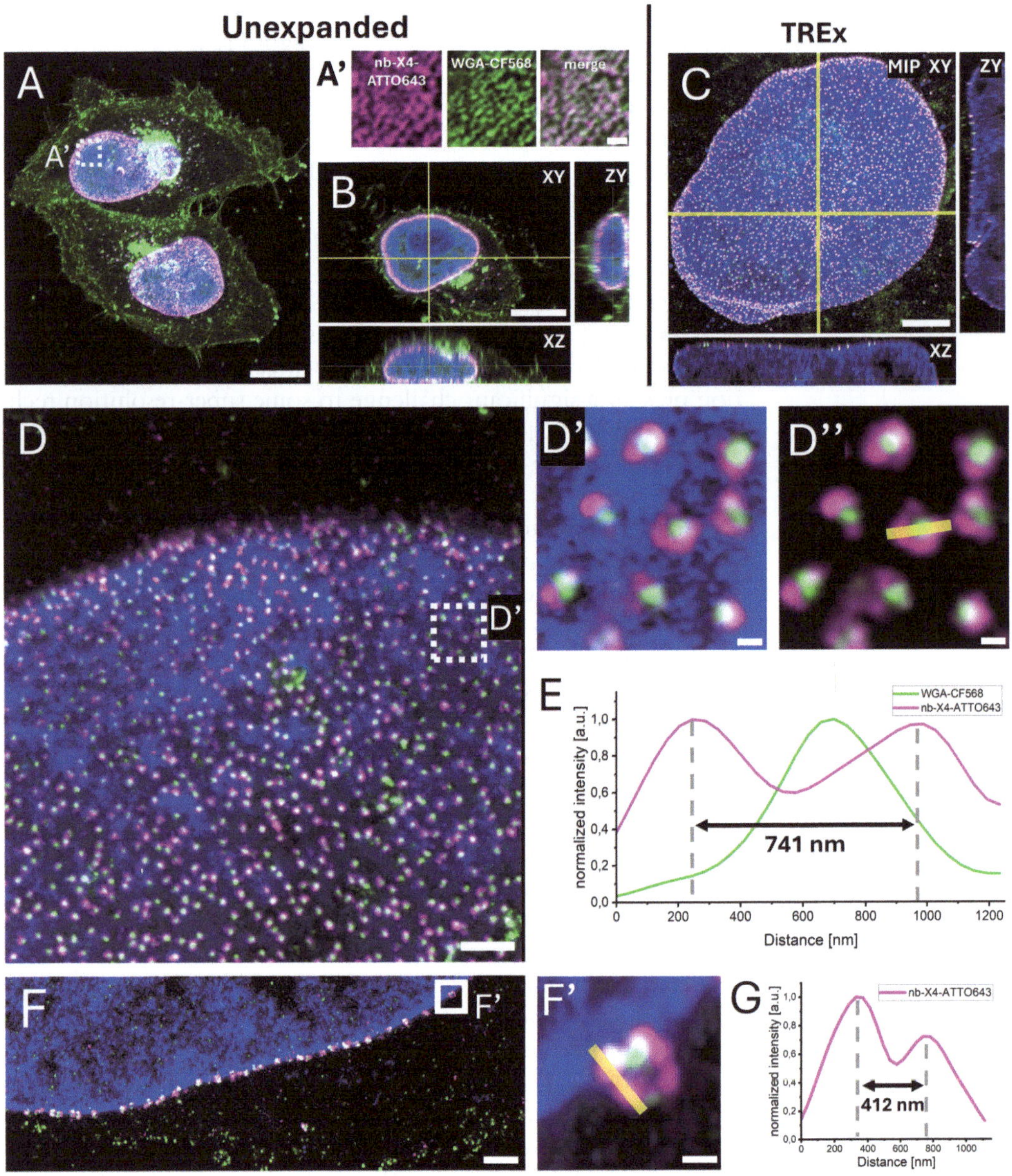

Fig. 1 TREx microscopy in combination with post-gelation immunostaining enables the visualization of the nuclear pore organization. (**A**) AiryScan imaging of an unexpanded HeLa-NUP107-GFP stained by nanobody-anti-GFP-X4-ATTO643 (magenta) and WGA-CF568 (green), to visualize the nuclear pores, along with a DAPI staining (blue). The system was configured for 3D, three-color imaging in AiryScan super-resolution mode with optimal parameters as suggested by the ZEN 3.5 blue microscope control software. A z-stack of ~4 μm was captured with optimal z-step intervals to enable reliable 3D processing. Post-processing of the data was performed in ZEN blue using the *3D Airyscan processing* function. (**A′**) Zoom-in view of the outlined region in A, showing channels for anti-GFP-nanobody-X4-ATTO643 and WGA-CF568 staining, along with a merge of both channels. (**B**) Orthogonal projections of the nucleus of unexpanded HeLa-NUP-GFP cells stained as described in A. (**C**) Orthogonal projection of TREx-expanded HeLa-NUP107-GFP nucleus, stained by using the post-staining protocol outlined in Subheading 3.2 and imaged by Airyscan confocal microscopy, as described for A.

achieve consistent staining of NPCs with anti-GFP nanobodies (Fig. 1). In these images, NPCs appear as cylindrical structures with a central minimum representing the nuclear pore. The pore is visualized by a pre-expansion staining with wheat germ agglutinin conjugated to CF568 (WGA-CF568), a lectin known to block nuclear transport, by binding inside the pore [17, 21].

1.5 Protocol Optimization and Validation of Expansion Microscopy

Over the last decade, ExM has proven its capability in offering sub-diffraction limit imaging to a broad audience of researchers. However, the variety of available expansion protocols suggests that extensive adjustments to the protocol may be required to obtain reliable staining and expansion of different target proteins or samples (i.e., cultured cells, isolated proteins, or tissues) [2, 4, 22, 23]. Therefore, while the protocol outlined in this chapter may work well for expanding NUPs, modifications to crosslinking or homogenization methods may yield better results for other target proteins. We observe an anisotropic expansion of NUPs when considering the intensity profiles in Fig. 1e, g. Here it is evident that the expansion across the long axis of the NPC is reduced compared to the expansion across their depth. This discrepancy is likely due to varying linker densities. The long axis of NPCs is protein dense, providing numerous anchor points to crosslink the proteins into the gel. In contrast, the depth of NPCs is less protein dense, and the nuclear lamina, which is not functionalized with anchor points for integration into the gel, separates NUPs on either side of the pore, thus leading to increased expansion in this direction. While a longer digestion step can improve the isotropy of expansion by more thoroughly cleaving peptide bonds, this might cause reduced signal in the post-staining with anti-GFP nanobodies, as its binding epitope might be degraded.

Relating to this point, we must note that, in our experience, extensive iterations and modifications to the employed expansion protocol are necessary for different targets to obtain the best results. As minor modifications to the expansion protocol can have large effects on the imaging data, it is necessary to validate its accuracy, i.e., by determining the true expansion factor of the

Fig. 1 (continued) (**D′**) Zoom-in of the region outlined in D, revealing distinct and separated nuclear pore complexes with WGA-CF568 signal in the center of the complex. (**D″**) The same region, showing only the staining for NUP107-GFP by nanobody-anti-GFP-X4-ATTO643 and the WGA-staining. The yellow bar indicates the location for the intensity profile provided in (**E**). The peak-to-peak distance of 741 nm of the nanobody-X4-ATTO643 intensity plot indicates an expansion factor of 6.8-fold, as the unexpanded diameter of NUP107-GFP is expected to be approximately 107 nm. (**F**) Nuclear lamina at the equatorial plane of the cell, providing a side-profile of nuclear pore complexes. (**F′**) Zoomed-in view of the boxed region in F, resolving the nuclear pore complex depth. The yellow line indicates the location of the intensity profile plotted in (**G**). The peak-to-peak distance of 412 nm suggests an expansion factor of approximately 9.8-fold, as the unexpanded depth of a nuclear pore complex is expected to be 45 nm. Scale bars: A, B, C, D and F 5 μm, A′, D′, D″ and F′ 500 nm

structure of interest. While it may be sufficient to determine the expansion factor of a gel by simply measuring the size of the fully expanded gel compared to the gel size after gelation, it is more reliable to image a structure with known dimensions, such as NPCs, or to image the same structure before and after expansion to accurately assess the true expansion of the sample. Experiments implementing the latter method can also benefit from the use of elastic transformations, which can highlight local distortions introduced in the expanded sample [12, 24, 25]. In our case, Airyscan confocal imaging in the far-red spectrum can achieve lateral resolutions of approximately 200 nm. Given the 6.8-fold expansion with respect to the long axis of the NPCs, we can estimate a maximal lateral resolution of around 30 nm, although higher resolutions are attainable with respect to the depth of NPCs.

2 Materials

2.1 Cell Culture

Unless otherwise specified, all buffers are prepared using Milli-Q or ultrapure water.

2.1.1 Culture of HeLa-NUP107-GFP Cells

1. *Culture flasks:* Nunclon Delta T25 flasks for cell culture of adherent cells.
2. *Culture medium:* DMEM high glucose, 10% FCS, 1% Penicillin/Streptomycin, 1 mM sodium pyruvate.
3. *Trypsin solution:* 0.05% trypsin in 1× PBS.
4. *Culture dish:* 4-well plastic culture chambers with No. 1.5 12 mm high-precision glass coverslip.

2.1.2 Staining of Nuclear Pore Complex in HeLa-NUP107-GFP Cells

1. *Cytoskeleton Buffer (CB):* 10 mM MES, 150 mM NaCl, 5 mM EGTA, 5 mM $MgCl_2$, and 5 mM glucose in ddH_2O adjusted to pH 6.1 (*see* **Note 1**).
2. *Fixation solution:* 2.4% Formaldehyde in CB.
3. *Blocking and Permeabilization Buffer:* 4 mM EDTA, 5% Bovine Serum Albumin (BSA), 0.02% Tween-80, 0.05% Sodium Azide and 0.25% Triton X-100 in 1× PBS pH 7.4.
4. *WGA staining solution:* 5% BSA and 7.5 μg/mL Wheat Germ Agglutinin (WGA) CF568 in 1× PBS pH 7.4.
5. *Nanobody staining solution:* 100 nM nanobody anti-GFP-X4-ATTO643 (N0304, NanoTag Biotechnologies) in 5% BSA in 1× PBS pH 7.4.

2.2 Expansion Microscopy

2.2.1 Materials

1. *Gelation chamber:* Polystyrene box with ice, parafilm, 4-well plastic dish, laboratory wipe, spatula, tweezers.

2. *Digestion:* Platform shaker with heating function, 1.5 mL reaction tubes.
3. *Washing and expansion:* 5 mL reaction tubes, round polystyrene petri dishes.

2.2.2 Solutions and Buffer

1. *AcX stock:* Acryloyl X-SE stock solution (5 mg) dissolved in anhydrous DMSO to a concentration of 10 mg/mL, portioned in 0.1 mg aliquots in 1.5 mL centrifuge tubes, desiccated and kept at −20 °C.
2. *AcX crosslinking solution:* 0.1 mg AcX is dissolved in 10 μl DMSO, then suspended in 100 mM $NaHCO_3$ in 1× PBS to a final concentration of 0.1 mg/mL.
3. *TREx monomer solution:* 1.1 M (w/w) sodium acrylate, 2.0 M (w/w) acrylamide, 90 ppm (w/w) N,N-′-methylenebisacrylamide (bis), dissolved in 1× PBS. 0.15% TEMED and 0.15% APS are added right before gelation (*see* **Note 2**). Stock solution should be kept in 485 μL aliquots at −20 °C.
4. *TEMED stock solution:* 10% TEMED in ddH_2O, kept at −20 °C (*see* **Note 2**).
5. *APS stock solution:* 10% APS in ddH_2O, kept at −20 °C.
6. *Digestion buffer:* 50 mM Tris, 0.5% Triton X-100, 0.8 M guanidine, 2 mM $CaCl_2$, in ddH_2O and adjusted to pH 8.0. Aliquots should be stored at −20 °C. Prior to the use of the buffer, Proteinase K is added at 8 units/mL.
7. *Washing buffer*: 0.1% Tween-20 in 1× PBS.
8. *Nucleus post-staining:* 1:500 solution of DAPI (1 mg/mL) in 1× PBS.

2.3 Imaging Chamber for Expanded Gels

1. *Coverslips:* 24 mm high-precision coverslips.
2. *Coverslip cleaning solutions:* 1 M potassium hydroxide, absolute EtOH.
3. *APTES solution*: 10 μL APTES in a solution of 8 mL EtOH, 0.5 mL ddH_2O and 0.5 mL anhydrous acetic acid.
4. *Imaging chamber:* Metal imaging chamber for 24 mm coverslips.

2.4 Airyscan Imaging of Expanded Gels

1. *Imaging system:* Zeiss LSM 900 with Airyscan 2, equipped with a 63×/1.4 NA oil and a 40×/1.2 NA water immersion objective for imaging of expanded samples and 4 laser lines: 405 nm (30 mW), 488 nm (30 mW) and 639 nm (25 mW) diode lasers and a 561 nm (25 mW) DPSS laser.
2. *Software:* ZEN blue 3.5 for processing of Airyscan imaging data, Fiji (ImageJ) for further processing.

3 Methods

3.1 Cell Culture

The HeLa-NUP107-GFP cell line was generated by the laboratory of Prof. Dr. Jungmann and colleagues according to published procedures [20].

HeLa-NUP107-GFP cells are maintained in DMEM cell culture medium, supplemented with 10% FCS and 1% Penicillin/Streptomycin. Cells are passaged three times a week and are used between passage numbers 5 and 30. For passaging adherent cells are transferred into suspension by exchanging the culture medium to 0.05% trypsin in 1× PBS.

3.2 Expansion and Post-staining of Nuclear Pore Complex in HELA-NUP107-GFP

3.2.1 Permeabilization, Fixation and WGA-Staining

1. Plate 10,000 to 20,000 cells per well in cell culture medium on 12 mm high-precision round coverglasses (No. 1.5) within 4-well culture plates, allowing the cells to proliferate for 24 h before fixation.
2. Fix cells with 2.4% Formaldehyde solution in CB-Buffer for 30 min at room temperature.
3. Wash the samples three times with 1× PBS for 5 min each.
4. Block and permeabilize cells with 0.25% Triton X-100 in blocking buffer for 60 min at room temperature.
5. Wash the samples three times with 1× PBS for 5 min each.
6. To stain the sample with WGA-CF568, add 5% BSA in 1× PBS together with 7.5 μg/mL WGA-CF568 into the 4-well culture plate and incubate for 2 h at room temperature under gentle agitation.
7. Wash the samples three times with 1× PBS for 5 min each.

3.2.2 Crosslinking

1. Prepare a fresh 1:100 dilution of AcX solution in 100 mM $NaHCO_3$ in 1× PBS to achieve a final concentration of 0.1 mg/mL. First, resuspend the dried stock in 10 μL of anhydrous DMSO. Then, dilute this solution with 990 μL 100 mM $NaHCO_3$ in 1× PBS, mixing thoroughly by pipetting up and down multiple times. Vortex the solution to ensure it is well-mixed.
2. Dispense 500 μL of the diluted AcX solution into each well of a 4-well chambered culture plate. Incubate the plate for 2 h at room temperature in a humidified chamber.

3.2.3 Gelation

1. Prepare a gelation chamber on ice. Cover the lid of the culture-well plate with parafilm, ensuring it is smooth and free of creases and place the plate on ice. Place pre-wetted paper towels in the corner of the chamber to prevent drying of the sample during gelation. (Fig. 2a).

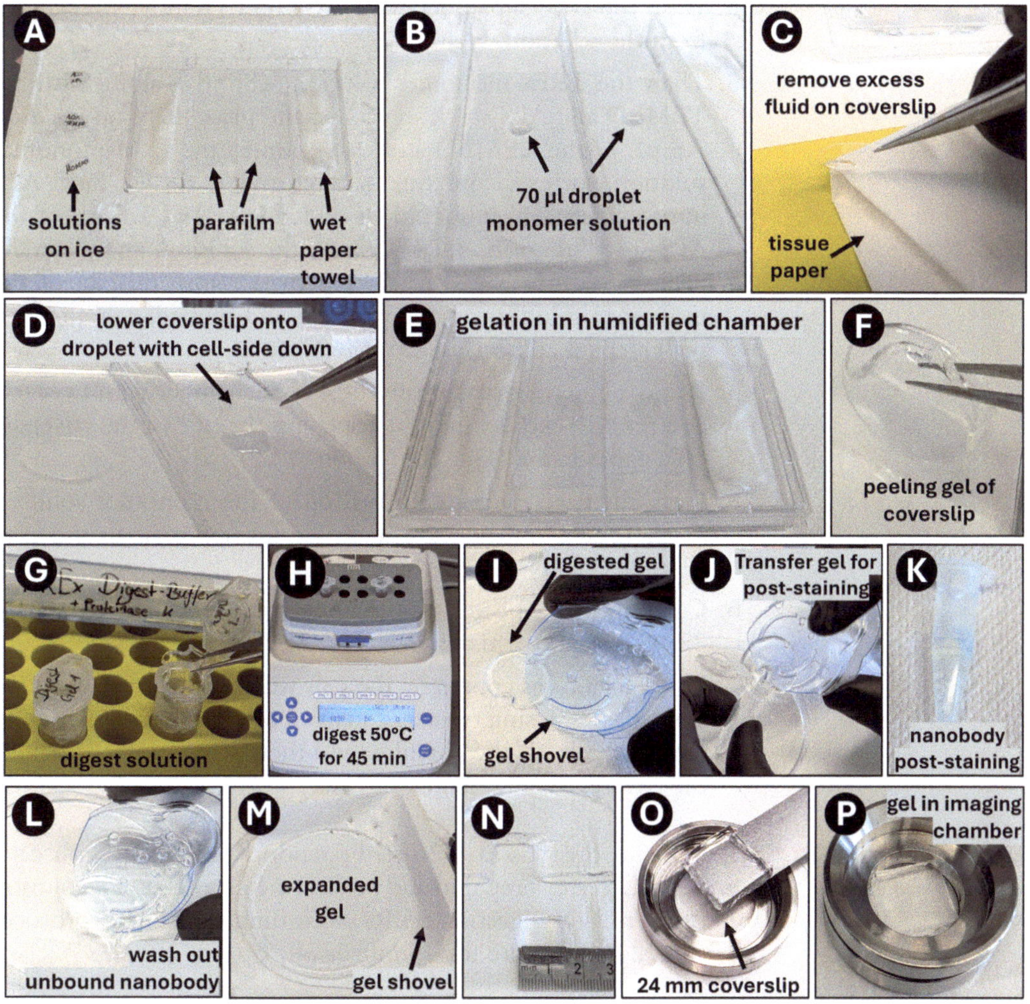

Fig. 2 Workflow and handling of gelation, digestion and expansion of TREx gels. (**A**) Preparation of gelation chamber and monomer solution on ice. A four-well plastic plate with a lid serves as a humidified chamber for gelation. A layer of parafilm is placed in the middle of two wells to provide a smooth and hydrophobic surface for gelation, while wet paper towels are placed in the outer two wells. (**B**) After initialization of gelation by adding TEMED and APS to the monomer solution, 70 μL drops of gelation solution are pipetted onto the parafilm. (**C**) Coverslips are retrieved from 4-well culture plates, and the excess liquid is removed by gently touching the corner of the coverslip to a piece of tissue paper. (**D**) Once the excess liquid is removed, the coverslips are lowered onto the droplets of monomer solution. Lowering the coverslip edge-first onto the drop creates a meniscus and ensures an even distribution of the solution below the coverslip. (**E**) Gelation is performed in the chamber with a closed lid, first on ice to allow the monomer solution to diffuse through the sample, and then at 37 °C. (**F**) After gelation, the coverslip and gel are carefully removed from the parafilm. The gel can then be peeled off the coverslip with tweezers. Although TREx gels are robust, careful handling of the gels is essential to avoid stretching of the polymerized gel. (**G**) The peeled-off gel is transferred into a 2 mL reaction tube filled with 1.5 mL pre-warmed digestion buffer. (**H**) The tubes are placed in a heating block at 50 °C for 45 min. (**I**) After digestion, the gel size increases to approximately twice its original size and should be washed several times in 1x PBS to remove the digestion buffer. Gel shovels, like those shown in the image, can be used to ease handling during the washing steps and ensure non-damaging transfer of the gels. Prior to

2. Wash the coverslips once with 1× PBS to remove excess AcX from the solution.
3. Thaw the TREx monomer solution (485 μL), along with the TEMED and APS stock solutions, and place them on ice. Add 7.5 μL of the TEMED stock solution to the ExM monomer solution, vortex the mixture, and return it to the ice. Since APS initiates gelation, it should be added last. Add 7.5 μL of the APS stock solution, quickly vortex the solution, and place the gelling solution back on ice. Pipette a 70 μL droplet of the monomer solution on the parafilm and immediately proceed with **steps 4** and **5** (Fig. 2b) (*see* **Note 2**).
4. Remove the coverglass from the 4-well chamber with tweezers and remove excess liquid by touching the side of the coverslip on a paper towel (Fig. 2c).
5. Place them cell-side down on top of the monomer solution droplet. Afterwards, leave it on ice for approximately 5 min (Fig. 2d) (*see* **Note 3**).
6. Carefully place the gelation chamber in a preheated chamber at 37 °C for approximately 1 h (Fig. 2e).
7. Once gelation is complete, the gels will remain attached to the coverglasses. Using tweezers, gently peel the coverglasses along with the gels off the parafilm, and then gently peel the gel off the coverglass (Fig. 2f).

3.2.4 Homogenization

1. Transfer the gels in a 2 mL Eppendorf tube filled with each 1.5 mL freshly prepared and pre-warmed Proteinase K solution (8 U/mL) in digestion buffer, according to the 10× protocol (Fig. 2g). Incubate for 45 min at 50 °C (Fig. 2h).

Fig. 2 (continued) post-staining with antibodies or nanobodies, a final washing step in 5% BSA in 1× PBS should be performed. (**J**) For post-staining the gels are transferred into 1.5 mL reaction tubes and supplemented with 5% BSA in 1× PBS to a total volume of approximately 1 mL. (**K**) Anti-GFP nanobody is added to the solution at a final concentration of 100 nM. Incubation is carried out at 4 °C overnight, or for 2 h at room temperature, under constant agitation (e.g., on a shaker plate or a slowly rotating wheel). (**L**) Post-stained gels are washed three to four times in 1× PBS to remove unbound nanobodies. (**M**) Expansion is performed in 15 cm plastic petri dishes filled with ddH_2O. Water should be exchanged at least three times, or until the gel no longer increases in size. We recommend using gel shovels for the washing steps in water to facilitate handling of the expanding gels. Shown here is a shovel cut to size from a disposable weighing dish with small drainage holes punched through on the side to allow excess water to drain while keeping the gel protected. Fully expanded gels can be measured for their total expansion. (**N**) To fit gels into the imaging chambers, small 1.5 × 1.5 cm squares are cut out of the fully expanded gel. We found that the cell side of the gel can be easily identified by its rough surface compared to the side exposed to the parafilm during gelation. We also observed that gels slide more easily on a plastic surface with the cell-side up. (**O**) The gel sections are carefully lowered cell side-down onto APTES-coated coverslips, to avoid the formation of air bubbles between coverslip and gel. (**P**) The imaging chamber is then assembled and ddH_2O is added to the gels to prevent shrinkage of the gels during imaging

2. After digesting, place the gels in a petri dish with an 8 cm diameter and add 25 mL 1× PBS (Fig. 2i).
3. Exchange 1× PBS 3–4 times every 20–30 min to rinse out the Proteinase K solution (*see* **Note 4**).

3.2.5 Post-staining of Nuclei with DAPI

Staining of the nucleus can still be performed after gelation and digestion of the hydrogels. To do this the gels, or sections of the gels, are transferred into a new plastic dish or reaction tube and submerged in 1× PBS. DAPI in a 1 mg/mL stock solution in DMSO is added 1:500 to the submerged gels and incubated at room temperature under gentle agitation for 1 h. After staining the gels are washed 3–5 times in 1× PBS to wash out unbound DAPI.

3.2.6 Post-staining of GFP by Anti-GFP Nanobody-X4-ATTO643

For the post-staining of the gel with the anti-GFP nanobody-X4-ATTO643, the section of the gel is submerged in 1× PBS with 5% BSA and 100 nM of anti-GFP nanobody-X4-ATTO643 (Fig. 2j, k). The sample is then incubated at 4 °C overnight, or 2 h at room temperature, under gentle agitation. After staining, the gels are washed 3–4 times in 1× PBS to wash out unbound nanobody (Fig. 2l). The post-stained hydrogel is then transferred into a round petri dish (150 mm) filled with 250 mL ddH_2O. The water is exchanged every 20–30 min until the maximum expansion factor of the gel is reached (Fig. 2m). This can also be carried out overnight (*see* **Note 5**).

3.3 Preparation of Imaging Chamber and Sample Mounting

3.3.1 Cleaning of 24 mm Coverglasses

1. Rinse the coverglasses twice with ddH_2O.
2. Sonicate the coverglasses in absolute ethanol for 15 min.
3. Rinse the coverglasses again twice with ddH_2O.
4. Sonicate the coverglasses in 1 M potassium hydroxide for 15 min.
5. Rinse the coverglasses 3–4 times with ddH_2O.
6. The cleaned coverglasses can be stored in a closed container to avoid dust contamination.

3.3.2 APTES Coating

1. Prepare the APTES solution (*see* Subheading 2.3) under a chemical hood.
2. Evenly distribute the APTES solution on the cleaned 24 mm coverglasses and leave them under a chemical hood for approximately 1 h until the solution has fully evaporated.
3. Rinse the coverglasses twice with absolute ethanol and allow them to air-dry in a coverslip rack. Store the coverglasses in a dust-free glass container.

3.3.3 Sample Mounting and Imaging

1. Cut the sample gel into a 1.5 cm × 1.5 cm rectangle and place it cell-side-down slowly on a silanized coverglass. Start at one side of the rectangle and slowly move on to the other, to avoid formation of air bubbles in between the sample and the coverglass (Fig. 2n, o).
2. Put the coverglass with the sample in the imaging chamber, screw the top half on with enough pressure to prevent leakage (Fig. 2p).
3. Add 500–750 μL ddH_2O into the chamber to prevent the gel from shrinking during the measurement.
4. Place the sample on the Zeiss LSM 900 Airyscan 2, using a 40×/1.2 NA water immersion objective for imaging in super-resolution mode.
5. For each fluorophore an appropriate imaging channel and setup is used, based on the suggested imaging setting provided by the microscope control software ZEN blue 3.5.
6. The Airyscan detector is calibrated based on the manufacturer's suggested protocol. The pixel dwell time is adjusted to values between 1 and 5 μs depending on the signal-to-noise ratio of the sample.
7. Z-stacks of the sample are captured with z-intervals set to *optimal* in ZEN.
8. The raw Airyscan imaging data is processed in ZEN blue 3.5 using the integrated *Airyscan processing* function, using the 3D Airyscan processing. The strength of Airyscan processing can be adjusted per channel based on the signal-to-noise ratio.

4 Notes

1. EGTA is insoluble in water at a pH below 9. We recommend creating a separate stock of 250 mM EGTA in ddH_2O. To fully dissolve EGTA in this stock, the pH must be raised above 9. Once EGTA is fully dissolved, the pH can then be lowered to 7.4 by titration with HCl.
2. Acrylamide is potentially carcinogenic, and TEMED is harmful if inhaled. Be sure to read the safety data sheets for all chemicals, wear protective gloves, and work under a chemical fume hood.
3. This is a somewhat time-critical step. While the monomer solution should not polymerize while on ice, it is recommended to place the coverslips onto the droplets of monomer solution as quickly as possible.

4. Insufficiently washed-out proteinase K and the digestion buffer can lead to a partial digestion of antibodies or nanobodies.
5. The maximal expansion factor is reached if the gel no longer increases in size after exchange of ddH_2O.

Acknowledgments

We thank Lisa Behringer-Pließ, Elke Maier, and Ivan Simeonov for excellent technical assistance. This work was supported by the European Union's Horizon 2020 research and innovation program (grant agreement No 835102 and 951275).

Conflict of Interest The authors of this chapter declare no conflict of interest.

References

1. Schermelleh L, Ferrand A, Huser T et al (2019) Super-resolution microscopy demystified. Nat Cell Biol 21(1):72–84. https://doi.org/10.1038/s41556-018-0251-8
2. Chen F, Tillberg PW, Boyden ES (2015) Expansion microscopy. Science 347(6221): 543–548. https://doi.org/10.1126/science.1260088
3. Truckenbrodt S, Maidorn M, Crzan D et al (2018) X10 expansion microscopy enables 25-nm resolution on conventional microscopes. EMBO Rep 19(9). https://doi.org/10.15252/embr.201845836
4. Damstra HGJ, Mohar B, Eddison M et al (2022) Visualizing cellular and tissue ultrastructure using ten-fold robust expansion microscopy (TREx). elife 11. https://doi.org/10.7554/eLife.73775
5. Gambarotto D, Zwettler FU, Le Guennec M et al (2019) Imaging cellular ultrastructures using expansion microscopy (U-ExM). Nat Methods 16(1):71–74. https://doi.org/10.1038/s41592-018-0238-1
6. Klimas A, Gallagher BR, Wijesekara P et al (2023) Magnify is a universal molecular anchoring strategy for expansion microscopy. Nat Biotechnol 41(6):858–869. https://doi.org/10.1038/s41587-022-01546-1
7. Chang J-B, Chen F, Yoon Y-G et al (2017) Iterative expansion microscopy. Nat Methods 14(6):593–599. https://doi.org/10.1038/nmeth.4261
8. Louvel V, Haase R, Mercey O et al (2023) iU-ExM: nanoscopy of organelles and tissues with iterative ultrastructure expansion microscopy. Nat Commun 14(1):7893. https://doi.org/10.1038/s41467-023-43582-8
9. Park J, Khan S, Yun DH et al (2021) Epitope-preserving magnified analysis of proteome (eMAP). Sci Adv 7(46):eabf6589. https://doi.org/10.1126/sciadv.abf6589
10. Ku T, Swaney J, Park JY et al (2016) Multiplexed and scalable super-resolution imaging of three-dimensional protein localization in size-adjustable tissues. Nat Biotechnol 34(9): 973–981. https://doi.org/10.1038/nbt.3641
11. Eilts J, Reinhard S, Michetschläger N et al (2023) Enhanced synaptic protein visualization by multicolor super-resolution expansion microscopy. Neurophotonics 10(4):044412. https://doi.org/10.1117/1.NPh.10.4.044412
12. Sachs S, Reinhard S, Eilts J et al (2024) Visualizing the trans-synaptic arrangement of synaptic proteins by expansion microscopy. Front Cell Neurosci 18:1328726. https://doi.org/10.3389/fncel.2024.1328726
13. Gao M, Maraspini R, Beutel O et al (2018) Expansion stimulated emission depletion microscopy (ExSTED). ACS Nano 12(5): 4178–4185. https://doi.org/10.1021/acsnano.8b00776
14. Beck M, Hurt E (2017) The nuclear pore complex: understanding its function through structural insight. Nat Rev Mol Cell Biol 18(2): 73–89. https://doi.org/10.1038/nrm.2016.147

15. Peters R (2009) Functionalization of a nanopore: the nuclear pore complex paradigm. Biochim Biophys Acta 1793(10):1533–1539. https://doi.org/10.1016/j.bbamcr.2009.06.003
16. von Appen A, Kosinski J, Sparks L et al (2015) In situ structural analysis of the human nuclear pore complex. Nature 526(7571):140–143. https://doi.org/10.1038/nature15381
17. Löschberger A, van de Linde S, Dabauvalle MC et al (2012) Super-resolution imaging visualizes the eightfold symmetry of gp210 proteins around the nuclear pore complex and resolves the central channel with nanometer resolution. J Cell Sci 125(Pt 3):570–575. https://doi.org/10.1242/jcs.098822
18. Szymborska A, de Marco A, Daigle N et al (2013) Nuclear pore scaffold structure analyzed by super-resolution microscopy and particle averaging. Science 341(6146):655–658. https://doi.org/10.1126/science.1240672
19. Bianchini P, Pesce L, Diaspro A (2021) Expansion microscopy at the nanoscale: the nuclear pore complex as a fiducial landmark. Methods Cell Biol 161:275–295. https://doi.org/10.1016/bs.mcb.2020.07.001
20. Schlichthaerle T, Strauss MT, Schueder F et al (2019) Direct visualization of single nuclear pore complex proteins using genetically-encoded probes for DNA-PAINT. Angew Chem Int Ed Engl 58(37):13004–13008. https://doi.org/10.1002/anie.201905685
21. Finlay DR, Newmeyer DD, Price TM et al (1987) Inhibition of in vitro nuclear transport by a lectin that binds to nuclear pores. J Cell Biol 104(2):189–200. https://doi.org/10.1083/jcb.104.2.189
22. Steib E, Tetley R, Laine RF et al (2022) TissUExM enables quantitative ultrastructural analysis in whole vertebrate embryos by expansion microscopy. Cell Reports Methods 2(10): 100311. https://doi.org/10.1016/j.crmeth.2022.100311
23. Sheard TMD, Hurley ME, Colyer J et al (2019) Three-dimensional and chemical mapping of intracellular signaling nanodomains in health and disease with enhanced expansion microscopy. ACS Nano 13(2): 2143–2157. https://doi.org/10.1021/acsnano.8b08742
24. Seehra RS, Warrington SJ, Allouis BHK et al (2023) Geometry-preserving expansion microscopy microplates enable high-fidelity nanoscale distortion mapping. Cell Rep Phys Sci 4(12):101719. https://doi.org/10.1016/j.xcrp.2023.101719
25. Zhu C, Wang A, Chen L et al (2021) Measurement of expansion factor and distortion for expansion microscopy using isolated renal glomeruli as landmarks. J Biophotonics 14(7): e202100001. https://doi.org/10.1002/jbio.202100001

Chapter 3

Super-resolution Spinning-Disk Confocal Microscopy Using Optical Photon Reassignment (SoRa)

Sevannah A. Steeves, Saber H. Saber, Mo Chen, and Frédéric A. Meunier

Abstract

Super-resolution spinning-disk confocal microscopy with optical photon reassignment (SoRa) is an advanced imaging technique that extends fluorescence microscopy resolution beyond the diffraction limit without the need for specialized sample preparation or intensive computational processing. By enhancing both lateral and axial resolution while preserving the rapid imaging speed of conventional spinning-disk systems, SoRa is particularly well suited for live-cell imaging and high-throughput studies. In contrast to other super-resolution approaches, SoRa enables fast, 3D imaging with low phototoxicity, making it ideal for capturing dynamic cellular events in real time. This chapter highlights the use of SoRa for two key applications: (1) high-throughput screening, specifically for assessing drug effects on neuronal processes such as dendritic outgrowth, and (2) high-resolution 3D imaging of live-cells including imaging mitochondrial trafficking in neurons. SoRa's versatility, speed, and compatibility with standard dyes make it a powerful tool for a wide range of biological and drug discovery research.

Key words Super-resolution, Confocal microscopy, Live imaging, Neurons, Mitochondrial trafficking

1 Introduction

Super-resolution spinning-disk confocal microscopy with optical photon reassignment (SoRa) is a confocal imaging technique that enhances spatial resolution to ~120 nm laterally and ~300 nm axially, producing sub-diffraction-limited images [1, 2]. This approach uses optical and computational photon reassignment to sharpen the effective point spread function (PSF), enabling real-time super-resolution imaging with minimal post-processing. Importantly, SoRa maintains the high-speed capabilities of a traditional spinning-disk microscope, allows super-resolution imaging, and is compatible with several labelling techniques, including traditional dyes, fluorophores, tags, and DNA-PAINT probes [2, 3]. This has made SoRa a valuable tool for high-throughput and fast live-cell super-resolution imaging.

Frédéric A. Meunier (ed.), *Super-Resolution Microscopy: Methods and Protocols*, Methods in Molecular Biology, vol. 3034,
https://doi.org/10.1007/978-1-0716-5268-8_3,

1.1 Advances in Super-resolution Spinning-Disk Confocal Microscopy

Fluorescence confocal microscopy is widely used for its optical sectioning capabilities, employing a pinhole to eliminate out-of-focus light and achieve lateral resolution of ~200 nm and axial resolution of ~500 nm [4, 5]. Traditional spinning-disk confocal microscopy uses a rapidly rotating disk with multiple pinholes to simultaneously excite and detect fluorescence across a sample, enabling faster acquisition rates, improved signal-to-noise ratios, and reduced phototoxicity compared to line-scanning confocal microscopy [4–6]. However, conventional spinning-disk microscopy remains limited by diffraction, as the PSF causes emitted photons to blur and overlap, reducing spatial precision.

SoRa builds upon traditional spinning-disk confocal microscopy by introducing a second microlens array into the detection path and computationally reassigning each detected photon to its most probable point of origin [1, 2] (*see* Fig. 1). This microlens array reimages and slightly magnifies the emission from each pinhole, effectively shifting detected photons toward their most probable source location. This is then refined computationally using a

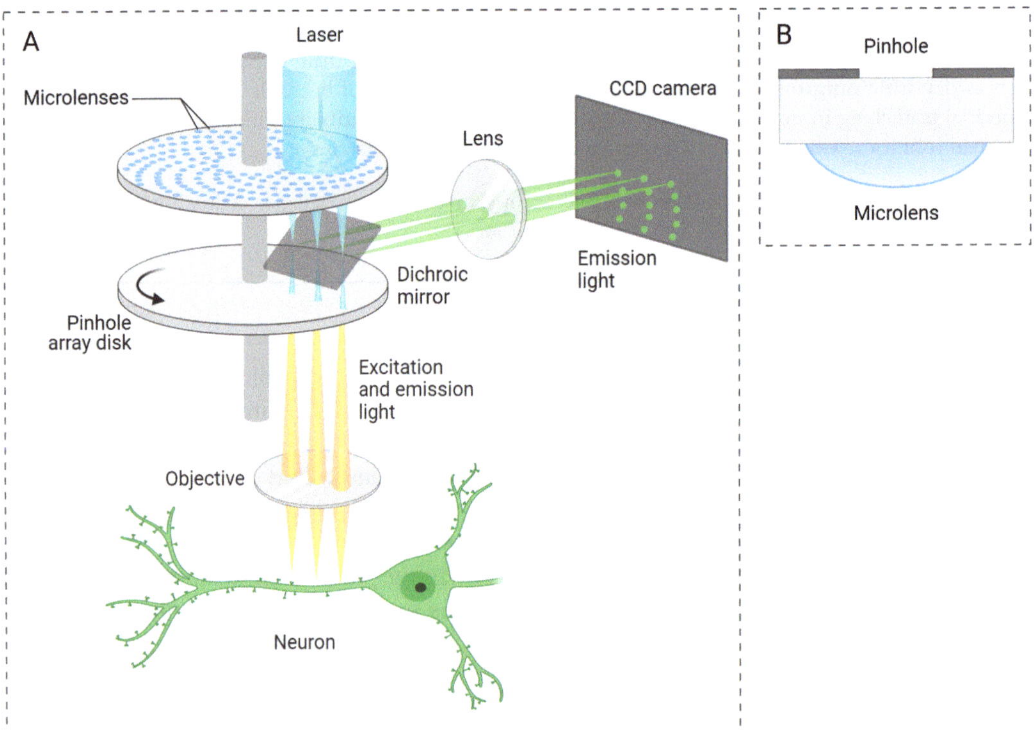

Fig. 1 Schematic of the optical layout of the SoRa spinning-disk confocal microscope. (**a**) Diagram illustrating the optical configuration of the SoRa module integrated into a spinning-disk confocal system. The microlens array is positioned in front of the confocal pinhole disk to concentrate excitation light into each pinhole, enhancing illumination efficiency and resolution. (**b**) Schematic of the pinhole-microlens arrangement used for sample illumination and fluorescence detection

reassignment algorithm based on the PSF and microscope geometry to reassign each detected photon to its most probable point of origin. This effectively redistributes light intensity toward the true emission coordinates in real-time, enabling super-resolution reconstruction with minimal post-processing.

A key advantage of SoRa super-resolution microscopy is its ability to image at high acquisition speeds. Other confocal super-resolution microscopy techniques include Airyscan, which achieves super-resolution imaging by combining confocal detection with a detector array to improve resolution and signal-to-noise ratio [7]. However, Airyscan utilizes line-scanning, which is significantly slower and more phototoxic than SoRa. Overall, SoRa is advantageous for high-speed, low-phototoxicity imaging, making it ideal for live-cell and high-throughput applications.

In this chapter, we outline the application of SoRa for (1) high-throughput super-resolution imaging and (2) live-cell super-resolution imaging.

1.2 High-Throughput Super-resolution Imaging

SoRa microscopy has also become a valuable tool for high-throughput drug screening. Neurite outgrowth is a key indicator of neuronal development, regeneration, and response to external stimuli, making it a valuable readout in drug screening for neuroprotective compounds [8]. Accurate quantification of neurite length, branching, and growth cone morphology requires high-resolution imaging to discern the subtle morphological changes to inform drug efficacy. When imaging dendrite and dendritic spine outgrowth, SoRa microscopy offers enhanced spatial resolution and rapid acquisition rates while minimizing photodamage, enabling detailed visualization of fine neural structures over time. This makes SoRa ideally suited for high-throughput drug screening applications, allowing precise and efficient analysis of neurite outgrowth in physiologically relevant conditions. High-throughput dendritic spine analysis is of increasing interest in drug screens to identify compounds that modulate synaptic plasticity, which is critical for learning, memory, and neurodegenerative diseases. Dendritic spines are small, dynamic structures where synapses form, and their morphology and density can change in response to neurological drugs, making them an ideal target for screening [9, 10]. However, accurately capturing the subtle alterations in spine morphology requires high-resolution imaging. SoRa's super-resolution 3D imaging capabilities allow researchers to visualize and measure fine neuronal processes with unprecedented detail and speed. This application of SoRa offers a powerful approach for drug discovery (*see* Fig. 2).

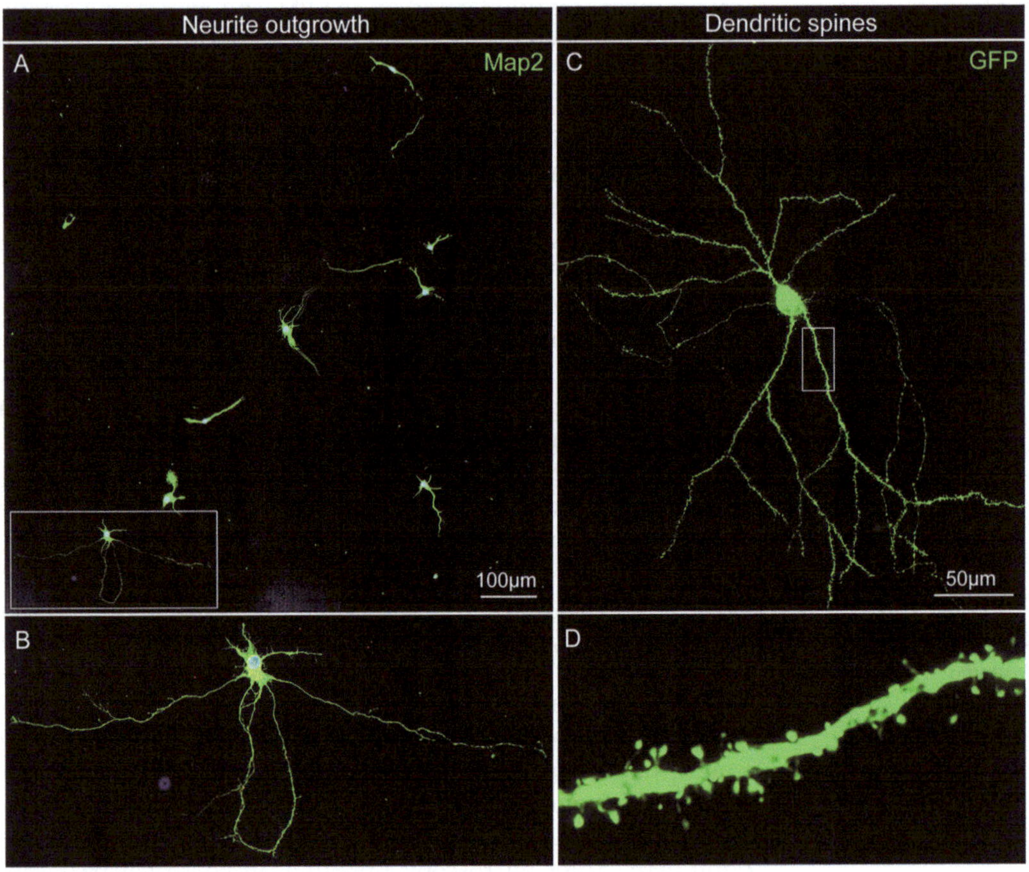

Fig. 2 SoRa images of neurites and dendritic spines. (**a**) Example 3D Z-stack maximum projection image of DIV4 neurons stained with Map2 and imaged using the SoRa automated imaging setup described below. (**b**) Zoom in on neuron. (**c**) Example 3D Z-stack maximum projection image of DIV21 neurons transfected with GFP using the SoRa automated imaging setup described below. (**d**) Zoom on dendrite and spines

1.3 Super-resolution Live-Cell Imaging of Mitochondria

Mitochondria are essential organelles in neurons, serving as the primary source of ATP through oxidative phosphorylation [11–14], while also regulating calcium buffering [13, 15–17], redox signaling [17, 18] and apoptosis [19, 20]. In neurons, which are highly polarized and energetically demanding cells, mitochondrial function is particularly critical to sustain synaptic activity [11–14], and plasticity [21]. Mitochondria in neurons are highly dynamic, undergoing constant fission, fusion, biogenesis, and degradation, collectively termed mitochondrial dynamics [22, 23]. These processes enable mitochondria to adapt to local metabolic needs, segregate damaged components, and maintain network integrity. Additionally, mitochondrial transport along microtubules is vital for delivering these organelles to energy-demanding regions such as growth cones and synaptic terminals [24, 25]. Disruption in mitochondrial dynamics or function is closely associated with neurodegenerative disorders, including Parkinson's disease [25, 26], Alzheimer's disease [27, 28], and hereditary spastic paraplegia

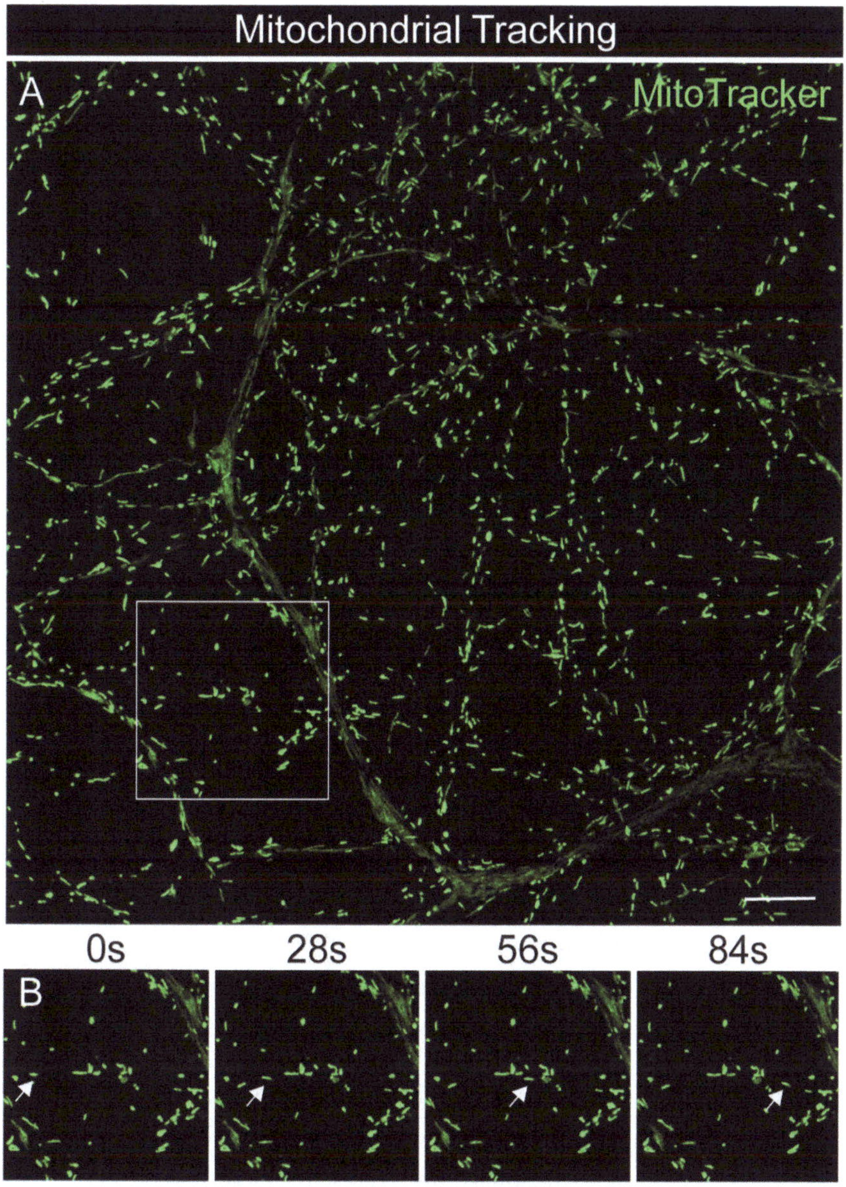

Fig. 3 SoRa images of mitochondrial trafficking. (**a**) Example image of DIV18 neurons stained with MitoTracker and imaged using the SoRa live-cell imaging setup described below. (**b**) Time-lapse images showing tracking of mitochondria (arrow)

[14], highlighting the importance of studying mitochondrial behavior in live neuronal systems.

Live-cell mitochondrial imaging provides critical insights into the dynamics, morphology, and function of mitochondria in neurons. Using high-resolution spinning-disk microscopy equipped with a SoRa unit enables enhanced lateral resolution while maintaining minimal phototoxicity, making it particularly suitable for delicate neuronal preparations [1] (*see* Fig. 3).

2 Materials

2.1 Primary Hippocampal Neuronal Culture

1. Primary hippocampal neurons isolated from C57BL/6J embryonic day (E) 16 mice (*see* **Note 1**).
2. Cell culture carbon dioxide incubator.
3. 10 mL syringes.
4. 0.22 μm polyethersulfone syringe filters.
5. Glass coating buffer: 1 mg/mL poly-L-lysine hydrobromide (PLL) in 0.1 M borate buffer (pH 8.4). 0.1 M borate buffer is made up of 0.2 M sodium tetraborate and adjusted to a pH of 8.4 with 0.2 M boric acid, topped up 1:1 with milli-Q water. Can be mixed up in advance and stored at 4 °C. Check pH is at 8.4 before use and filter immediately before use using the 10 mL syringe and 0.22 μm filter.
6. Plating neurobasal medium: 1% Glutamax (Gmax), 2% B27, 50 U/mL penicillin/streptomycin (pen/strep), 5% heat-inactivated fetal bovine serum (FBS) in neurobasal medium (*see* **Note 2**). Filter immediately before use using the 10 mL syringe and 0.22 μm filter.
7. Culturing neurobasal medium: 1% Gmax, 2% B27, 50 U/mL pen/strep, in neurobasal medium (*see* **Note 2**). Filter immediately before use using the 10 mL syringe and 0.22 μm filter.
8. Imaging medium: DMEM/F-12, HEPES, no phenol red.
9. 96-well high content screening microplates with glass bottom.
10. 29 mm Glass bottom dish with 10 mm micro-well #1.5 cover glass.
11. Reagent reservoirs (recommended).
12. Multichannel pipette (recommended).

2.2 Buffers, Media, and Antibodies

1. Compounds of choice (*see* **Note 3**).
2. 16% paraformaldehyde (PFA) (*see* **Note 4**).
3. 1× phosphate-buffered saline (PBS).
4. Permeabilizing solution: 0.1% Triton X-100 in PBS.
5. Blocking solution: 5% bovine serum albumin (BSA) in PBS.
6. Antibodies of choice (e.g., guinea pig anti-MAP2, mouse anti-beta actin, goat anti-guinea pig Alexa Fluor 488, goat anti-mouse Alexa Fluor 561).
7. Plasmids (e.g., eGFP).
8. Lipofectamine® 2000.
9. ProLong gold anti-fade mounting media (optional).
10. MitoTracker™ Red CMXRos.

2.3 Microscope

1. Objectives: UPLXAPO10X (NA 0.4, WD 3.1 mm), UPLXAPO20X (NA 0.8, WD 0.6 mm), UPLXAPO60X (NA 1.42, WD 0.15 mm), and UPLXAPO100X (NA 1.45, WD 0.13 mm).
2. Lasers: Coherent OBIS 488 nm LS (150 mW), Coherent OBIS 561 nm LS (150 mW).
3. Spinning-disk confocal microscope (SpinSR10; Olympus) built around an Olympus IX3 body with a SoRa disk (*see* Fig. 4).
4. Two ORCA-Fusion BT sCMOS cameras.
5. Metal microtiter plate holder.

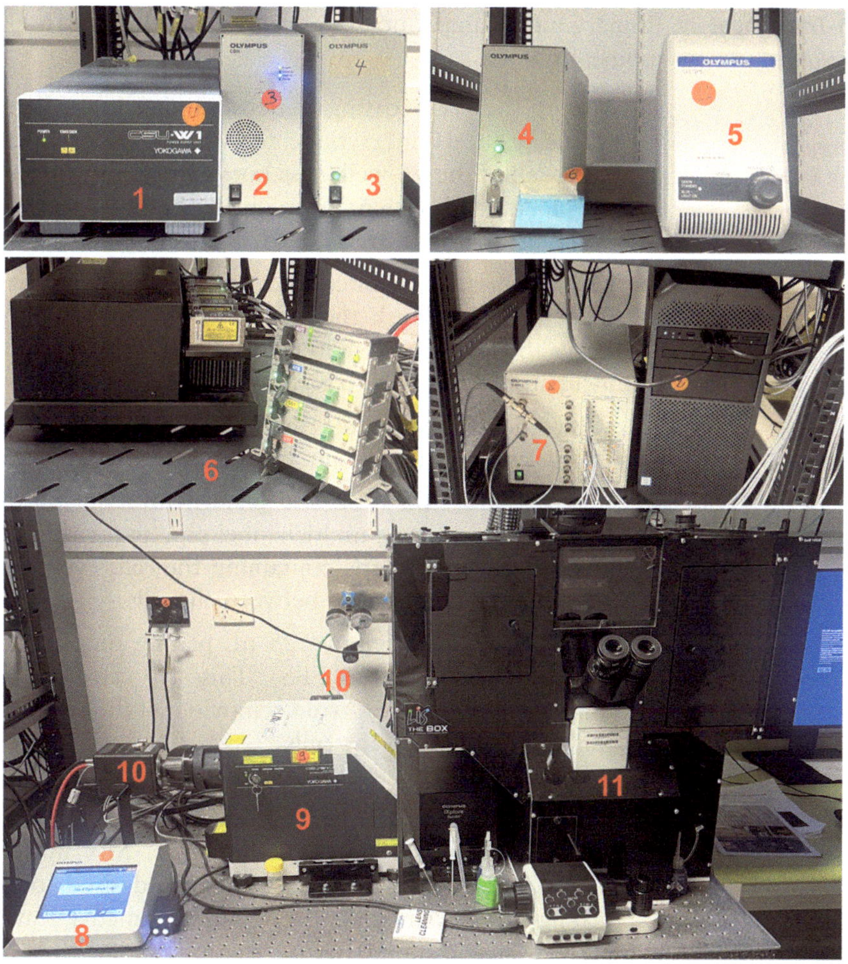

Fig. 4 SoRa spinning-disk microscope setup. (**a**) (1) Yokogawa CSU-W1 power supply and control unit, (2) Olympus CBH control box, (3) Olympus power supply. (**b**) (4) Olympus power supply 2, (5) Olympus U-LGPS epifluorescence. (**c**) (6) OBIS coherent lasers. (**d**) (7) Olympus U-RTCE real-time controller. (**e**) (8) Olympus LCD touch panel screen controller, (9) Yokogawa CSU-W1 high-speed spinning-disk confocal unit, (10) ORCA-Fusion BT sCMOS cameras (11) Olympus IX3 inverted microscope

6. Olympus cellSens software.
7. Fluorescence immersion oil.
8. Environmental chamber set at 37 °C, 5% CO_2.

2.4 Analysis Software

1. Python (v3.11) (https://www.python.org/downloads/).
2. PyCharm (2025.25) (https://www.jetbrains.com/pycharm/).
3. Python-based mitochondria motility workflow (GitHub: https://github.com/momo19931988/MitoDiffTrack/).

3 Methods

3.1 High-Throughput Super-resolution Imaging

3.1.1 Primary Hippocampal Neuron Culture and Treatment

Primary hippocampal neurons are obtained from E16 C57BL/6J mouse embryos.

1. Mix up the glass coating buffer and filter immediately before use using the 10 mL syringe and 0.22 μm filter. Add 50 μL to each well of the 96-well glass bottom plate using a multichannel pipette and reagent reservoir (*see* **Note 5**). Leave the plate in the incubator at 37 °C, 5% CO_2 overnight.
2. The following day, filter PBS using the 10 mL syringe and 0.22 μm filter. Wash the wells 3× with freshly filtered PBS. Leave the last PBS wash in the well and return the plate to the incubator until ready to plate.
3. Filter the plating media immediately before use using the 10 mL syringe and 0.22 μm filter and add the cells (*see* **Note 1**).
4. Remove final PBS wash and plate neurons in 50 μL plating media. Do not use a multichannel pipette for this step. While plating, invert the tube containing the cells every ~5 wells to ensure consistent density between all wells.
5. Three hours later remove 25 μL plating media add 125 μL culture media (without FBS). This step can be done with a multichannel pipette. Avoid removing all the plating media as this will dry and kill the cells.
6. 3–20 days later mix up the compounds in fresh culture media (*see* **Note 3**). 50 μL will be added to each well and each compound will be tested in duplicate. Mix up 100 μL of media and add compounds at twice the desired concentration (20 μg/mL). Remove 100 μL of plating media, leaving 50 μL in the well, and add 50 μL of the compound mixture to give the final desired concentration (10 μg/mL). Return the plate to the incubator.
7. 24 h later mix up ~5 mL 8% PFA in PBS (*see* **Note 4**). Remove 50 μL media, leaving 50 μL media and immediately add 50 μL

8% PFA using a multichannel pipette and reagent reservoir (final concentration of 4% PFA). Fix the cells for 20 min at room temperature. Wash the cells 3× with PBS.

3.1.2 Immunocytochemistry Neurite Outgrowth Assay

1. Permeabilize cells with 0.01% Triton X-100 in PBS for 5 min (*see* **Note 6**) and wash 3× in PBS.
2. Block cells with 5% BSA in PBS for 20 min at room temperature, then add primary antibody at the appropriate concentration (e.g., MAP2 1:2000, beta Actin 1:1000) in 5% BSA in PBS overnight at 4 °C or room temperature.
3. Wash cells 3× in PBS and incubate with secondary antibody at the appropriate concentration (e.g., Alexa Fluor 488 and 561 1:2000) for 2 h at 4 °C or room temperature.
4. Wash cells 3× in PBS. Either image in PBS or consider using mounting media such as ProLong Gold Anti-Fade.

3.1.3 Transfection Dendrite and Spine Outgrowth Assay

1. Transfect primary neurons (with eGFP to visualize dendrites and spines) at DIV 14 using Lipofectamine® 2000, as per manufacturer's instructions.

3.1.4 Microscopy

1. Start cellSens and select all options (*see* **Note 7**). Open the following windows: *Camera Control*, *Well Navigator*, *Stage Navigator*, *Live Imaging Window*, *Microscope Control*, *Process Manager*, and *Adjust Display*.
2. Place the 96-well plate on the metal microtiter plate holder, ensuring it is completely flat (*see* **Note 8**).
3. Using the *Well Navigator*, load the 96-well plate template. First focus on a well of interest using the 10× objective and brightfield with 20 ms exposure, then start calibrating the dish. Calibration will start with well A1, and ensure the plate has moved to the correct well. Move the image to align the central target to each of the calibration spots flashing in green. Move to the next well, A12, repeat the 3-point calibration, and move to the final calibration well, H12, and repeat calibration (*see* **Note 9**).
4. Using the *Define Pattern* function in the *Well Navigator*, select *Full Well In* to image the entire well as a tile-image or image regions of interest (ROIs) (*see* **Note 10**).
5. Using the *Apply Pattern* function in the *Well Navigator*, select the wells you wish to image in the order you wish to image them.
6. For *dendrite and spine outgrowth assay*: if you wish to image specific ROIs of the plate, consider using the *Stage Navigator* function. First take an overview image of the plate using the 10× objective and your channel of interest. This can take

~30 min. Then, using the *Rectangle Scan Area with Mouse* or *Polygon Scan Area with Mouse* function in the *Stage Navigator*, manually select your ROIs from each well, going through the overview image to identify your GFP-expressing cells of interest.

7. Swap to the desired objective and add in the magnifier if required. 20× with 1× magnification is sufficient for confocal high-throughput imaging. 60× oil objective with 3.2× magnification is recommended for super-resolution imaging (*see* **Note 11**).
8. Select the *Multichannel* function in the *Process Manager*, add the appropriate channels (e.g., SoRa488 and SoRa561) and set the exposure time for each channel individually by going into live mode and assessing the maximum intensity using *auto contrast*.
9. Select the *Multi-Position* function in the *Process Manager*. Set up the *Focus Map* using the three-point function in your brightest channel (*see* **Note 12**).
10. Select the *Z-stack* function if wanting to perform 3D imaging and set up the Z-stack using the *Range* function (*see* **Note 13**). Set the range to include your whole sample (for neurons we recommend 5–10 μm and apply the recommended step size).
11. In *Acquisition Settings*, set up *Naming—Process* and *Saving—Process*. Ensure saving is directed to a drive with sufficient storage as acquisition will not work if there is not enough space on the drive.
12. Final checks before starting acquisition: Ensure that the SoRa disk is selected, the correct magnification is selected for each channel, the laser power is below 20% for each channel to avoid bleaching and the binning is at the highest resolution (2304 × 2304).

3.2 Mitochondrial Dynamics: Live Imaging in Primary Neurons

Primary hippocampal neurons are prepared as described above; however, for this assay cells are plated in 29 mm/10 mm glass-bottom imaging dishes (*see* **Note 1**). Primary hippocampal neurons are ready for live imaging at DIV18–21 when they have well-formed axons and dendrites.

3.2.1 Mitochondrial Labelling

1. Prepare MitoTracker Dye for mitochondrial staining by diluting MitoTracker stock solution (1 mM in DMSO) to 200 nM in pre-warmed imaging medium and protect from light (*see* **Note 14**).
2. Remove the culture medium from each well and gently add the MitoTracker working solution. Incubate the cells at 37 °C for 20–30 min to allow dye uptake.

3. Wash the neurons 2–3 times with pre-warmed imaging medium to remove excess dye.
4. Set up the SoRa spinning-disk confocal microscope and environmental chamber for live-cell imaging. Maintain the chamber at 37 °C and 5% CO_2 during imaging.

3.2.2 Microscopy

1. Start cellSens and open the following windows: *Camera Control*, *Stage Navigator*, *Live Imaging Window*, *Microscope Control*, *Process Manager*, and *Adjust Display*.
2. Using the *Stage Navigator* function, first take an overview image of the dish using the 10× objective and your channel of interest. Then, using the *Rectangle Scan Area with Mouse* or *Polygon Scan Area with Mouse* function in the *Stage Navigator*, select your ROIs going through the overview image to identify your cells of interest.
3. Swap to the 100× oil objective with 3.2× magnification for live mitochondrial imaging.
4. Select the *Multichannel* function in the *Process Manager*, add the appropriate channel (MitoTracker Red CMXRos: 561 nm), and set the exposure time (100–300 ms) (*see* **Note 15**).
5. Select the *Z-stack* function in the *Process Manager* to perform 3D imaging. Set up the Z-stack using the *Range* function and set the step size to 0.3 μm.
6. Select the *Timelapse* function in the *Process Manager* to perform live imaging. Set the frame interval to 10 s and cycles to >40.
7. Final checks before starting acquisition: Ensure that the SoRa disk is selected, the correct magnification is selected for each channel, the laser power is below 20% for each channel to avoid bleaching (*see* **Note 16**), and the binning is at the highest resolution (2304 × 2304).

3.2.3 Analysis

1. Run MitoDiffTrack_V01.py from an IDE (such as PyCharm) by clicking *Run*, or in the IDE's integrated terminal using "*python MitoDiffTrack_V01.py*," which will launch the full workflow with all panels automatically loaded.
2. In the *load_czi* panel, select a.czi or.vsi file using the *Load CZI/VSI* button. The movie will appear as a new layer.
3. In the *calibration* panel, set *Pixel size (μm/px)* and *Frame interval (s)*, then press *Apply* (*see* **Note 17**).
4. In the *seg_params* panel, choose the Detection channel, adjust preprocessing parameters such as *BG sigma* (removes large-scale background), *Gaussian sigma* (smooths noise before thresholding), and *CLAHE clip* (enhances local contrast),

select a thresholding method, and set the Minimum and Maximum area to filter out unrealistically small or large objects (*see* **Note 18**).

5. In the *process_export (seg)* panel, set the segmentation *Channel* and time range, then press *Run detection on all frames.*
6. In the *track_mito* panel, configure *Max disp (px/frame)* (the maximum allowed displacement for normal movement), *Fast disp (px/frame)* (an upper bound for unusually rapid jumps), *Fast penalty factor* (adds extra cost to fast jumps to reduce false associations), *Max area ratio*, and *Max intensity ratio* (reject matches if object size or brightness changes implausibly between frames) (*see* **Note 19**). Press *Run LAP tracking from detections.*
7. In the *process_export (CSV)* panel, set an *Output prefix* and select export options such as tables or a snapshot. Press *Export CSV/snapshot.*
8. (Optional) Perform motion-based analysis in the *flow_motion_-filaments* and *fast_tracks_skeleton* panels by adjusting *Channel* (the imaging channel used for motion estimation), *T start* and *T end* (the start and end frames) (*see* **Note 20**).
9. (Optional): Save or load GUI settings using the *Save GUI config* and *Load GUI config* buttons.

4 Notes

1. For the *neurite outgrowth assay* plate ~300 neurons per well in a 96-well glass-bottom plate. It is ideal to have no overlapping processes so be sure to plate the cells drop-wise around the well to avoid clumping in the center. For the *dendrite and spine outgrowth assay* plate ~10,000 neurons per well in a 96-well glass-bottom plate. For the *mitochondrial dynamics live imaging,* plate ~45,000 neurons per 10 mm/29 mm glass-bottom dish.
2. Mix up plating media and culture media the day before, filter and place in the incubator overnight to equilibrate.
3. For the *neurite outgrowth assay* add compounds on day in vitro (DIV) 4. For the *dendrite and spine outgrowth assay* add compounds on DIV 20.
4. All steps using PFA should be conducted in a fume hood with proper PPE and all waste removed into appropriate chemical waste containers.
5. Please note that on some microscopes, automated imaging is only possible for the inner 60 wells as the metal plate holder hits the objective when imaging the very outer wells. Consider only using the inner 60 wells.

6. Permeabilizing for too long will result in loss of intracellular proteins and cell lysis.
7. CellSens will give you the option to execute stage cleaning. If you are having issues with long automated acquisitions of multi-well dishes, you may need to execute stage cleaning before you start.
8. The 96-well plates sit significantly higher than other plates, dishes or slides. If you are having trouble focusing on the dish, you may need to extend the Z-stage range in *Calibration* to the maximum range.
9. If there are issues with the objective hitting the metal plate holder, calibrate at the following wells: B2, B11 and G11.
10. *Full Well In* is best suited for low-resolution imaging of whole cells. 20× magnification with 1 channel acquisition in 2D will take ~1 h to acquire with this function. Using these parameters with the 60× objective and 3.2× magnifier will take substantially longer; therefore, consider using *Part Well In* to set up ROIs instead.
11. 100× objective reduces the signal significantly and is only recommended for very bright samples. The 60× and 100× objectives are oil objectives and require immersion. It is safe to remove the plate after calibration and to add the immersion. Oil will dry up on the objective, so it can help to add oil to the underside of the plate. You may need to pause the acquisition to add oil if you find it is drying up.
12. It is important to set up the *Focus Map* using the objective you plan to use for your imaging as the *Focus Map* for 20× and 60× will differ.
13. It is important to use the *Range* setting for 3D imaging of an entire multi-well plate as this will use the *Focus Map* to determine the center point of each Z-stack.
14. Avoid over-staining with MitoTracker as longer incubations or high concentrations can lead to mitochondrial dysfunction.
15. If the mitochondrial signal appears weak, either increase the MitoTracker concentration to improve labelling efficiency (while avoiding mitochondrial dysfunction) or increase the exposure time (while avoiding phototoxicity).
16. If photobleaching is observed, reduce the laser power or the frame rate (while preserving sufficient temporal resolution).
17. Correct pixel size and frame interval are essential, because all velocities are reported in µm per second.
18. As general guidance, *BG sigma* values around 8, *Gaussian sigma* values around 1.0, and *CLAHE clip* values around

0.015 work well for most mitochondrial datasets; users may fine-tune these values depending on image noise, contrast, and mitochondrial density.

19. In practice, values such as *Max disp* = *5 px*, *Fast disp* = *5–10 px*, *Fast penalty factor* = *5*, *Max area ratio* = *3*, and *Max intensity ratio* = *3* work well for typical mitochondrial datasets and can be increased slightly for recordings with fast-moving or low-contrast mitochondria.
20. Motion-based analysis uses pixel-level displacement between consecutive frames to highlight regions where mitochondria exhibit faster movement, allowing rapid trajectories to be extracted based on changes in pixel positions rather than intensity alone.

References

1. Azuma T, Key T (2015) Super-resolution spinning-disk confocal microscopy using optical photon reassignment. Opt Express 23: 15003–15011. https://doi.org/10.1364/OE.23.015003
2. Verth F, Fairn GD (2023) Super-resolution spinning-disk confocal microscopy using optical photon reassignment (SoRa) to visualize the actin cytoskeleton in macrophages. In: Botelho RJ (ed) Phagocytosis and phagosomes. Methods in molecular biology, vol 2692. Humana, New York, pp 79–90
3. Zaza C, Joseph MD, Dalby OPL et al (2025) Super-resolution imaging in whole cells and tissues via DNA-PAINT on a spinning disk confocal with optical photon reassignment. Nat Commun 16:4991. https://doi.org/10.1038/s41467-025-60263-w
4. Paddock SW, Eliceirin KW (2014) Laser scanning confocal microscopy: history, applications, and related optical sectioning techniques. In: Paddock S (ed) Confocal microscopy. Methods in molecular biology, vol 1075. Humana Press, New York, pp 9–47
5. Conchello JA, Lichtman JW (2005) Optical sectioning microscopy. Nat Methods 2:920–931. https://doi.org/10.1038/nmeth815
6. Oreopoulos J, Berman R, Browne M (2014) Spinning-disk confocal microscopy: present technology and future trends. Methods Cell Biol 123:153–175
7. Huff J (2015) The Airyscan detector from ZEISS: confocal imaging with improved signal-to-noise ratio and super-resolution. Nat Methods 12:i–ii. https://doi.org/10.1038/nmeth.f.388
8. Martínez-Mármol R, Chai Y, Conroy JN et al (2023) Hericerin derivatives activates a pan-neurotrophic pathway in central hippocampal neurons converging to ERK1/2 signaling enhancing spatial memory. J Neurochem 165:791–808. https://doi.org/10.1111/jnc.15767
9. Hering H, Sheng M (2001) Dendritic spines: structure, dynamics and regulation. Nat Rev Neurosci 2:880–888. https://doi.org/10.1038/35104061
10. Fiala JC, Spacek J, Harris KM (2002) Dendritic spine pathology: cause or consequence of neurological disorders? Brain Res Brain Res Rev 39:29–54. https://doi.org/10.1016/S0165-0173(02)00158-3
11. Nicholls DG, Budd SL (2000) Mitochondria and neuronal survival. Physiol Rev 80:315–360. https://doi.org/10.1152/physrev.2000.80.1.315
12. Mandal A, Drerup CM (2019) Axonal transport and mitochondrial function in neurons. Front Cell Neurosci 13:373. https://doi.org/10.3389/fncel.2019.00373
13. Misgeld T, Schwarz TL (2017) Mitostasis in neurons: maintaining mitochondria in an extended cellular architecture. Neuron 96: 651–666. https://doi.org/10.1016/j.neuron.2017.09.055
14. Saber SH, Yak N, Yong XLH et al (2025) DDHD2 provides a flux of saturated fatty acids for neuronal energy and function. Nat Metab. https://doi.org/10.1038/s42255-025-01367-x

15. Nicholls DG (2005) Mitochondria and calcium signaling. Cell Calcium 38:311–317. https://doi.org/10.1016/j.ceca.2005.06.011
16. Parekh AB (2003) Mitochondrial regulation of intracellular Ca2+ signaling: more than just simple Ca2+ buffers. News Physiol Sci 18: 252–256. https://doi.org/10.1152/nips.01458.2003
17. Bygrave FL (1978) Mitochondria and the control of intracellular calcium. Biol Rev Camb Philos Soc 53:43–79. https://doi.org/10.1111/j.1469-185x.1978.tb00992.x
18. Horn A, Van der Meulen JH, Defour A et al (2017) Mitochondrial redox signaling enables repair of injured skeletal muscle cells. Sci Signal 10:eaaj1978. https://doi.org/10.1126/scisignal.aaj1978
19. Wang C, Youle RJ (2009) The role of mitochondria in apoptosis. Annu Rev Genet 43:95–118. https://doi.org/10.1146/annurev-genet-102108-134850
20. Wang X (2001) The expanding role of mitochondria in apoptosis. Genes Dev 15:2922–2933
21. Mattson MP (2007) Mitochondrial regulation of neuronal plasticity. Neurochem Res 32:707–715. https://doi.org/10.1007/s11064-006-9170-3
22. Bertholet AM, Delerue T, Millet AM et al (2016) Mitochondrial fusion/fission dynamics in neurodegeneration and neuronal plasticity. Neurobiol Dis 90:3–19. https://doi.org/10.1016/j.nbd.2015.10.011
23. Dorn GW (2019) Evolving concepts of mitochondrial dynamics. Annu Rev Physiol 81:1–17. https://doi.org/10.1146/annurev-physiol-020518-114358
24. Melkov A, Abdu U (2018) Regulation of long-distance transport of mitochondria along microtubules. Cell Mol Life Sci 75:163–176. https://doi.org/10.1007/s00018-017-2590-1
25. Cheng XT, Sheng ZH (2021) Developmental regulation of microtubule-based trafficking and anchoring of axonal mitochondria in health and diseases. Dev Neurobiol 81:284–299. https://doi.org/10.1002/dneu.22748
26. Büeler H (2009) Impaired mitochondrial dynamics and function in the pathogenesis of Parkinson's disease. Exp Neurol 218:235–246. https://doi.org/10.1016/j.expneurol.2009.03.006
27. Bonda DJ, Wang X, Perry G et al (2010) Mitochondrial dynamics in Alzheimer's disease: opportunities for future treatment strategies. Drugs Aging 27:181–192. https://doi.org/10.2165/11532140-000000000-00000
28. Su B, Wang X, Bonda D et al (2010) Abnormal mitochondrial dynamics – a novel therapeutic target for Alzheimer's disease? Mol Neurobiol 41:87–96. https://doi.org/10.1007/s12035-009-8095-7

Chapter 4

Improving STED-Based Super-resolution-Imaging with Fluorescence Lifetimes

Vasileios Moysidis, Marina Mikhaylova, and Daniela Hacker

Abstract

The neuronal cytoskeleton and synaptic organization play crucial roles in neuronal information processing. The cytoskeleton is composed of individual filaments that differ in their posttranslational modifications, which often reflect differences in stability or dynamics. Similarly, chemical synapses consist of multiple functionally specialized compartments within which proteins are organized into nanodomains. Understanding the spatial organization of these components requires imaging techniques with sufficient resolution to visualize multiple marker proteins simultaneously and in z-stacks that can be used to display the three-dimensional organization. Super-resolution methods such as Stimulated Emission Depletion (STED) microscopy are essential to achieve this purpose. STED overcomes the diffraction-limited resolution of conventional fluorescence microscopy by narrowing down the area of detected fluorescence. As an extension of that, tauSTED uses the information of fluorescence lifetime to further enhance lateral, axial, and temporal resolution, thus offering several advantages for biological imaging. These advanced microscopy methods, however, require sample preparation with additional considerations compared to regular fluorescent microscopy. Thorough planning is critical to successfully acquire high-quality super-resolution images using STED microscopy. In this chapter, we place a particular focus on tauSTED nanoscopy. We address key aspects, including sample preparation, understanding the underlying theoretical principles and finally the selection of suitable imaging parameters.

Key words tauSTED, STED, FLIM, Cytoskeleton, Synapses, Dissociated cultures, Hippocampus

1 Introduction

1.1 Biological Discoveries Using Super-resolution Microscopy

To study the morphology of cells, microscopy has become the most important tool at the disposal of cellular scientists. However, since the development of Golgi-stainings, which outline the basic morphology of unselectively stained neurons [1], the demands for labelling specific sub-cellular structures and obtaining images with a higher resolution have increased. Many subcellular structures are only a few nanometers long, making our understanding of their organization hinge on the obtainable resolution beyond the diffraction limitation.

Frédéric A. Meunier (ed.), *Super-Resolution Microscopy: Methods and Protocols*, Methods in Molecular Biology, vol. 3034, https://doi.org/10.1007/978-1-0716-5268-8_4,

Historically, super-resolution microscopy has often focused on cell lines, since these cultures are generally the least complex and easiest to culture. A prominent structure within cells are the nuclear pore complexes, which are highly symmetrical and distances between individual components have been determined, e.g., by cryo-electron microscopy [2], making them the perfect point of comparison for new developments in the field of super-resolution microscopy.

Another interesting target for super-resolution microscopy are organelles, which thus far have been difficult to resolve with a lot of detail. They are typically several hundred nanometers to only a few micrometers long and can contain specific internal structures such as mitochondrial cristae. They have been challenging to resolve using fluorescent microscopy, which was made possible to easily resolve through super-resolution microscopy [3, 4], revealing differences in their structures during fission [3].

For neurons, one of the most prominent features is their distinct organization into a soma (~20 μm diameter, for typical pyramidal neurons in rodents [5]), a dendritic arbor and a single axon separated from the soma by the axon-initial segment. This structural specialization requires corresponding adaptations of the cytoskeleton, which ensure that the post-mitotic neurons are both stable and adaptable to, e.g., plasticity-induced changes during learning and memory. Thus unsurprisingly, the cytoskeleton has received a considerable amount of attention, as super-resolution microscopy allows for the resolution of individual filaments. For instance, the periodic organization of the actin-spectrin ring-structure [6] could only be determined using super-resolution microscopy as the distances between individual spectrin filaments are around 180–190 nm and thus slightly below the diffraction limit. Its presence has since been confirmed for many different cell types [7] and subcellular structures such as axons [6], dendrites [8], and dendritic spine necks [9].

The neuronal microtubule cytoskeleton is particularly difficult to resolve, due to the dense accumulation of microtubule filaments (Fig. 1). Single microtubules have an outer diameter of 25 nm and are composed of only 4 nm thick protofilaments [10]. These filaments are often organized into bundles with 20–70 nm distance between individual filaments [11–13] and their stability or dynamicity can be identified by visualizing their posttranslational modifications [14]. STED microscopy has helped to understand that stable microtubules are enriched in the core of the dendritic shaft while dynamic microtubules are found closer to the membrane and the number of filaments belonging to either group could be identified [11].

The most notable subcellular part of neurons are chemical synapses which are often located on dendritic protrusions called spines. Due to their critical functional role and small size of

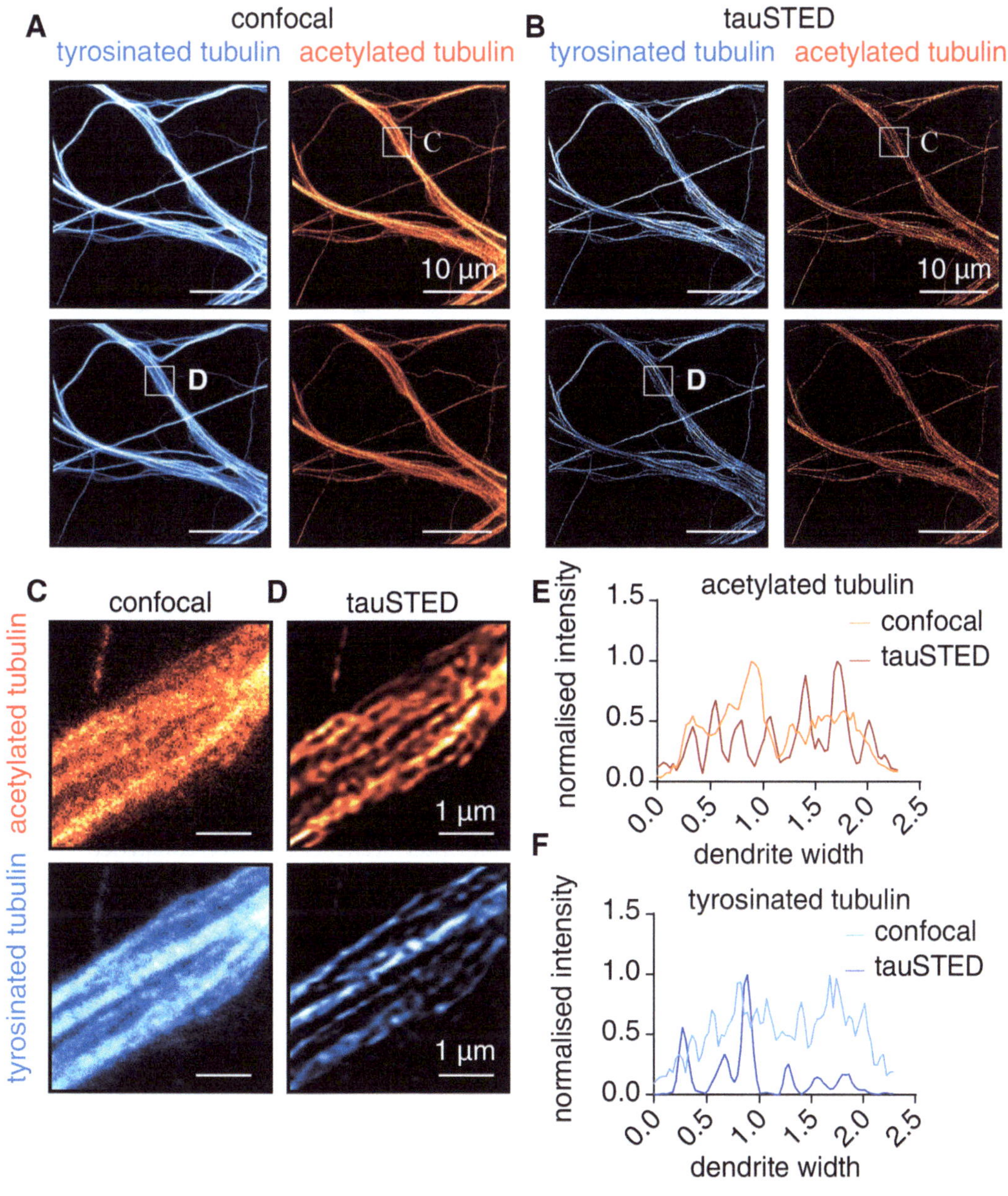

Fig. 1 2D confocal and tauSTED stack of stable (acetylated) and dynamic (tyrosinated) microtubule populations of dissociated hippocampal neurons. (**a** and **b**) Confocal (**a**) and tauSTED (**b**) images of tyrosinated (cyan) and acetylated (red) microtubule populations in dendrites, depicted are single planes of 2D-STED z-stacks. Scale bar 10 μm. (**c** and **d**) Zoom-ins of confocal and tauSTED images shown in (**a**) and (**b**), respectively, indicated by letters next to regions of interest (ROIs). Scale bar 1 μm. (**e** and **f**) Line profiles of normalized fluorescent intensities of tyrosinated and acetylated tubulin in dendrite cross-sections shown in (**c**) and (**d**), respectively

0.5–2 μm, these structures have been of large interest in super-resolution microscopy [15–18]. In the recent years, it has become clear that even within these small subcellular structures, proteins organize into nanodomains [19, 20] and determining the position of a protein of interest in a specific nanodomain can give information on its involvement in certain pathways. The field of super-resolution microscopy thus aims to steadily improve the axial as well as lateral resolution with which we can observe the cytoskeleton and synaptic organization of neurons. One method that has been especially successful in uncovering the subcellular organization of neurons is stimulated emission depletion (STED) microscopy (Fig. 2).

To obtain the ideal samples for super-resolution microscopy, a variety of details need to be considered for cell preparation, fixation, antibody selection, and microscopy settings (summarized in Fig. 3).

In the following sections, we will first explore the fundamental physical principles and key advantages of STED microscopy and its extension tauSTED (also denoted τSTED).

1.2 The Diffraction Barrier

The spatial resolution of conventional fluorescence microscopy is limited by the diffraction of light. Diffraction is a physical property of light owed to its wave-nature, which means light sourced from a punctual spot "spreads out" when it hits small obstacles or passes through small openings. This can lead to a shift in the phase of light waves, and when these diffracted light waves interact with each other, we speak about interference. Due to constructive- and destructive- interference (*see* **Note 1**), patterns are formed, called diffraction or interference patterns [21].

In the case of microscopes and punctual light sources we want to observe, this behavior of light creates spatially extended diffraction patterns at the image plane with a peak intensity in the center and concentric dark and bright rings around it [21]. These patterns are called Airy discs [21]. In microscopy we mostly refer to them as the point spread function (PSF). The PSF also considers all other imperfections, like optical aberrations that can affect the diffraction pattern.

The theoretical maximum resolution of our system, the minimum distance between two points (e.g., fluorophores or in general punctual light sources) that can still be resolved, is given by the Rayleigh Criterion. Discrimination of two points is possible when the central maximum peak of the Airy disc produced by one point lies on the first minimum of the Airy disc produced by the second point [21].

$$d = \frac{1.22\lambda}{2n\ \sin\alpha} = \frac{1.22\lambda}{2\mathrm{NA}}$$

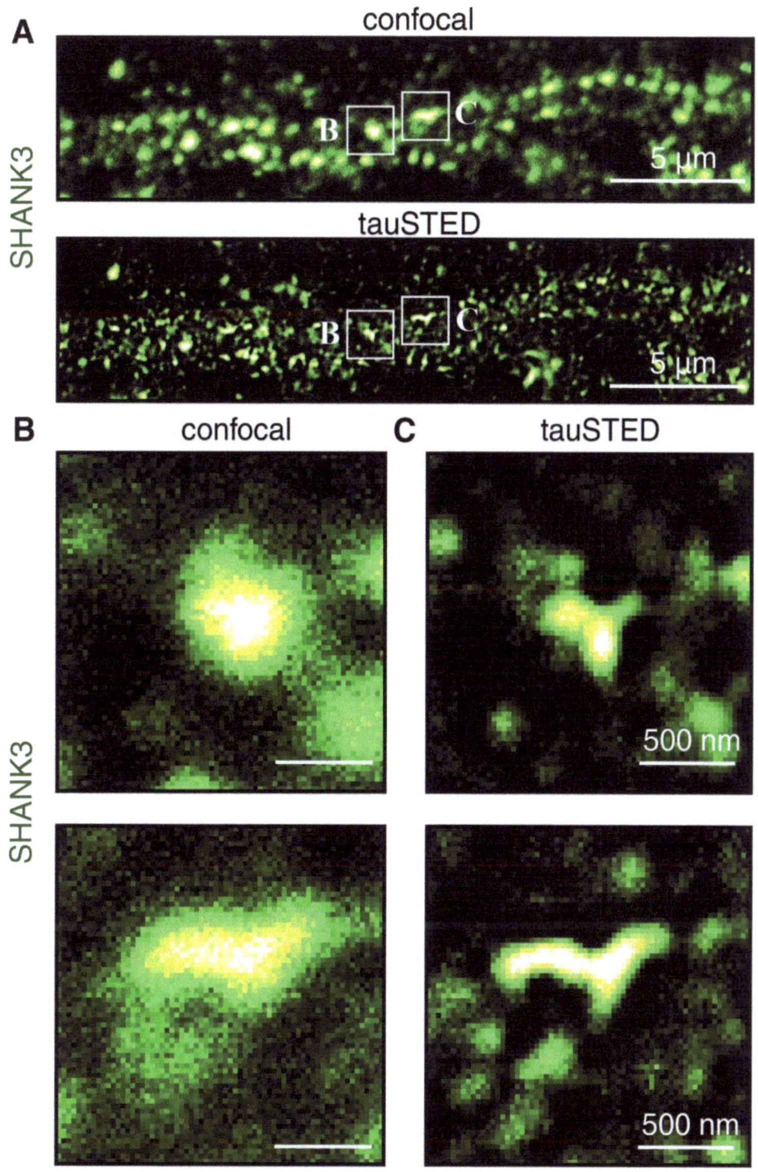

Fig. 2 Confocal and tauSTED images of postsynaptic densities of dissociated hippocampal neurons. (**a**) Confocal (upper panel) and tauSTED (lower panel) images of SHANK3 in dendrites. Scale bar 5 µm. (**b** and **c**) Zoom-ins of confocal (**b**) and tauSTED (**c**) images. Scale bar 500 nm

So, this minimal distance d is determined by the wavelength (λ) of the light and the system's numerical aperture NA (refractive index n of the optical medium times the sine of the half angular aperture α of the objective's lens).

Also applicable in microscopy and very similar is Abbe's diffraction limit [22]:

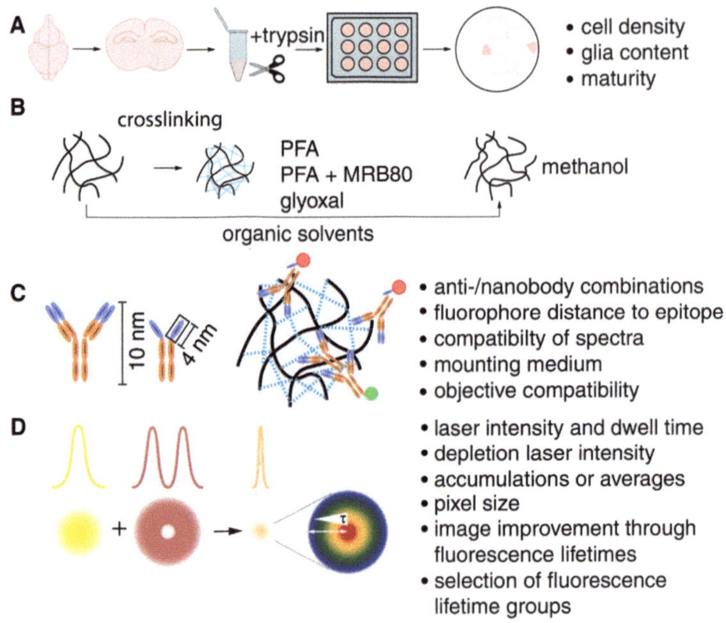

Fig. 3 Steps to consider when obtaining tauSTED microscopy images of hippocampal, dissociated neurons. (**a**) Hippocampal, dissociated culture preparation. (**b**) Selection of suitable fixation methods. (**c**) Immunocytochemistry with antibodies conjugated to fluorescent dyes. (**d**) tauSTED microscopy considerations for imaging settings

$$d = \frac{\lambda}{2n \sin \alpha} = \frac{\lambda}{2\mathrm{NA}}$$

According to this equation, in microscopy, the rule of thumb is established that the best possible resolution is half the wavelength of the light used.

So, the only direct ways to increase the diffraction-limited resolution are, on the one hand, to increase the numerical aperture with high-quality objectives combined with immersion oil of a high refractive index, and on the other hand, to use shorter wavelengths. The last way is how, for instance, cryo-electron microscopy (cryo-EM) achieves resolutions down to the atomic level by using highly accelerated electrons with much shorter wavelengths than those of photons from the visible spectrum (VIS).

Here we are just dealing with the VIS light, since it is more versatile for most biological questions due to multicolor and live cell imaging, for example, faster in sample preparation, and generally more accessible than cryo-EM.

The diffraction limit, though for light of the VIS according to the Rayleigh Criterion, if we, for instance, use yellow light of 580 nm and an NA of 1.4, would just allow us to theoretically resolve two points in a proximity of around $d \approx 250$ nm.

1.3 Fluorescence Microscopy and Its Limits

A common tool in modern cell biology is fluorescence microscopy, which utilizes immunohistochemistry and genetically encoded fluorescent proteins to tag virtually all cellular structures and components of interest with fluorophores. In general, fluorophores are molecules with extended delocalized π-electron systems, able of absorbing light of a specific wavelength in the process of excitation, and spontaneously after some time, releasing a red-shifted (lower energy) photon in a process called emission [23] (*see* **Note 2**) (Fig. 4b). The emission of a red-shifted photon is called Stokes shift [23] and is the foundation of all fluorescence microscopy techniques.

For many applications, the diffraction-limited resolution of around 200–250 nm, that can be obtained by simple epifluorescence microscopy, would be sufficient. Considering however that the width of a microtubule is ~25 nm [24], it is impossible to distinguish single microtubules in the dense cytoskeleton of neurons as mentioned above (Fig. 1). For such studies the transition to diffraction-unlimited super-resolution microscopy methods is the only option.

1.4 Super-resolution Microscopy and the STED Approach

During the past decades, different fluorescence microscopy methods have been developed to overcome the diffraction limit, ranging from widefield methods like Structure Illumination Microscopy [25] (SIM; *see* Chapter 1), stochastic methods like single molecule localization microscopy (SMLM) with PALM [26] and STORM [27] (*see* Chapters 5, 6, and 9) to extensions of confocal microscopy. Examples of these are Stimulated Emission Depletion (STED) [28], gSTED, tauSTED nanoscopy [29] and even the SMLM-variant MINFLUX [30].

The diffraction-unlimited STED microscopy can be seen as an extension of confocal microscopy (*see* **Note 3**). It is based on the same scanning principle, with the main difference being the addition of a depletion (or deexcitation) laser, capable of stimulating the earlier emission of photons identical to the stimulating ones [22, 31] (*see* **Note 4**) (Fig. 4a, b). In simpler terms this is a way of deactivating excited fluorophores.

The physical principle of stimulated emission is used in 2D-STED to create a much narrower effective PSF and, with that, to reach a higher lateral resolution [28]. This is achieved thanks to the "doughnut" shape of the STED laser beam, which is coaxially aligned with the excitation beam, and has a null (or zero) intensity center and high intensity in the periphery (*see* **Note 5**), letting only fluorophores from that sub diffraction small central spot to emit fluorescence (Fig. 4d).

According to the founder of STED, Stefan W. Hell, the resolution is not limited by the diffraction of light anymore but theoretically only determined by the intensity of the STED laser beam

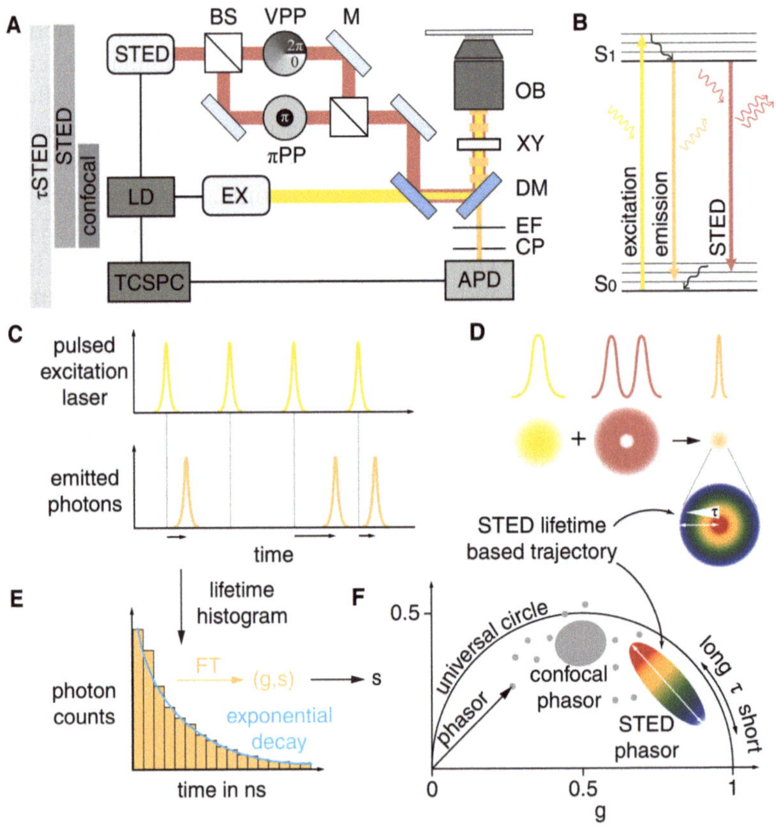

Fig. 4 tauSTED microscopy and underlying principles. (**a**) Confocal microscope setup consisting of excitation laser (EX), laser driver (LD), a scanning unit (XY), confocal pinhole (CP), detector, e.g., avalanche photo diode (APD), objective (OB), dichroic mirrors (DM), and emission filter (EF). Additions for STED microscopes: deexcitation laser (STED), vortex phase plate (VPP), π phase plate (πPP), beam splitters (BS), and mirrors (M). Additions for τSTED microscopes: time correlated single photon counting electronics (TCSPC). (**b**) Jablonski diagram showing excitation of a fluorophore and subsequent emission of light of a longer wavelength. Emission can be suppressed with longer-wavelength light (STED laser). During this depletion process the energy of the excited state is released in form of identical photons to that of the depletion photons, and no fluorescence. (**c**) Pulsed excitation laser with light pulses in equal time intervals (upper panel). Detection time of spontaneously emitted photons after each pulse (lower panel). (**d**) Schematic depicting intensity profiles of the diffraction limited excitation (yellow) and STED (red) laser, resulting in sub-diffraction signal (orange). Different fluorescent lifetimes (τ) within the resulting signal are depicted. (**e**) Lifetime histogram showing photon-distribution according to detector-arrival time after excitation. Approximation of fluorescence lifetime by fitting an exponential decay function to the histogram for each pixel. Fit-free alternative approach with computation (FT) of phasor coordinates (g,s) for each histogram (or pixel). (**f**) Phasor plot displaying lifetime information from each pixel of the original image in form of points or phasors. Phasors closer to (1,0) represent shorter lifetimes, and closer to (0,0) longer lifetimes. Conventional confocal microscopy leads to uniform distribution of phasors around a small central spot (confocal phasor). The STED effect becomes visible in form of a lifetime-based trajectory (STED phasor)

[28]. This is reflected in the following formula as an extension of Abbe's diffraction limit formula:

$$d \approx \frac{\lambda}{2n\,\sin\alpha\sqrt{\frac{1+I}{I_s}}} = \frac{\lambda}{2\mathrm{NA}\sqrt{\frac{1+I}{I_s}}}$$

I is the intensity of the STED laser and I_s is the saturation intensity of the fluorophore, thus the characteristic intensity at which half of the fluorescence is depleted. Clearly, with increasing STED intensity, the root term tends toward infinity, resulting in d approaching zero.

1.5 The Effect of STED on Fluorescence Lifetime

The conventional STED approach for increasing the spatial resolution is based solely on the information coming from the fluorescence intensity of fluorophores originating from a subdiffractional confined spot as described above. The phenomenon of fluorescence though, is not only characterized by wavelength and intensity. A very important property and information source is the fluorescence lifetime (τ), which is characteristic for each fluorophore [23]. The fluorescence lifetime of a fluorophore is the average time that it spends in the excited state before it emits a photon and returns to the ground state, this happens typically in the range of 10 ns [23] (Fig. 4b).

The lifetime of ordinary fluorescence processes is given by the reciprocal of the total fluorescence decay rate k_{tot}, which is composed of the rate for the radiative decay k_r, and the rate for the non-radiative decay k_{nr}, [29, 32], thus fluorescent photon emission and non-radiative energy release as heat, respectively.

$$\tau = \frac{1}{k_{\mathrm{tot}}} = \frac{1}{k_r + k_{\mathrm{nr}}}$$

In the case of STED microscopy, fluorescence lifetime is also influenced by the decay constant for stimulated emission k_{STED}, which is directly proportional to the intensity of the STED laser beam [32].

$$\tau_{\mathrm{STED}} = \frac{1}{k_r + k_{\mathrm{nr}} + k_{\mathrm{STED}}}$$

That means the higher the STED intensity, the shorter the fluorescence lifetime, or with other words the STED process competes with the fluorescence emission. The STED intensity profile, inversely corresponds to the fluorescence lifetime, building a fluorescence lifetime gradient with long lifetimes in the middle and decreasing lifetimes towards the periphery of the "doughnut"--shaped STED beam (Fig. 4b, d).

One of the first approaches for taking advantage of the fluorescence lifetime in STED microscopy was gated STED (gSTED) microscopy [33, 34], where a preset time gate (or threshold) excludes early detected photons, hence short lifetime fluorescence, normally coming from the depleted periphery of the spot and would decrease the effective resolution (Fig. 4d).

1.6 tauSTED as a Combination of STED and FLIM

A more elegant way of using the fluorescence lifetime information without losing photons (intensity) in STED microscopy is realized in combination with a FLIM (*see* **Note 6**) approach.

The tauSTED nanoscopy is a composite imaging method that integrates conventional STED microscopy (based on intensity information), with fit-free time-domain phasor FLIM (based on fluorescence lifetime information), along with novel computational tools such as complex wavelet filtering (CWF) and tau trajectory analysis [35]. This approach offers numerous advantages over previously mentioned methods, for instance, improved resolution and contrast even at lower excitation and STED intensities, an expanded palette of fluorescent dye combinations due to lifetime-based separation of spectrally overlapping dyes, and phasor FLIM capabilities (Fig. 4).

1.7 tauSTED and Phasor Plots

As mentioned previously, the FLIM approach is the core of tauSTED nanoscopy, providing fluorescence lifetime information. While in conventional FLIM, fluorescence lifetime data are usually analyzed by fitting an exponential function to the lifetime histogram at each pixel [36] (Fig. 4c, e), tauSTED employs fit-free time-domain phasor FLIM [29, 35, 37] (Fig. 4c, e, f). The phasor plot offers a simple graphical representation of the distribution of fluorescence lifetimes in an image and requires no fitting. During image acquisition, the lifetime histogram of each individual pixel, which represents the corresponding fluorescence decay, is transformed into Fourier space [38, 39]. This results in a point (phasor) on a two-dimensional phasor plot, with $g_{i,j}(\omega)$ and $s_{i,j}(\omega)$, as the x and y coordinates, respectively (Fig. 4e, f).

$$g_{i,j}(\omega) = \int_0^\infty I_{i,j}(t)\cos(\omega t)\,dt \Big/ \int_0^\infty I_{i,j}(t)dt$$

$$s_{i,j}(\omega) = \int_0^\infty I_{i,j}(t)\sin(\omega t)dt \Big/ \int_0^\infty I_{i,j}(t)dt$$

$I_{i,j}(t)$ represents the fluorescence decay corresponding to pixel (i,j) and ω is the angular frequency of the pulsed excitation laser. All possible single-exponential (mono-exponential) decays lie on the so-called universal circle, a semicircular curve with centre at (0.5, 0) and a radius of 0.5. Phasors located further to the right, closer to (1, 0), correspond to shorter lifetimes, while those further to the left, closer to (0, 0), represent longer lifetimes. Phasor points inside the universal circle indicate multiexponential decays which may result from mixtures of single- or multiexponential components (Fig. 4f). For example, a multiexponential decay can be a linear combination (or vector sum) of two single- or multiexponential lifetimes. In a case like that, if two fluorophore species are located in the same pixel, the resulting phasor will lie on the line connecting their respective phasor positions. The ratio between the two species is then reflected in the relative position on this line. More complex combinations, including three or more lifetimes, are also possible [37]. Phasors outside the universal circle typically indicate reactions occurring in the excited state, such as Förster Resonance Energy Transfer (FRET).

In tauSTED, the concept of fit-free time-domain phasor FLIM analysis enables the use of the STED effect to fluorescence lifetime (as described above) to further increase the resolution and contrast, in a simple graphical way. A conventional (single-channel) confocal-FLIM image would result in a uniform cluster of points representing similar lifetimes in a phasor plot (confocal phasor) (Fig. 4f). The STED effect becomes visible in the phasor plot in the way that the confocal cluster of similar lifetimes turns into a lifetime gradient, the so-called STED phasor or STED (lifetime-based) trajectory. This trajectory appears as a wide straight-line cluster that contains longer lifetimes, originating in the center of the doughnut, towards the left side of the phasor plot, and the short lifetime components, from the STED beam's periphery, towards the right side of the phasor plot (Fig. 4f).

As indicated previously, in tauSTED the mapping of the fluorescence lifetimes from each pixel of an image to the phasor plot works bidirectional. This means that, by processing the phasor plot, it is possible to reconstruct the image and make visible the effects induced by changes in the phasor plot. The power of tauSTED lies exactly in that concept. In commercial systems the enhancement of the images works mostly in an automated manner.

For instance, filters are applied like the complex wavelet filter (CWF) [35] that enable the automatic identification of the STED trajectory and noise filtering, resulting in better resolution even with low photon counts. This can be particularly important for live cell imaging. Suppression of background and noise can also be performed automatically, by selectively removing phasors outside of the expected STED Trajectory (TauBackground Suppression [29]). Also, automatic assignment of weights to the pixels according to their corresponding lifetime can lead to a more efficient use of photons and higher resolutions (TauStrength [29]). This is realized by taking into account that longer lifetimes are more important and the shorter lifetimes around them serve with an assistive role for narrowing down the PSF (Fig. 4d, f). An additional feature of tauSTED nanoscopy thanks to its FLIM nature, is the capability of easily distinguishing a plethora of fluorophores according to their lifetime, even if their emission spectra overlap. This separation can also be easily performed by creating regions of interest (ROIs) of the respective clusters in the phasor plot. Besides these advantages contributing to enhanced subdiffractional resolution, tauSTED is also capable of FLIM applications with phasor analysis.

2 Materials

2.1 Buffers, Media, and Ligands

All buffers should be prepared using MilliQ water, unless specified. Storage at 4 °C, −20 °C or at room temperature is indicated after each buffer.

1. Hanks' Balanced Salt Solution (HBSS), 4 °C.
2. Trypsin (0.25% EDTA), −20 °C.
3. Trypan blue, 4 °C.
4. Full medium: Dulbecco's Modified Eagle Medium (DMEM), 10% heat-inactivated fetal bovine serum (FBS) (FBS: heat-inactivate in a water bath at 56 °C for 30 min, invert 1–2 times during the incubation (*see* **Note 7**), −20 °C), 1% penicillin/streptomycin, −20 °C.
5. 4% PFA, 4% sucrose in PBS, −20 °C
6. Microtubule reaction buffer 80 (MRB80) (80 mM PIPES, 1 mM EGTA, 4 mM $MgCl_2$, pH 6.8), 4 °C.
7. 4% PFA, 4% sucrose in MRB80, −20 °C
8. MRB80 with 0.3% Triton ×-100, 0.1% glutaraldehyde.
9. Glyoxal, 4 °C.
10. Methanol with 1 mM EGTA, −20 °C.
11. 20× Phosphate-buffered saline (PBS): 3 M NaCl, 161 mM Na_2HPO_4, 39 mM KH_2PO_4, RT
12. PBS permeabilization buffer (0.25% Triton ×-100).
13. Blocking buffer: PBS containing 10% heat-inactivated horse serum and 0.1% Triton-X, −20 °C.
14. Mowiol: 9.6 g granulated mowiol, 24 g glycerol, and 24 mL water are stirred for approximately 2 h, optional: add 2.4 g of the anti-bleaching agent triethylenediamine (also known as DABCO). Add 48 mL of 0.2 M Tris-HCl (pH 8.5). Heat the mixture to 50 °C in a water bath for 2 h and stir every for around 2 min for each 20 min incubation period to allow for even heating. When the majority of mowiol has been dissolved, centrifuge the mixture for 5000 g for 15 min to remove any excess mowiol. Prepare small 1.0 mL aliquots and store at −20 °C.

3 Methods

3.1 Hippocampal Dissociated Cultures

The hippocampus is a structure crucial for learning and spatial memory. Hippocampal neurons can be investigated using acute or organotypic hippocampal slices as well as with cryosections, in which the endogenous distribution of proteins is well preserved in their native state. Despite the possibility of cutting cryosections with 5 μm thickness, they still contain a high density of biological material that can make the confident resolution of small structures challenging. When aiming to investigate differences on the nanometer scale, dissociated hippocampal cultures become increasingly important. They can be prepared with a lower cellular density—relating to both neuronal and glial cells—and are thus not as limited when it comes to detecting the subcellular positioning of endogenous proteins. The cell density should be as high as possible to ensure healthy synaptic connections between neurons, and as low as possible to allow for the accurate detection of proteins without background signal e.g. from other cells. Here, we briefly describe the protocol needed to prepare dissociated hippocampal cultures from embryonic day 18.5 rats.

1. Anesthesia pregnant rats carrying E18.5 (from embryonic day 18 and 19) embryos using the method of choice, e.g., isofluorane inhalation.
2. When fully anesthetized, decapitate pregnant rats, e.g., using a guillotine.
3. Remove embryos in the uterus; quickly place them in a sterile, cooled container.
4. On ice, decapitate the embryos, collect their heads in 4 °C HBSS.
5. Quickly remove their brains and hippocampi.
6. Collect hippocampi in 4 °C HBSS.
7. Carefully wash 5× with 4 °C HBSS.
8. Dissociate by incubating with 37 °C trypsin (0.25% EDTA) for 15 min.
9. Carefully wash 5× with 37 °C HBSS.
10. Slowly triturate the hippocampi with a 20G and subsequently with a 26G needle, to separate the cells.
11. Remove cell clumps by passing through EasyStrain filter (100 μm).
12. Determine the number of live cells by mixing the cell solution with trypan blue and counting in a Neubauer chamber.

13. Dilute cells according to the desired cell density in 37 °C full medium (DMEM with 10% heat-inactivated FBS and 1% P/S) (*see* **Note 8**).
14. Plate on poly-L-lysine-coated 18 mm or 24 mm glass coverslips. Ideally use high-precision coverslips that have a more even thickness and reduced autofluorescence than regular coverslips.
15. After cells have fully attached to the surface of the coverslips, full medium is gently replaced with neuronal culture medium such as BrainPhys or NeuroBasal, including supplements.
16. Let cultures mature at 37 °C 5% CO_2 atmosphere until reaching the desired maturity (*see* **Note 9**).
17. Counteract evaporation by removing 10% of the culturing medium with double the amount of freshly prepared culturing medium once a week.

3.2 Selecting Suitable Fixation Protocols for STED Microscopy

Since lower laser powers are required for tauSTED nanoscopy than in STED, it is therefore more suitable for live imaging as it reduces cell damage through phototoxicity. This gives it a considerable advantage in time-lapse imaging over extended periods of time as it ensures cells remain viable and the signal remains strong. It further yields considerable advantages for imaging z-stacks. To maximize the resolution that can be harnessed with STED, fixed samples are ideal, since they allow for the reduction of acquisition speed (dwell time) and the use of fluorescent dyes optimized for STED imaging, increasing the resolution. Fixation methods can influence how accessible antigens are and how well structures are preserved, thus substantially influencing sample quality. Although it is possible to improve antigen-accessibility post-fixation, these methods should only be used if necessary, and if possible, a suitable fixation protocol should be selected based on the expected-to-be-used antibodies before resorting to post-fixation troubleshooting.

1. For all described fixation methods: first replace the growth medium. Only remove medium from 1 to 2 coverslips at a time, to avoid them from drying out.
2. Immediately, but carefully add the fixative to each coverslip (*see* **Notes 10** and **11**). Do not pipette the fixative directly onto the coverslips but start by pouring it onto the well plate wall. This is simple to achieve when initially tilting the plate perpendicular to its orientation and with increasing volume, returning it to its original, horizontal orientation.
3. After 10 min, remove the fixative from the cells. From this point on, the cells are not considered fragile anymore and more coverslips can be treated at the same time. Add PBS to each coverslip and incubate at room temperature for around 10 min.

4. Repeat the PBS wash **step 2** times.
5. (Optional): Employ a post-fixation treatment such as antigen-retrieval. They often rely on a partial removal of the fixative and can be employed to improve antigen accessibility.
6. Following the washes after the fixation (and optionally any post-fixation treatment), the store coverslips at 4 °C until they are used in a staining. Fixed coverslips should only be stored for a few weeks in PBS. If long-term storage is intended, add sodium azide to the PBS to avoid growth of contaminants.

3.3 Staining Procedure

Typically, immunocytochemistry (ICC) staining protocols include similar steps, such as the fixation of samples, blocking of unspecific binding of antibodies, primary and secondary antibody binding, multiple wash-steps in-between and the mounting of the sample onto glass slides.

Here, we present a standard protocol used for STED microscopy. If not described otherwise, steps are done at room temperature.

1. Fix cells using the selected fixation technique from Table 1: Fixation techniques for STED microscopy.
2. Remove fixative by washing samples three times for 10 min each with phosphate-buffered saline (PBS).
3. Permeabilize samples to prepare for antibody penetration by replacing PBS with PBS permeabilization buffer containing 0.25% of the detergent TritonX-100.
4. Remove permeabilization buffer by washing twice for 5 min with PBS.

Table 1
Fixation techniques for STED microscopy

Fixation reagent	Steps required	Utility
PFA	1. 10 min fixation with 4% PFA (including 4% sucrose)	Standard protocol, suitable for most approaches
PFA	1. Wash cells prior to fixation with 37 °C warm MRB80 buffer, 0.3% triton ×-100, 0.1 glutaraldehyde 2. 10 min fixation with 37 °C warm 4% PFA/4% sucrose in MRB80 buffer	Tubulin imaging
Glyoxal	An extensive description of glyoxal fixation for STED microscopy exists; please refer to this reference [45].	Alternative to the PFA standard protocol
Methanol	1. 10 min fixation with −20 °C cold methanol (including 1 mM EGTA)	Tubulin imaging

5. Block the samples to prevent unspecific antibody binding with blocking buffer (PBS containing 10% heat-inactivated horse serum (*see* **Note 12**) and 0.1% Triton-X).
6. Dilute primary antibody in blocking buffer.
7. Prepare a humified staining chamber: Use a staining box in which slides can be placed horizontally on top of the holders, creating a surface parallel to the ground. Fill this created chamber below with water to reduce evaporation and prevent coverslips from sticking to the staining chamber. Place parafilm upside down on the slides. Can be marked on the side that does not face the coverslips to easily identify samples.
8. Add one drop of ~50 μL antibody solution for a 12-well plate coverslip (18 mm).
9. With fine forceps, take the coverslip out of its well plate, tip gently on a clean tissue to remove excess liquid, and place the antibody-drop cells facing downwards.
10. Incubate samples with primary antibody overnight at 4 °C (*see* **Note 13**).
11. Transfer samples as in 9. into well plates with fresh PBS; this time, cells should face upwards.
12. Wash samples three times for 10 min each with PBS to remove unbound primary antibody.
13. In the staining chamber, place another layer of parafilm on top of the already existing one.
14. Add one drop of ~50 μL secondary antibody solution diluted 1:500 in blocking buffer for a 12-well plate coverslip (18 mm). To remove aggregating secondary antibodies that would decrease the SNR, quickly spin down the antibody solution before using it on samples using a tabletop centrifuge.
15. Wash samples three times for 10 min each with PBS to remove unbound secondary antibody.
16. Optional: When using a prelabelled antibody that has the same species as another primary antibody, the protocol can be repeated in **steps 10–12** using the prelabelled antibody. This way, antibodies that were raised in the same species and even anti−/nanobodies with the same isotype, can be used in the same sample.
17. Optional: incubate the sample with additional dyes such as DAPI, although it is not recommended. If it, or a similar dye, is still used, wash 2–3 times with PBS afterwards (*see* **Note 14**).
18. To mount coverslips onto slides, mounting medium such as mowiol can be used. For easier handling, mowiol can be pre-heated to up to 50 °C to reduce acute viscosity (*see* **Note 15**).

19. Place a drop of ~10 μL of the polyvinyl alcohol mowiol onto a cleaned glass slide. Dry the sample at room temperature.
20. After 3–4 h, the coverslips may be firmly attached to the slide.
21. Image the samples or store at 4 °C until doing so (*see* **Note 16**). Light exposure of the samples should be minimalized to prevent photobleaching.
22. Before taking the samples to a STED microscope, quickly check them at, e.g., a widefield LED microscope to confirm a suitable staining quality and rule out any detection issues caused by the excitation laser of the STED microscope (*see* **Note 17**).

3.4 Determining Mixture of Antibodies for STED Microscopy

For most immunostainings, the combination of multiple antibodies is required to visualize different targets. Compared to regular confocal microscopy, more care has to be taken when selecting suitable antibodies. Although an antibody-stained sample may look acceptable under regular confocal microscopy, in super-resolution microscopy, it becomes particularly evident when primary and secondary antibodies are not suitable to detect the protein of interest or the primary antibody, respectively. The higher the resolution, the more important the distance between the protein of interest and the fluorophore used to detect it becomes. In a classical staining, a primary antibody is used to detect an epitope on the protein of interest and a fluorescently labelled secondary antibody that binds to the primary antibody makes its location visible. However, this is no direct labelling of the protein of interest and the distance between the epitope and the fluorophore used to detect it can be in the range of 10–20 nm, resulting in an area of uncertainty of up to 40 nm. Using classical microscopy methods, this is negligible as the diffraction limitation of ~200 μm is not able to resolve these distances. In super-resolution microscopy however, even small distances between protein of interest and fluorophore become highly relevant. Thus, nanobodies that originate from heavy-chain-only antibodies produced in camelids and chondrichthyes [40] are preferable to regular antibodies. Ideally, prelabelled primary nanobodies or even endogenous tags can be used to visualize the proteins of interest. In reality, this is often too expensive for practical reasons such as the loss of anti−/nanobody in the labelling process or the establishment of new mouse lines. Unfortunately, the ideal antibodies that need to be used in a combined staining, often happen to have the same host-species. To still be able to use the most specific primary anti−/nanobodies, the use of immunoglobin (Ig) isotype-specific secondary nanobodies comes in handy. For example, different isotypes of mouse Igs exist: IgA, IgD, IgE, five IgGs, and IgM. These isotypes can be targeted individually by nanobodies conjugated to a fluorophore.

3.5 Determining a Combination of Fluorophores for STED Microscopy

Antibodies need to be as small as possible for super-resolution microscopy; each combination (or single antibody) detecting an epitope/protein also needs to bear a fluorophore that allows its detection. For regular STED as in fluorescent microscopy generally, when interested in more than one target, it is crucial to ensure fluorophores used are compatible with each other; it especially needs to be ensured that their emission (and ideally their excitation) spectra do not overlap or not completely. This can be done by using readily available databases such as https://www.fpbase.org/spectra/ or manually, relying on manufacturer's webpages. For tauSTED nanoscopy, additional options can be considered that grant various advantages over regular STED. While regular STED is limited in the number of targets that can be imaged at the same time, FLIM can help differentiate fluorophores with overlapping spectra such as mCherry and abberior STAR 580, allowing for multiplexing. It should, however, be avoided to image structures that are overlapping completely, although theoretically possible. When interested in multiple targets, the ones for which the best resolution is required should be detected using STED-compatible dyes that, when excited, emit light in the red to far red (625–800 nm) spectrum. When the to-be-used microscope is equipped with a 590–600 nm STED depletion laser, most regular dyes emitting green light can be super-resolved as well (*see* **Note 18**). When this STED depletion laser is not present, using long stoke shift dyes can be used a as a workaround. They can enable multicolor-STED using the same depletion laser as for regular red or far-red emitting fluorophores. Fluorophores that we have successfully used for tauSTED nanoscopy are: Atto647N, STAR 635P emitting in far red, and STAR 580 emitting in red. For an extensive list of suitable fluorophores, supplier webpages should be consulted (*see* **Note 19**).

3.6 Choosing Imaging Settings

In the following, we will describe imaging settings that are based on the Leica STELLARIS 8 STED Falcon and the STELLARIS STED microscope. Settings might deviate slightly when using other systems.

When planning to image a sample at a (tau)STED microscope, one should allocate enough time not just to image, but also to start the microscope prior to imaging, to allow the setup to "warm up" and stabilize (*see* **Note 20**).

Before the sample is placed on the stage, a suitable objective has to be selected. Typically, 93× glycerol or 100× immersion oil objectives with high numerical apertures (NA) of 1.3–1.6, resulting in a high spatial resolution are used for thin, fixed samples in STED microscopy. The choice of immersion medium should be tied to the mounting medium used. For the glycerol-based mounting-medium mowiol, a glycerol objective would be most suitable.

As a rule, first, it should be tested whether the individual channels display the signal as intended in confocal mode, and use the FLIM mode if fluorophores with significant overlap in their emission spectra are present in the same sample. This may even be useful when not using strongly overlapping fluorophores since it can increase signal specificity and reduce background signal. Test images of every fluorophore present in the sample should be taken.

After successfully imaging each fluorophore in, e.g., confocal and FLIM mode, the settings for super-resolution imaging can be considered. For STED microscopy in general, it needs to be ensured that both depletion and excitation lasers are aligned. When interested in multicolor STED, it is additionally crucial that all excitation and depletion lasers used in STED resolution are calibrated. Depending on the setup, this is done every few hours to every few weeks and in varying degrees of automation. To determine the ideal settings for super-resolution microscopy, several factors need to be considered: exposure time, laser power, and wavelength of both excitation and depletion lasers for individual channels, which detectors to use and how to set their sensitivity and the nanometer range of the emission they collect, the number of accumulations or averaging and pixel size settings. The region of interest should be placed in the center of the field of view. When possible, acquisition speed (dwell time) should be set as low as possible to maximize the resolution. As a rule of thumb, samples that are excited with and emit light in higher wavelengths, should be imaged first and the other channels in descending order. Imaging in order of decreasing wavelength ensures the emitted light of one fluorophore does not excite and possibly bleaches another fluorophore.

As a starting point for each fluorophore, their excitation and emission spectra can be determined using databases such as FPbase (https://www.fpbase.org/spectra/). When using, e.g., a tunable white light laser as an excitation laser, it can be adjusted carefully to increase the signal-to-noise ratio (SNR). The excitation laser should ideally be set to excite a fluorophore as much as possible while exciting other fluorophores as little as possible.

Laser power and exposure time for individual fluorophores needs to be tested manually. It should be high enough to get a decent SNR, but low enough to not bleach the fluorophores. The wavelength of the depletion laser is determined by the emission of the used fluorophores and should be red-shifted compared to its maximum. Depletion lasers generally have a fixed wavelength and cannot be adjusted as carefully. The higher their intensity, the better the resolution will get. The maximum intensity that can be set is determined by the photostability of the fluorophore. The measured fluorescent intensity also depends on the sensitivity of the detectors used. For tauSTED nanoscopy, FLIM-compatible detectors need to be used.

When taking images, the same area should be imaged multiple times. It thus should be decided whether the pixel values of single images are added up (accumulations) or whether the average is used (averaging). When intending to compare intensities between different groups, averaging needs to be selected. Accumulations increase the SNR but do not grant a linear increase in signal intensities. The same area is usually set to accumulate/average 4–16 line scans.

Pixel size is generally set to oversample (i.e., the pixel size is set smaller than theoretically achievable) to improve the SNR.

Z-stacks can be set as they would be for confocal microscopy, which is rarely possible with regular STED microscopy. When interested in 3D reconstructions, the STED "doughnut" can be adapted to a three-dimensional shape. This improves axial resolution at the cost of lateral resolution. An extensive overview of STED methods in which axial and lateral resolutions are compared can be found [41]. Not included in the overview is tauSTED nanoscopy, which is more suitable for volumetric imaging than regular STED, owing mainly to its lower use of laser intensities.

3.7 Post-imaging Analysis

Like most fluorescence microscopy methods, tauSTED nanoscopy does not end with the acquisition of images. Thanks to the collection of vast amounts of metadata, including multiple scans of the same area and acquisition of fluorescence lifetime information for each pixel, many post-acquisition analyses are possible. While confocal images are often deconvolved, which is based on fluorescence intensity, tauSTED data can be deconvolved based on the lifetime that is measured for each pixel, which further improves spatial resolution. Another analysis method based on fluorescence lifetime information involves the selection of specific populations of fluorophores from phasor plots. For further reading, a number of scientific articles are available [43, 44]. It should, however, be noted that the acquisition of quantitatively valuable datasets is possible, but settings need to be carefully considered during acquisition with a particular focus on averaging or accumulations, which can result in linear or exponential fluorescence intensity level changes, respectively.

4 Notes

1. *Interacting light waves.* Depending on the phase shift of interacting light waves, relative to each other, we observe constructive interference when two waves meet in phase (crests-crests and troughs-troughs) so an increase in the amplitude of light, and destructive interference when they meet out of phase (crests-throughs), leading in cancelation of each other (zero amplitude of the resulting light wave) [21].

2. *Excitation-emission process.* Illuminating a fluorophore with light can lead to its excitation, the transition of an electron from the energetic ground state S_0 to a higher vibrational state mostly of the first excited state S_1 ($10^{-15}s$) [23]. This happens when the wavelength of the excitation light corresponds to the energy difference between the ground and excited state of the fluorophore. In a process called vibrational relaxation the fluorophore makes almost immediately a transition to the lowest vibrational state of S_1 under energy release in form of heat (also transition from the second excited state S_2 to S_1 ($10^{-12}s$), termed internal conversion, is possible). After some time ($\sim 10^{-9}s$) the fluorophore spontaneously returns to its ground state (spontaneous relaxation) by emitting a photon of a higher wavelength (red-shifted photon). The graphical representation of this process is normally given in form of a Jablonski diagram (Fig. 4b).
3. *Confocal microscopy.* Conventional confocal microscopy is a scanning method [42, 43], in which the excitation laser illuminates individual points line by line, elevating the fluorophore (s) of the current spot (point) to a higher (excited) electronic energy state. The red-shifted fluorescence light of the spontaneous emission is then collected from the objective and travels backwards the same pathway as the excitation light until it reaches the dichroic mirror and emission filter, where only the higher wavelength (fluorescence) is let to pass through and reach the detector (mostly a photomultiplier tube (PMT) or an avalanche photodiode (APD)) after the out of focus light, is blocked by the confocal pinhole (Fig. 4a).
4. *Deexcitation process. Stimulated emission.* Deexcitation is based on the principle of stimulated emission that was first described by Einstein [44]. In this process, a fluorophore in an excited electronic state can be forced to return to its ground state by stimulation with a photon of a typically longer wavelength (red-shifted) than the excitation wavelength. The result is the emission of a second photon, identical to the stimulating one and no fluorescence (Fig. 4b).
5. *Phase plates and the shape of the STED laser beam in 2D- and 3D-STED.* Technically the hole of the STED laser beam is created by destructive interference in the central focus of the beam, caused by a spiral wave plate (Vortex phase plate) on its pathway [22]. This is a plate of azimuthally increasing thickness (like a spiral staircase) from 0 to λ_{STED} (wavelength of the STED laser), which induces a variable phase shift from 0 to 2π, respectively (phase singularity in the central focus). With an additional phase plate of thickness of half the wavelength (π-step phase mask, or just π-phase plate) and a hole in the middle (or vice versa) placed in a part of the STED laser

beam, it is possible to deplete additional fluorophores above and under the focus plane to also increase the axial resolution, the method is then called 3D STED [18] (Fig. 4a). 3D STED does not refer to the volumetric capabilities of STED (z-stacks).

6. *FLIM microscopy and the TCSPC method.* The fluorescence lifetime of a fluorophore also depends on its nano-environment within the cell. Important influencing factors include pH, ion concentration, metabolite concentration, the presence of potential FRET partners, and more [37]. The study of such molecular properties using changes in fluorescence lifetime as indicators is the subject of fluorescence lifetime imaging microscopy (FLIM). FLIM images show the color-coded spatial distribution of fluorescence lifetime on the sample. Considering only the time-domain (there is also the frequency-domain) for measurement and calculation of the fluorescence lifetime in microscopy, the most common approach is the time-correlated single photon counting (TCSPC) method [23].

   TCSPC in confocal FLIM approaches requires the use of pulsed excitation lasers, thus excitation lasers of high repetition rates, like 80 MHz. While the laser scans the sample, over a thousand (depending on the dwell time) concentrated light pulses are applied to the fluorophores for each individual pixel. After each pulse, the time until an eventually emitted photon arrives at the detector (which is usually a PMT or an APD) is measured, and the single photons are counted (and recorded) according to their arrival times in the form of a histogram (lifetime histogram) for each individual pixel of the image (Fig. 4a, c, e). This histogram is typically well approximated by fitting an exponential decay function, where I_0 is the number of all counted photons, and τ is the fluorescence lifetime.

$$I(t) = I_0 e^{\frac{-t}{\tau}}$$

   Another method for extracting the lifetime information from these histograms, without the need to fit an exponential function, is based on phasor plots (Fig. 4e, f), which are later discussed in the text.

7. Aliquot serum into 50 mL reaction tubes prior to heat-inactivation to facilitate a thorough inactivation.

8. *Neuronal culture density.* When interested in synaptic stainings, cultures should be quite dense (60000–80,000 cells/mL) to ensure the development of mature synapses is possible. Denser cultures generally also contain healthier neurons. Most major components of the neuronal cytoskeleton are not exclusively found in neurons; thus a lower density of cells needs to be plated (30000–50,000 cells/mL). When a higher density is

used, the addition of Cytosine arabinoside (AraC) is recommended to avoid signal from glia cells to mask signal from neurons. Note that the density needs to be tested in each laboratory, as there are large variations even when different experimenters prepare cultures. The numbers presented here should rather serve as rough guidelines.

9. *Presence of glia cells in neuronal cultures.* When growing neuronal cultures for an extended time period, the presence of glial cells can become a problem, which can cause a decrease in the SNR. To counteract this, 1:500 AraC can be added to the full medium to prevent cell divisions. AraC should only be supplemented when needed, as it also decreases the viability of neurons, both by itself and by reducing the number of glia cells supplementing surrounding neurons. Whether AraC also depends on the day in vitro cells are grown and may be less critical for younger cultures than for older ones.
10. *MRB80 fixation.* The only exception is the fixation protocol shown in the second line of Table 1 in which the cells are washed once before the fixative is added.
11. *Fixation techniques.* As an alternative to PFA, glyoxal can be used, which yields better results for many targets [45]. All mentioned fixatives also permeabilize the cell membrane, making post-hoc surface stainings impossible and necessitating their performance prior to fixation. Many fixation protocols, including PFA-based ones, can alter the ultrastructure of samples [46].
12. *Blocking agent alternatives.* As alternatives to horse serum, BSA, goat serum, FBS, etc., can be used. The selection depends on which species the primary antibodies have been raised in and should not be the same to avoid detection by secondary antibodies.
13. *Primary antibody incubation step.* Dyes such as phalloidin to visualize actin may also be added during this process or later, with the secondary antibodies. The staining process can be sped up by incubating for 2–3 h at room temperature, but it may reduce binding efficiency.
14. *DAPI.* Do not incubate with DAPI when intending to use a ~ 600 nm STED depletion laser.
15. *Mounting medium.* The main function of the mounting medium is to fix samples on coverslips to glass slides. For super-resolution microscopy, especially, it needs to have minimal autofluorescence and to protect samples from photobleaching when stored (at 4 °C). It should also decrease light scattering artefacts to a minimum.

16. *For long-term storage*, the sample should be placed at −20 °C. The sample can also be stored at room temperature, although this should not be done for extended periods of time (weeks). If, in addition to antibodies, dyes have been used, samples should be imaged as soon as the mounting medium has dried, not more than a few days after.

    Samples should be labelled with non-reflective markers, or to avoid accidental removal of any labelling through rigorous cleaning prior to imaging, a pencil can be used on frosted slides.
17. *Prior to imaging*, coverslips should be cleaned gently with ethanol or isopropanol to reduce impurities.
18. *Dye efficiency.* Although it is possible to achieve STED resolutions of green-emitting dyes using a ~ 600 nm STED depletion laser, it is far more efficient to use red or far-red-emitting dyes instead.
19. *Owed to lower laser powers used in tauSTED* compared to regular STED, even usually unsuitable dyes may be super-resolved, although with a lower success rate.
20. *Microscope surroundings.* It should, however, also be ensured that the area surrounding the microscope is well conditioned so the temperature does not drastically increase due to laser activity. This may be more difficult in live imaging, which is often done at physiological temperatures such as 37 °C. With a STED microscope, it can, however, still be challenging to keep the temperature constant.

Acknowledgments

This work was supported by the Deutsche Forschungsgemeinschaft (DFG) SFB1315 A03 and the Excellence Strategy—EXC-2049–390688087.

We would like to thank the Interdisciplinary Center Life in Time and Space (IZ LIST) for awarding a travel grant to support a stay in Heidelberg. https://izlist.hu-berlin.de/travel-grant-report-daniela-hacker-heidelberg.

We would further like to thank the EMBL Imaging Centre and especially Timo Zimmermann and Dietrich Walsh for access to the STELLARIS 8 STED Falcon and the STELLARIS STED microscope.

Conflict of Interest The authors of this chapter declare no conflict of interest.

References

1. Golgi C (1885) Sulla fina anatomia degli organi centrali del sistema nervoso. U. Hoepli, Milano
2. Akey CW (1989) Interactions and structure of the nuclear pore complex revealed by cryo-electron microscopy. J Cell Biol 109:955–970
3. Yang X, Yang Z, Wu Z et al (2020) Mitochondrial dynamics quantitatively revealed by STED nanoscopy with an enhanced squaraine variant probe. Nat Commun 11:1–9
4. Stephan T, Roesch A, Riedel D et al (2019) Live-cell STED nanoscopy of mitochondrial cristae. Sci Rep 9:1–6
5. Fiala JC, Harris KM (2012) Dendrite structure. In: Dendrites. Oxford University Press Inc/Children's Hospital, New York/ Boston
6. Xu K, Zhong G, Zhuang X (2013) Actin, spectrin, and associated proteins form a periodic cytoskeletal structure in axons. Science 339: 452–456
7. He J, Zhou R, Wu Z et al (2016) Prevalent presence of periodic actin-spectrin-based membrane skeleton in a broad range of neuronal cell types and animal species. Proc Natl Acad Sci USA 113:6029–6034
8. Han B, Zhou R, Xia C et al (2017) Structural organization of the actin-spectrin–based membrane skeleton in dendrites and soma of neurons. Proc Natl Acad Sci USA 114:E6678–E6685
9. Bär J, Kobler O, Van BB et al (2016) Periodic F-actin structures shape the neck of dendritic spines. Sci Rep 6:37136
10. Akhmanova A, Steinmetz MO (2015) Control of microtubule organization and dynamics: two ends in the limelight. Nat Rev Mol Cell Biol 16:711–726
11. Katrukha EA, Jurriens D, Pastene DS et al (2021) Quantitative mapping of dense microtubule arrays in mammalian neurons. elife 10: 367925
12. Mikhaylova M, Cloin BMC, Finan K et al (2015) Resolving bundled microtubules using anti-tubulin nanobodies. Nat Commun 6:1–7
13. Chen J et al (1992) Projection domains of MAP2 and tau determine spacings. Nature 360:674–677
14. Janke C, Magiera MM (2020) The tubulin code and its role in controlling microtubule properties and functions. Nat Rev Mol Cell Biol 21:307–326
15. Lauterbach MA, Guillon M, Desnos C et al (2016) Superresolving dendritic spine morphology with STED microscopy under holographic photostimulation. Neurophotonics 3:041806
16. Nägerl UV, Willig KI, Hein B et al (2008) Live-cell imaging of dendritic spines by STED microscopy. Proc Natl Acad Sci USA 105: 18982–18987
17. Arizono M, Idziak A, Nägerl UV (2025) Live STED imaging of functional neuroanatomy. Nat Protoc 20:1–25
18. Calovi S, Soria FN, Tønnesen J (2021) Super-resolution STED microscopy in live brain tissue. Neurobiol Dis 156:105420
19. Han Y, Cao R, Qin L et al (2022) Neuroligin-3 confines AMPA receptors into nanoclusters, thereby controlling synaptic strength at the calyx of held synapses. Sci Adv 8:1–14
20. Broadhead MJ, Horrocks MH, Zhu F et al (2016) PSD95 nanoclusters are postsynaptic building blocks in hippocampus circuits. Sci Rep 6:1–14
21. Giancoli DC (2010) Physik Lehr- und Übungsbuch. Pearson Deutschland GmbH, München
22. Masters BR (2020) Superresolution optical microscopy the quest for enhanced resolution and contrast. Springer International Publishing, Cham
23. Lakowicz JR (2006) Principles of fluorescence spectroscopy. Springer, Baltimore, p 21201
24. Akhmanova A, Steinmetz MO (2008) Tracking the ends: a dynamic protein network controls the fate of microtubule tips. Nat Rev Mol Cell Biol 9:309–322
25. Guerra JM (1995) Super-resolution through illumination by diffraction-born evanescent waves. Appl Phys Lett 66:3555–3557
26. Betzig E, Patterson GH, Sougrat R et al (2006) Imaging intracellular fluorescent proteins at nanometer resolution. Science 313:1642–1645
27. Huang B, Babcock H, Zhuang X (2010) Breaking the diffraction barrier: super-resolution imaging of cells. Cell 143:1047–1058
28. Hell SW (2007) Far-field optical nanoscopy. Science 316:1153–1158
29. Alvarez LAJ, Schwarz U, Friedrich L et al (2021) TauSTED: pushing STED beyond its limits with lifetime. Biomed Opt Express 12: 3463
30. Schmidt R, Weihs T, Wurm CA et al (2021) MINFLUX nanometer-scale 3D imaging and microsecond-range tracking on a common

fluorescence microscope. Nat Commun 12:1–12

31. Heineke J (2013) Screening for novel calcium-binding proteins that regulate cardiac hypertrophy: CIB1 as an example. Methods Mol Biol 963:279–301
32. Wang L, Chen B, Yan W et al (2018) Resolution improvement in STED super-resolution microscopy at low power using a phasor plot approach. Nanoscale 10:16252–16260
33. Hernández IC, Buttafava M, Boso G et al (2015) Gated STED microscopy with time-gated single-photon avalanche diode. Biomed Opt Express 6:2258
34. Moffitt JR, Osseforth C, Michaelis J (2011) Time-gating improves the spatial resolution of STED microscopy. Opt Express 19:4242
35. Wang P, Hecht F, Ossato G et al (2021) Complex wavelet filter improves FLIM phasors for photon starved imaging experiments. Biomed Opt Express 12:3463
36. Datta R, Heaster TM, Sharick JT et al (2020) Fluorescence lifetime imaging microscopy: fundamentals and advances in instrumentation, analysis, and applications. J Biomed Opt 25:1
37. Ranjit S, Malacrida L, Jameson DM et al (2018) Fit-free analysis of fluorescence lifetime imaging data using the phasor approach. Nat Protoc 13:1979–2004
38. Digman MA, Caiolfa VR, Zamai M et al (2008) The phasor approach to fluorescence lifetime imaging analysis. Biophys J 94:L14–L16
39. Weber G (1981) Resolution of the fluorescence lifetimes in a heterogeneous system by phase and modulation measurements. J Phys Chem 85:949–953
40. Frecot DI, Froehlich T, Rothbauer U (2023) 30 years of nanobodies – an ongoing success story of small binders in biological research. J Cell Sci 136:jcs261395
41. Wu Z, Xu X, Xi P (2021) Stimulated emission depletion microscopy for biological imaging in four dimensions: a review. Microsc Res Tech 84:1947–1958
42. Minsky M (1988) Memoir on inventing the confocal scanning microscope. Scanning 10:128–138
43. Amos WB, White JG (2003) How the confocal laser scanning microscope entered biological research. Biol Cell 95:335–342
44. Einstein A (1916) 7. Zur Quantentheorie der Strahlung, 209–228
45. Richter KN, Revelo NH, Seitz KJ et al (2018) Glyoxal as an alternative fixative to formaldehyde in immunostaining and super-resolution microscopy. EMBO J 37:139–159
46. Hobro AJ, Smith NI (2017) An evaluation of fixation methods: spatial and compositional cellular changes observed by Raman imaging. Vib Spectrosc 91:31–45

Chapter 5

Single-Molecule Localization Microscopy with Fixed Photoactivatable Fluorescent Proteins and Direct Stochastic Optical Reconstruction Microscopy

Rumelo Amor, Anusha Malapaka, Alex J. McCann, Frédéric A. Meunier, and Shanley F. Longfield

Abstract

Fluorescence microscopy has become one of the most widely used tools in modern cell biology. By selectively labelling molecules of interest with fluorescent tags, researchers can visualize the structural organization, distribution, and clustering of proteins within their native cellular environment. Single-molecule localization microscopy (SMLM) has extended fluorescence imaging by enabling spatial resolution far beyond the diffraction limit of light. Among SMLM techniques, fixed Photoactivated Localization Microscopy (PALM), Stochastic Optical Reconstruction Microscopy (STORM), and Direct Stochastic Optical Reconstruction Microscopy (dSTORM) provide access to subcellular organization below the diffraction limit that is otherwise difficult to achieve through conventional or even intermediate super-resolution methods, such as Structured Illumination Microscopy (SIM) or Stimulated Emission Depletion microscopy (STED). These techniques attain super-resolution by temporally separating fluorophore emission events, localizing individual molecules with high precision, and assembling their positions into a reconstructed image. Despite the recent advances of newer super-resolution modalities, fixed PALM and (d)STORM remain widely used, due to their molecular specificity, accessibility, and broad applicability across diverse biological systems. This chapter briefly introduces the core principles of SMLM, situates PALM and (d)STORM within the evolution of super-resolution microscopy, and provides detailed protocols for the image acquisition and data analysis of fixed-cell PALM and dSTORM imaging in monolayer cultures, such as PC12 cells and primary hippocampal neurons. The workflow described here is compatible with standard SMLM platforms and uses Zeiss ZEN Black 2012 and Abbelight NEO software for reconstruction and analysis.

Key words SMLM, Fixed PALM, dSTORM

1 Introduction

1.1 Core Principles of Fluorescence Microscopy

Fluorescence microscopy allows for the visualization of cellular structures and molecular interactions through the emission of light by specific probes or dyes. These probes or dyes, called fluorophores, usually contain a conjugated π-electron system in which

Frédéric A. Meunier (ed.), *Super-Resolution Microscopy: Methods and Protocols*, Methods in Molecular Biology, vol. 3034,
https://doi.org/10.1007/978-1-0716-5268-8_5,

electrons are delocalized across a series of alternating single and double bonds [1]. This arrangement lowers the energy gap between molecular orbitals and permits the electrons to be promoted to an excited electronic state when the molecule absorbs a photon of the appropriate wavelength [2]. Once in the excited state, the molecule typically undergoes rapid non-radiative relaxation through vibrational and rotational energy losses, ultimately returning to its ground state by emitting a photon. As some energy has been dissipated during the non-radiative relaxation steps, the emitted photon carries less energy than the absorbed photon and therefore has a longer wavelength. This difference in wavelengths is known as the Stokes shift [3]. It is this shift that allows optical filters to separate fluorescence emission from excitation light, thereby enabling high-contrast imaging in biological samples.

While the Stokes shift governs spectral separation, the wave nature of light in an optical system causes the emission from a point source to spread into a well-defined intensity distribution (blur) rather than remaining confined to a single location. This distribution is known as the Point Spread Function (PSF) [4] and is determined by the numerical aperture of the lens in the microscope and the wavelength of the emitted light. Due to this characteristic blurring, two fluorophores that lie closer together than 200 nanometers (nm) in the lateral plane cannot be distinguished as separate objects. This is the classical diffraction limit described by Abbe and Rayleigh in 1873 and 1879 [5, 6] and sets the fundamental resolution limit of conventional light microscopy.

1.2 Evolution of Conventional Fluorescence Microscopy to Nanoscale Imaging

Traditional fluorescence microscopy approaches such as - widefield imaging, laser scanning confocal microscopy, and spinning-disk confocal microscopy provide excellent molecular specificity but are fundamentally limited in spatial resolution by the diffraction limit of light. This limit obscures the fine details of many subcellular structures, most of which are composed of proteins ranging from 1 to 5 nm in size [7]. This mismatch between the biological complexity of cellular systems and the optical resolution of the microscopy techniques drove the development of super-resolution fluorescence microscopy approaches capable of bridging the gap.

These approaches can be broadly divided into methods that improve resolution by illuminating the sample in a structured or controlled manner (e.g., Structured Illumination Microscopy (SIM) or Stimulated Emission Depletion (STED) microscopy) and those that increase resolution by separating fluorescent detections in time. SIM provides a gentle, widefield approach that roughly doubles the resolution of standard microscopy [8, 9] and in contrast STED microscopy, employs a high-intensity doughnut-shaped depletion beam to confine fluorescence emission to a specific focal point [10, 11]. Both techniques highlighted important optical milestones, where SIM introduced accessible super-resolution imaging and STED demonstrated that resolution well below 100 nm was optically achievable.

1.2.1 The Emergence of PALM, STORM, and dSTORM

The next major conceptual leap came with Single-Molecule Localization Microscopy (SMLM). Rather than attempting to resolve all fluorophores simultaneously, SMLM techniques separate their emission in time. In each camera frame, only a sparse subset of fluorescent molecules is experimentally modulated to emit, producing isolated diffraction-limited spots that can be precisely localized by fitting their point-spread functions. Repeating this process over thousands of frames yields a large set of molecular coordinates which, when combined, form a super-resolved reconstruction of the underlying structure. This temporal separation strategy overcomes the diffraction limit not by altering the optics, but by redistributing detected photons across time.

The development of SMLM modalities emerged through a series of conceptual advances, beginning with Photoactivated Localization Microscopy (PALM; Fig. 1a; Table 1) [12, 13]. In PALM, photoactivatable or photoconvertible fluorescent proteins are held in a dark state until exposed to low-intensity ultraviolet (UV) light, which stochastically switches a small subset of molecules into a bright, irreversible fluorescent state. This controlled sparse activation ensures that only a limited number of emitters fluoresce in each frame, enabling accurate single-molecule localization. Stochastic Optical Reconstruction Microscopy (STORM; Fig. 1b, d; Table 1) [14] extended this principle by exploiting photo-switchable organic dye pairs. Controlled activation and rapid cycling between emissive and dark states allows for high-density molecular imaging with superior photon output. Building on this foundation, Direct Stochastic Optical Reconstruction Microscopy (dSTORM; Fig. 1c, e; Table 1) [15, 16] revealed that conventional organic fluorophores such as Alexa Fluor 647 can reversibly photo-switch in reducing and oxygen-scavenging buffer systems without the need for activator dyes. This simplification substantially broadened the accessibility of SMLM while preserving the high localization precision inherent to organic dyes.

1.3 Applications in Cellular and Molecular Imaging

PALM, STORM, and dSTORM have become indispensable for interrogating biological structures at nanometer resolution, enabling direct visualization of protein organization, molecular clustering, and subcellular architecture that lie far below the diffraction limit of light. In neuroscience, these methods have provided critical insights into synaptic nano-organization [17], cytoskeletal architecture [18], receptor distributions [19], and signaling assemblies [20], where spatial arrangements on the order of tens of nanometers directly influence function. PALM offers high molecular specificity, making it well-suited for resolving the

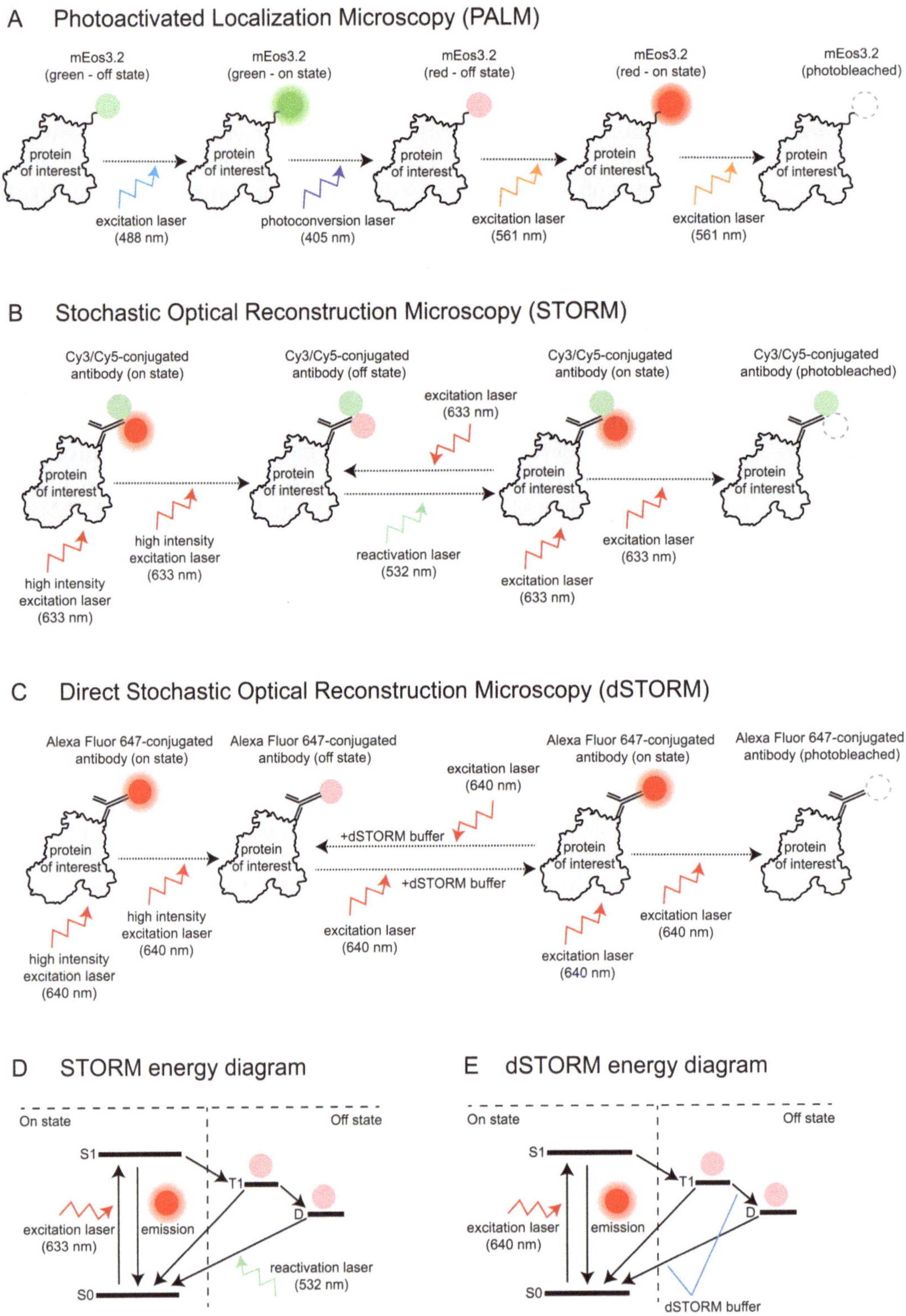

Fig. 1 Schematic overview of PALM and STORM. (**a**) Principle of PALM imaging. The protein of interest is genetically fused to an mEos3.2 photoconvertible fluorescent protein. Upon activation, mEos3.2 undergoes an irreversible switch from the green (on) state to the red (converted) state, allowing for sequential detection and

Table 1
Comparison of fixed-cell PALM, STORM, and dSTORM

Parameter	Fixed-cell PALM	STORM	dSTORM
Core principle	Uses photoactivatable (PA) or photo-convertible fluorescent proteins	Uses engineered activator–reporter dye pairs	Uses standard organic dyes that blink through redox driven OFF/ON transitions in chemical imaging buffers
Fluorophore examples	PA-GFP, PA-mCherry, mEos family, Dendra2	Activator: Cy3/Cy2/Alexa Fluor 405 Reporter: Cy5	Alexa Fluor 647, CF647, Cy5, ATTO647N
Blinking mechanism	Activation induces an irreversible ON → bleach cycle; requires controlled, sparse activation	Activator dye triggers reversible switching of the reporter dye; OFF–ON cycling governed by laser power	Thiol-rich, oxygen-scavenging buffers produce reversible radical states; blinking mainly buffer controlled
Labelling strategies	Genetic expression of the protein of interest fused with a PA-fluorophore	Immunolabelling with dye pairs	Immunolabelling with single organic dyes
Imaging requirements	Standard fixed-cell conditions; needs activation + imaging lasers, no special buffer	Activation + imaging lasers; may use mild buffer additives to support switching	Requires a dedicated switching buffer (MEA/GLOX) for reversible dye photoswitching
Best use cases	High-fidelity localization of genetically encoded proteins; structural studies with precise targeting	Highest control of photoswitching; ideal when labeling density is limited but precision is critical	Most common method for fixed-cell nanoscopy; broad applicability with standard dyes and antibodies

Fig. 1 (continued) precise localization of single molecules. (**b** and **c**) Principle of (**b**) STORM and (**c**) dSTORM imaging. The protein of interest is labelled with an antibody conjugated with either (STORM) an activator-reporter dye pair (e.g., Cy3 as the activator (green) and Cy5 as the reporter (red)), or (dSTORM) a single conventional fluorophore (e.g., Alexa Fluor 647). STORM and dSTORM: high-intensity excitation drives most fluorophores from the 'on' state (fluorescent/capable of fluorescing) into long-lived 'off' states (non-fluorescent). STORM: A reactivation laser (e.g., 532 nm for Cy3) excites a small subset of activator dyes, which, through a photochemical interaction, induces the nearby reporter dye to return from the long-lived 'off' state to the 'on' state. The reporter dye is then excited with an excitation laser (e.g., 633 nm for Cy5), allowing for the detection of single fluorescently-tagged molecules. dSTORM: fluorophores are reduced to a long-lived dark 'off' state using a thiol-containing buffer and return to the 'on' state spontaneously through buffer-mediated oxidation, enabling single-molecule detection. STORM and dSTORM: Individual fluorophores repeatedly cycle between 'off' (triplet (T_1) and long-lived dark (d)) states and 'on' (singlet ground (S_0) and singlet excited (S_1)) states until they ultimately enter an irreversible photobleached state. (**d** and **e**) A Jablonski energy-level diagram illustrates the transitions among S_0, S_1, T_1, and D states for (**d**) STORM and (**e**) dSTORM imaging techniques, highlighting the stochastic blinking behavior required for single-molecule localization

nanoscale distribution of defined neuronal proteins or examining structural plasticity across fixed preparations. STORM and dSTORM, which leverage the superior photon output and photostability of organic dyes, are particularly powerful for dense structures such as synapses, axonal segments, and cytoskeletal networks, where high localization precision is essential for accurate reconstruction of complex molecular patterns.

1.4 Experimental Considerations and Limitations

More recently, STORM has been used less routinely due to its greater labelling complexity, stringent laser requirements, and limited practical advantages over dSTORM. dSTORM provides comparable or superior performance with a substantially simpler implementation. For these reasons, fixed PALM and dSTORM are the more widely adopted fixed-cell SMLM modalities in biological and neuroscience research and form the focus of the protocol presented here.

Before outlining the protocols, it is important to consider the specific limitations and experimental requirements of these techniques. For PALM the behavior of photoactivatable fluorescent proteins (including variable activation efficiency, modest photon output, and susceptibility to premature photobleaching) demands precise regulation of the UV-activation light and sufficiently long imaging sequences to maintain sparse single-molecule emission. dSTORM is highly sensitive to the chemical environment; buffer composition, oxygen-scavenging capacity, and laser power strongly influence dye-switching kinetics, blinking behavior, and photostability.

Both techniques require robust correction of sample drift, which may accumulate over the thousands of frames acquired for SMLM. Drift must be addressed using fiducial markers or computational approaches to avoid systematic positional errors. Data analysis introduces additional challenges, including fluorophore blinking, which can generate multiple localizations of a single molecule; low-photon events, which degrade localization precision; and inappropriate PSF fitting or thresholding, which can introduce reconstruction artifacts. These factors highlight the need for rigorous optimization and validation at every stage of the workflow to ensure accurate and biologically meaningful nanoscale measurements.

In the sections that follow, we present detailed protocols for fixed-cell PALM and dSTORM imaging in monolayer cultures, including sample preparation, imaging conditions, and data-analysis procedures. The workflow described here has been implemented on an Abbelight SAFe 360 system for fixed-cell PALM and on a Zeiss Elyra PS.1 system for dSTORM but is compatible with any equivalent SMLM platform capable of PALM or dSTORM acquisition. Image reconstruction and analysis are performed using Zeiss ZEN Black 2012 and Abbelight NEO software.

2 Materials

2.1 Cell Culture Reagents

1. Primary hippocampal neurons derived from E18 C57BL/6 mice or rat pheochromocytoma cells (PC12; ATCC: CRL-1721™) seeded in glass-bottom dishes containing 10 mm micro-wells.
2. *Neuronal culture media:* Neurobasal™ media, B-27™ Supplement 50× (1/50), GlutaMAX™ Supplement (1/100), and 10,000 U/mL Penicillin-Streptomycin (1/100).
3. *PC12 culture media:* Dulbecco's Modified Eagle Medium (DMEM), Fetal bovine serum (FBS; final concentration: 10%), Horse serum (HS; final concentration: 10%), and GlutaMAX™ Supplement (final concentration: 1%), all from Thermo Fisher.

2.2 Cell Culture Transfection Reagents

1. For neuronal culture: Lipofectamine™ 2000.
2. Neuronal culture media.
3. For PC12 culture: Lipofectamine™ LTX and PLUS™ Reagent and Opti-MEM™ (Reduced Serum Media).
4. PC12 culture media.
5. DNA plasmid of interest (transfection quality).
6. 35 mm culture dish.
7. Poly-D-lysine hydrobromide (PDL).
8. UltraPure™ DNase/RNase-Free Distilled Water.
9. Glass-bottom dishes with 10 mm micro-wells.
10. 15 mL or 50 mL Falcon tubes.
11. 1.5 mL Microcentrifuge tubes.

2.3 PALM & dSTORM Fixation and Immunolabelling Reagents

1. Phosphate-Buffered Saline (PBS; 1×).
2. 4% Paraformaldehyde (PFA) in PBS.
3. 0.2% Triton X-100 in PBS.
4. *Blocking Buffer:* 1% Bovine Serum Albumin (BSA) and 2% Goat Serum in PBS.
5. Primary antibodies.
6. Parafilm.
7. Secondary antibodies.

2.4 dSTORM Buffers

Although dSTORM imaging buffers can be readily prepared in-house using standard reagents (see below for reagents and composition), a commercially available ready-to-use option is the Abbelight Smart Buffer Kit for adherent cells (Abbelight E-Shop). Prepare all buffers using Milli-Q or ultrapure water unless otherwise stated.

1. *Sample Storage Buffer (20 mL)*: 2.5 mL of 4% PFA buffer (final concentration: 0.5%), 80 μL of 5% sodium azide (final concentration: 0.02%) and 17.42 mL of PBS. Store at 4 °C for up to 1 month.
2. *Dilution Buffer (10 mL):* 0.1 mL of 1 M Tris-HCl (final concentration: 10 mM), 0.5 mL of 1 M NaCl stock solution (final concentration: 50 mM) and 9.4 mL PBS. Store at room temperature (RT; 20 °C–22 °C) for up to 6 months (*see* **Note 1**).
3. *Imaging Buffer Base (10 mL):* 1 g glucose (final concentration: 10%), 0.5 mL of 1 M Tris (final concentration: 50 mM) and 0.1 mL of 1 M NaCl stock solution (final concentration: 10 mM). Store at RT for up to 6 months.
4. *Enzyme Stock Solution with oxygen scavengers (250 μL):* 14 mg glucose oxidase, 200 μL Dilution Buffer, and 50 μL catalase (20 mg/mL). Store at 4 °C for up to 2 weeks.
5. *Cysteamine (MEA) Solution (1 mL):* 77 mg cysteamine (final concentration: 1 M) in 1 mL of 360 mM hydrochloric acid. Store solution at 4 °C for up to 1 month.
6. *dSTORM Imaging Buffer (1 mL):* 890 μL Imaging Buffer Base, 10 μL Enzyme Stock Solution, and 100 μL MEA Solution (final concentration: 100 mM). Make the solution immediately before imaging on ice and use it for up to 2 h at RT.

2.5 Imaging Reagents, Equipment, and Software

1. Class II biosafety cabinet.
2. Tissue culture incubator (set to 37 °C and 5% CO_2).
3. Fume hood.
4. Fluorescence-free immersion oil, $n = 1.518$ at 23 °C (Zeiss Immersol 518 F; Cat. No. 444960-0000-000).
5. SMLM system Zeiss Elyra PS.1 or SMLM system Abbelight SAFe 360 or any equivalent SMLM microscope.
6. Zeiss alpha Plan-Apochromat 100 ×1.46 NA oil-immersion objective or Nikon CFI Apochromat TIRF 100 × C1.49 NA oil-immersion objective.
7. Zeiss ZEN Black 2012 or Neo analysis software.

3 Methods

3.1 Fixed Photoactivated Localization Microscopy (PALM)

3.1.1 Preparation of Neuronal Cultures and Transfection

Primary hippocampal cultures should be prepared and maintained following previously published procedures [21, 22] and seeded onto glass-bottom dishes with 10 mm micro-wells at a density of 5.1×10^4 cells/cm^2. These neurons are considered classically mature at days in vitro (DIV) 21 and should be transfected prior to this stage. The appropriate transfection day may vary depending on promoter strength within the plasmid construct. In our hands, CMV-driven plasmids provide reliable expression when neurons are transfected 2–3 days prior to fixation.

Table 2
Composition of the Lipofectamine™ 2000 transfection reaction for primary hippocampal neurons

Components	Tube A	Tube B
Neurobasal media	100 μL	100 μL
Plasmid DNA (1–3 μg)		1–3 μg
Lipofectamine™ 2000 (1 μL per μg)	1–3 μL	

1. Prepare the transfection reaction tubes according to Table 2, with one reaction (tubes A + B) being sufficient for three glass-bottom dishes.
2. Mix the Lipofectamine™ 2000 (tube A) with the plasmid DNA (tube B) and incubate the mix for 25–30 min at RT (*see* **Note 2**).
3. Collect the media in which the neurons are cultured (conditioning media) into a 15 mL or 50 mL falcon tube and store the media in the incubator with the lid slightly ajar for CO_2 equilibration (*see* **Note 3**).
4. Add 67 μL of the transfection mixture dropwise to each 10 mm well and incubate the neurons at 37 °C and 5% CO_2 for 90 min.
5. After the incubation period, wash the neurons three times by ejecting warm culture media on the side of the well and allowing the media to gently flow over the neurons (*see* **Note 4**).
6. After the final wash, add back the conditioning media collected from the cells in **Step 3** and return them to the incubator.
7. Allow cells to incubate for the appropriate duration post-transfection to permit adequate expression of the transfected construct prior to fixation.

3.1.2 Preparation of PC12 Cultures and Transfection

1. Passage PC12 cells and seed them into a 35 mm tissue culture dish such that the culture reaches approximately 60% confluence the following day.
2. Prepare the transfection reaction tubes according to Table 3, with one reaction being sufficient for one 35 mm culture dish.
3. Add the Lipofectamine™ LTX to the plasmid DNA and PLUS reagent mixture and incubate for 25 min at RT.
4. Add the transfection mixture dropwise to each 35 mm culture dish and incubate at 37 °C and 5% CO_2.

Table 3
Composition of the Lipofectamine™ LTX transfection reaction for PC12 cells

Components	Single Reaction
Lipofectamine™ LTX	6.75 μL
Opti-MEM (reduced serum media)	500 μL
Plasmid DNA (1–3 μg)	1–3 μg
PLUS™ reagent (1 μL per μg)	1–3 μL

5. After 12–15 h or overnight, replace the transfection media with fresh PC12 culture media and incubate the cells while preparing the glass-bottom dishes to replate the cells.
6. Prepare a sterile 0.1 mg/mL PDL solution with ultra-pure water.
7. Add 150 μL of that solution to each 10 mm micro-well in the glass-bottom dishes and incubate for 30 min at 37 °C.
8. Wash the coated dishes 3 times with PBS and add 150 μL fresh PC12 culture media to each well. Store in the incubator at 37 °C until needed.
9. To harvest the transfected cells, replace the culture medium with 2 mL pre-warmed fresh medium and gently detach the cells by triturating 10 times using a 1 mL pipette.
10. Add this cell suspension to 8 mL warm PC12 culture media.
11. Redistribute the resulting cell suspension by replating into 10 glass-bottom dishes with the 10 mm micro-wells (*see* **Note 5**).
12. Allow cells to incubate for the appropriate duration post-transfection to permit adequate expression of the transfected construct prior to fixation.

3.1.3 Chemical Fixation of Transfected Primary Neurons and PC12 Cells

1. Retrieve the transfected cell cultures from the incubator and transfer them to a certified fume hood for subsequent processing.
2. Remove the conditioning media from the glass-bottom dish, leaving media within the 10 mm micro-well.
3. Add 1 mL 4% PFA to each dish and incubate for 30 min at RT (*see* **Note 6**).
4. Discard the PFA into an appropriate chemical waste container, then gently rinse each glass-bottom dish with 1 mL PBS. Incubate the cells in PBS for 5 min.
5. Repeat the PBS wash step three times (3 × 5 min).

6. After the final wash, leave a small volume of PBS covering the surface of the dishes. Store the samples at 4 °C until it is time to proceed with imaging.

3.1.4 Fixed PALM Image Acquisition Workflow

Custom-built SMLM setups require significant in-house expertise to build and maintain. Therefore, most SMLM experiments are performed on commercial systems such as the Zeiss Elyra 7, Nikon N-STORM, Leica SR GSD 3D, Abbelight SAFe MN360, Abberior Instruments MINFLUX, Bruker Vutara VXL, and ONI Nanoimager. The protocol described here was performed using an Abbelight SAFe 360 SMLM system with NEO Live Imaging acquisition software and should be applicable to most comparable platforms.

1. Apply immersion oil to the Nikon CFI Apochromat TIRF 100XC/1.49 NA objective, mount the specimen, and find the focus using the oculars.
2. In the NEO Live Imaging GUI, open *System* > *Hardware Configuration*, choose the appropriate saved configuration (e.g., *SAFe 360—Single color—Nanoscopy—2D*) and click *Validate* (Fig. 2; *see* **Note 7**).
3. Select *Imaging* > *Project* > *New Project* and create a new project folder (Fig. 3; *see* **Note 8**).
4. Click *Settings*. In the Camera tab, set the sensor size, region of interest (ROI) size, exposure, number of frames, and readout

Fig. 2 NEO Live Imaging software showing system hardware configuration and optical paths

Fig. 3 NEO Live Imaging Project window

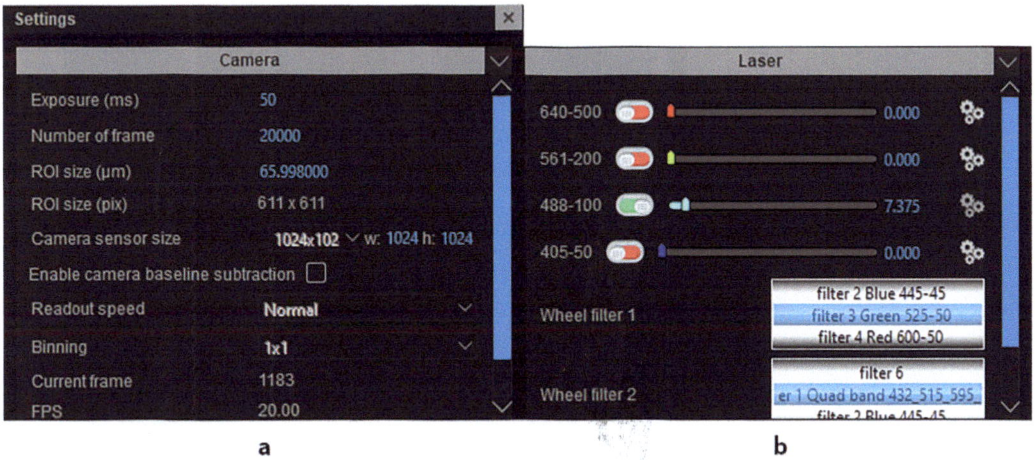

Fig. 4 Camera and laser settings windows in NEO Live Imaging. (**a**) Camera parameters, including exposure, number of frames in the stream, ROI size, sensor size, readout speed, and binning. (**b**) Laser tab showing available laser lines and respective power levels, and emission filters on the filter wheel for each camera

speed (Fig. 4a; *see* **Note 9**). In the Settings window's Laser tab, set the laser power for the acquisition line and choose the appropriate emission filter on the camera's filter wheel (Fig. 4b; *see* **Note 10**).

5. Click *Preview* on the left side of the GUI. Modify the ROI size by drawing a rectangle on the field of view with the mouse and adjust the exposure time and laser power as required (Fig. 5; *see* **Note 11**).
6. In the Field of Excitation tab, move the slider to ensure the illumination fills only the ROI without distortion. In the Beam Angle tab, select the required illumination mode (Epi-, HILO, TIRF) and set the illumination angle (Fig. 6a; *see* **Note 12**). In the Processing tab, set the detection threshold and single-molecule minimum distance. In the Saving tab, tick *ROI* and *Nano-2D* (Fig. 6b; *see* **Note 13**).
7. Before starting the stream acquisition, activate the laser wavelengths for activation and fluorescence excitation, and switch to the appropriate emission filter (Fig. 7; *see* **Note 14**).
8. Click *Stop Preview*, then click *Start* to begin the stream acquisition (Fig. 8a; *see* **Note 15**). Drag a camera view (e.g., *Camera, ROI or Nano-2D*) to monitor the acquisition (Fig. 8b).
9. Click the small bar graph icon and select either the Drift (Fig. 9a), FoV Hist (Fig. 9b), Localization count (Fig. 9c), or Nb photons per loc (Fig. 9d) tabs to observe the relevant live-stream acquisition parameters.

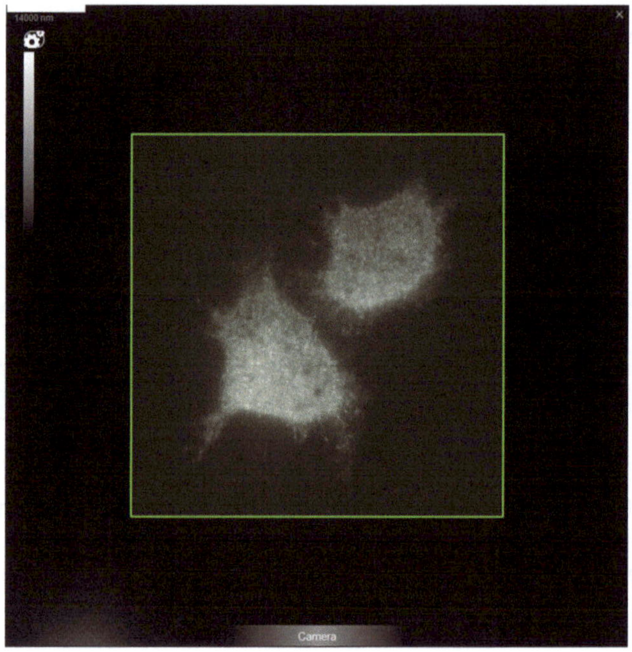

Fig. 5 Live camera view displaying the PC12 specimen and the defined region of interest

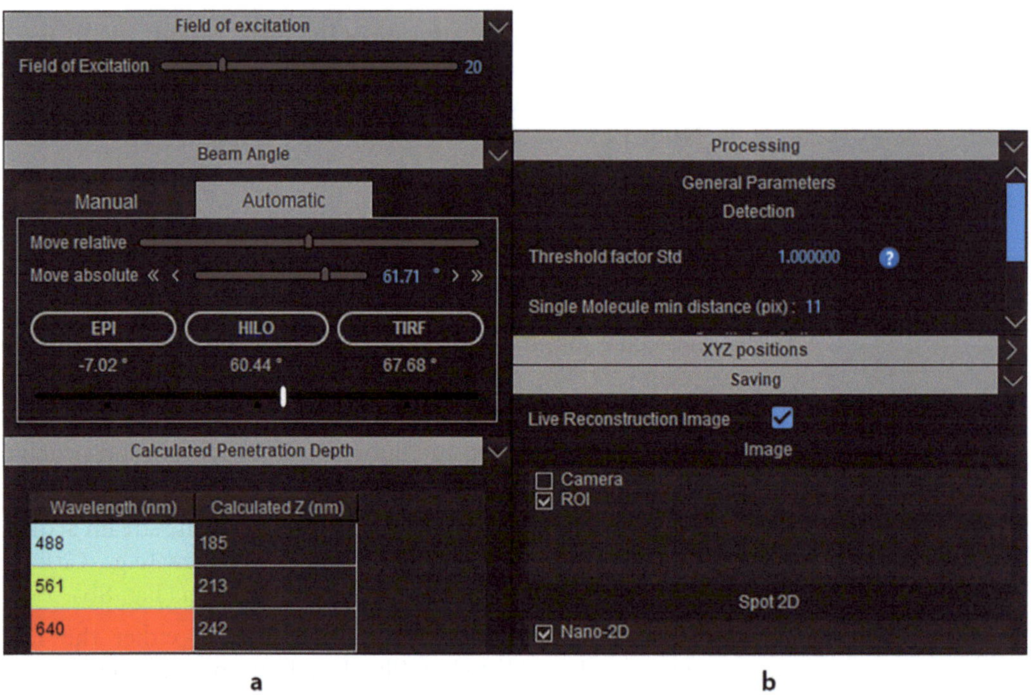

Fig. 6 Field of excitation, beam angle, processing, and saving tabs. (**a**) Field of Excitation tab, for setting the illumination region and Beam Angle tab, for Epi-, HILO, and TIRF illumination modes and associated penetration depths. (**b**) Processing tab for detecting single particles in real-time and Saving tab for saving the raw data and super-resolved reconstruction

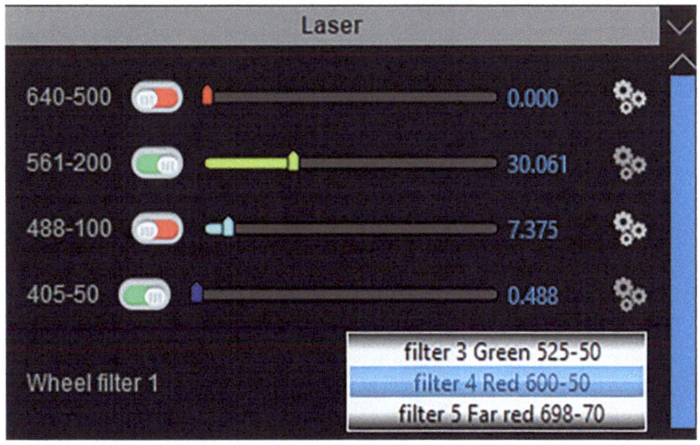

Fig. 7 Laser and filter settings showing parameters for activating and imaging mEos

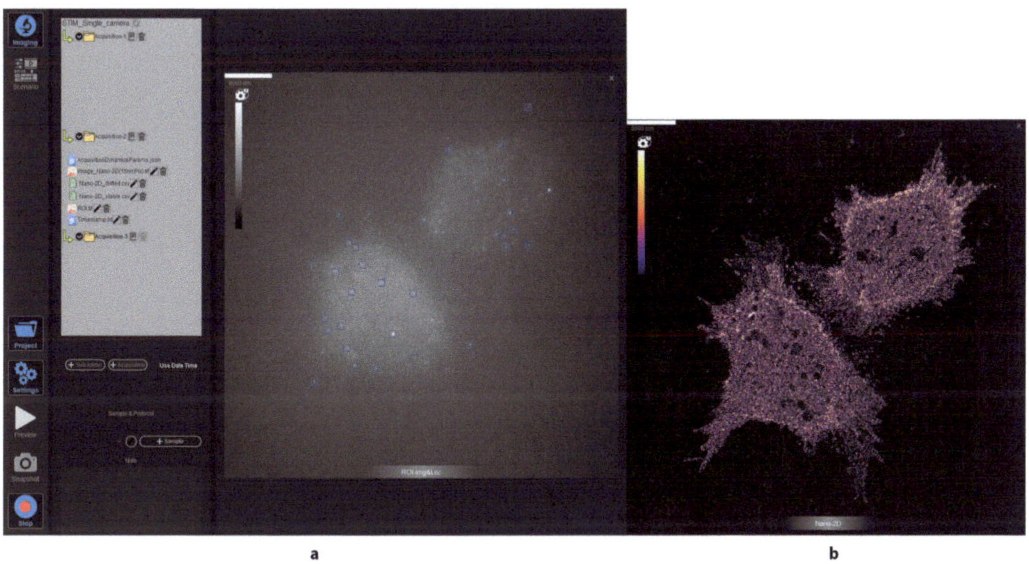

Fig. 8 Stream acquisition and camera views. (**a**) Stream acquisition showing localizations in real-time. (**b**) Stream acquisition showing super-resolution reconstruction view

10. When the acquisition is finished, the data is saved. To abort and save the frames already acquired, click *Stop* (*see* **Note 16**).

3.1.5 SMLM Data Analysis

We recommend processing the raw data using Abbelight NEO Analysis for a complete set of detection parameters (background removal, detection, and super-localization parameters, drift correction).

1. Open the NEO Analysis software and click *Batch*. Choose the appropriate camera mode (SAFe 180 for single-camera or SAFe 360 for dual-camera acquisition) and the desired imaging

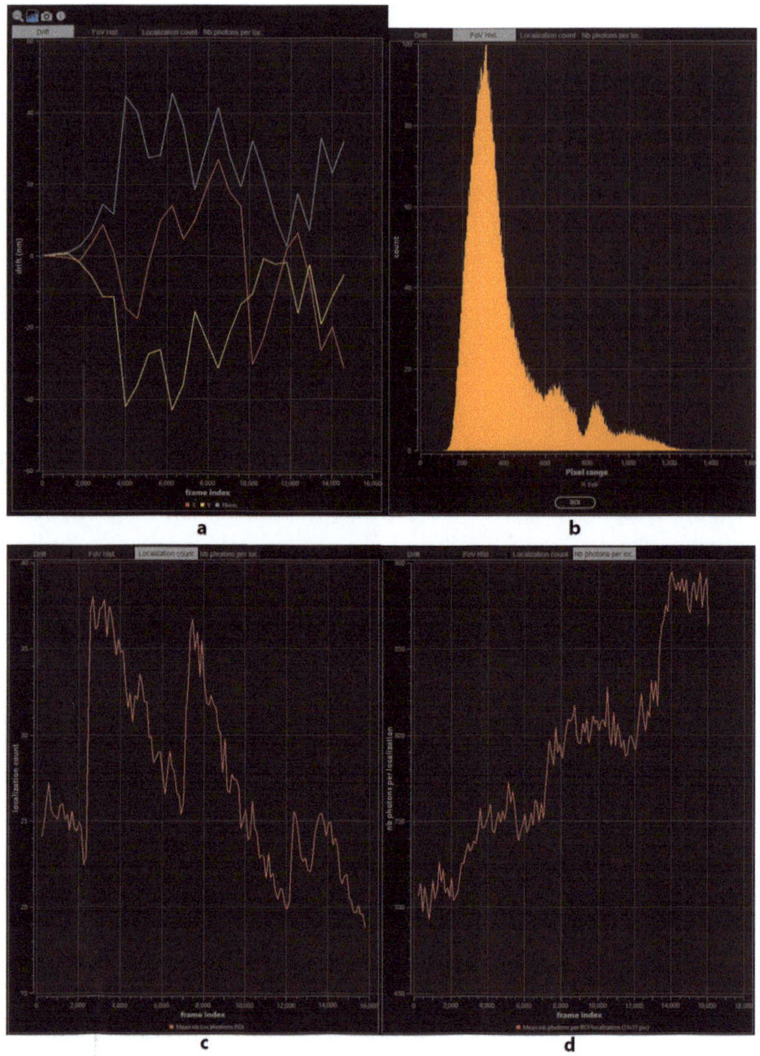

Fig. 9 NEO Live Imaging stream acquisition data parameters. (**a**) Drift curves. (**b**) Histogram curve. (**c**) Localization counts plot. (**d**) Number of photons per localization graph

modality. Enter the camera parameters, direct the software to the *ROI TIFF file* for processing, and assign a destination folder for the output (Fig. 10; *see* **Note 17**).

2. In the Background Removal tab, choose the methods for detection, super-localization, and photon-counting, and define the number of images for the temporal median (Fig. 11a; *see* **Note 18**). Set the detection threshold, local mean kernel size, minimum/maximum PSF areas, super-localization method, photon-counting method, mask size, and PSF-matching radius (Fig. 11b; *see* **Note 19**).

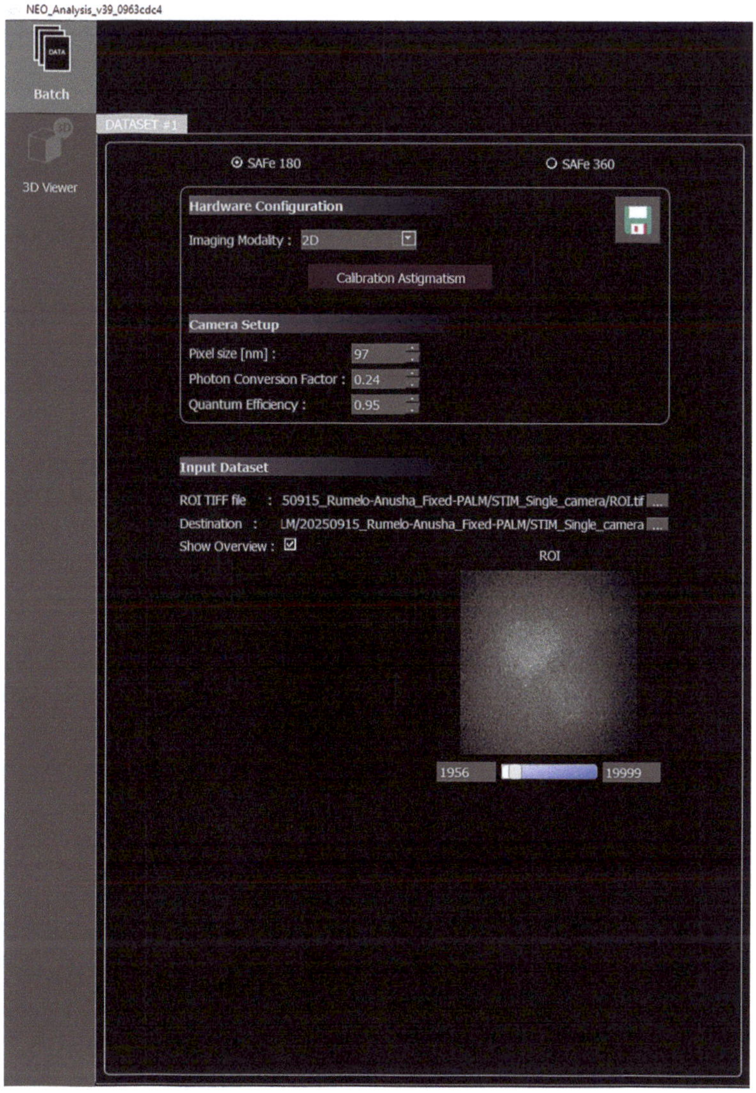

Fig. 10 NEO Analysis batch-processing window

3. In the Drift Correction tab, tick *Compute Drift Correction* or load a previously created drift-correction table (Fig. 12a; *see* **Note 20**). Click *Add Dataset* to include more files, then select *Start Batch* (Fig. 12b; *see* **Note 21**).
4. Exit batch processing by clicking *Batch* again, then click *3D Viewer*. Select *Load File* and open the coordinate table.csv file (*see* **Note 22**).
5. Within the General tab, modify the *Point Size* parameter to alter the visual size of localization points and adjust the *Alpha Value* to set their brightness. Enable or disable the scale bar and xyz marker using the respective checkboxes (Fig. 13; *see* **Note 23**).

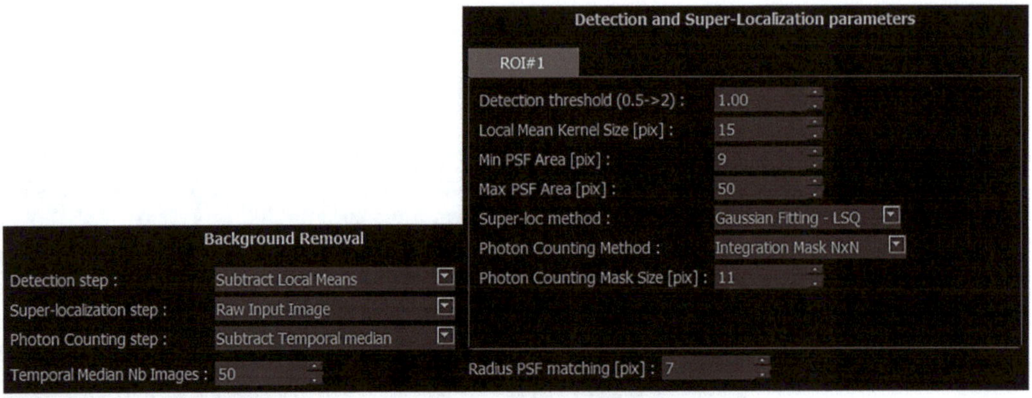

a b

Fig. 11 Batch-processing parameters. (**a**) Background-removal parameters, including the detection, super-localization, and photon-counting algorithms, and the temporal median number of images. (**b**) Detection and super-localization settings, including the detection threshold, local mean kernel size, minimum and maximum PSF areas, super-localization and photon-counting methods, photon-counting mask size, and radius for PSF matching

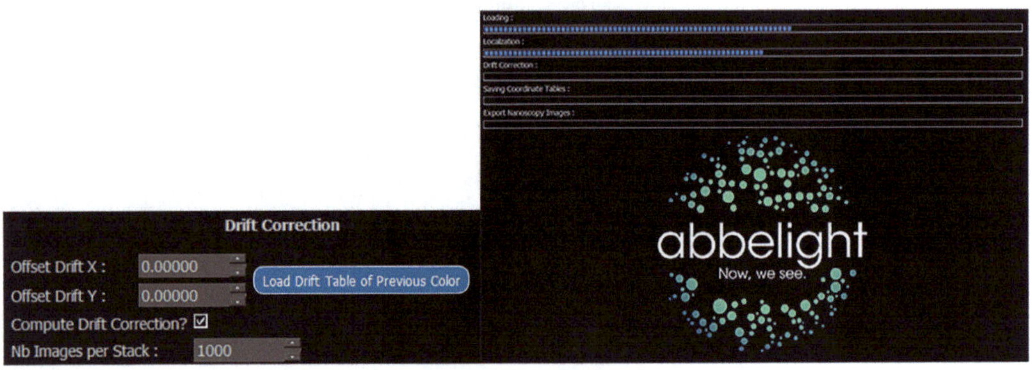

a b

Fig. 12 Drift correction and processing status. (**a**) Drift-correction parameters, including the facility to load a saved drift-correction table. (**b**) Batch-processing loading, localization, drift correction, and saving status bars

6. On the Spatial tab, use the Spatial Distribution sub-tab to view surface area, localization count, and density. Select *Clipping* to crop and view ROI properties; *Filtering* to filter data based on neighborhood settings and *Transformation* for translation, scaling, rotation, and multi-channel realignment (Fig. 14; *see* **Note 24**).
7. In the SMLM tab, use *Frame Detection* to inspect detections over time and filter by frame number; *Uncertainty* to view localization precision histogram and filter detections by precision range; *Intensity* to view photon count histogram and filter by photon number; *Blinking Spatial Distance* to view nearest-

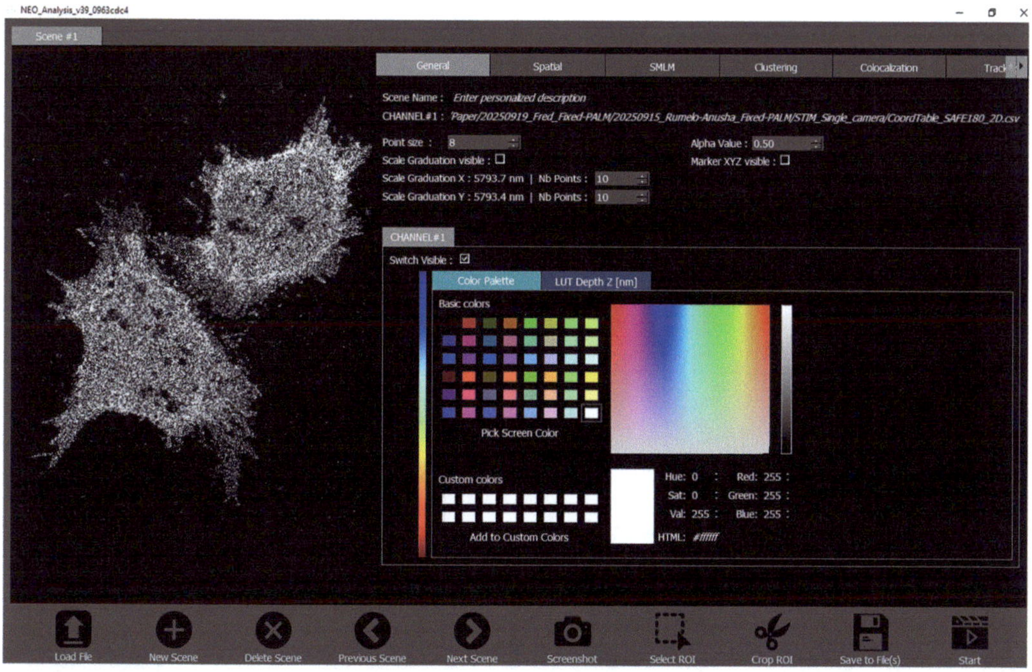

Fig. 13 NEO Analysis 3D Viewer window showing a localization-based reconstruction generated from accumulated detections (left) and the processing tabs (right)

neighbor distance and filter by minimum distance and *Blinking Temporal Analysis* to analyze blinking events and merge detections from the same molecule (Fig. 15; *see* **Note 25**).

8. In the Clustering tab, apply clustering algorithms (e.g., K-Ripley, DBSCAN, or Voronoi) to the localization dataset as needed. Identify specific clusters using their Cluster ID and filter/export to extract their associated metrics (Fig. 16; *see* **Note 26**).

3.2 Direct Stochastic Optical Reconstruction Microscopy (dSTORM)

3.2.1 Preparation of Neuronal Cultures

As above, primary hippocampal cultures should be prepared and maintained following previously published procedures [21, 22] and seeded onto glass-bottom dishes with 10 mm micro-wells at a density of 5.1×10^4 cells/cm^2 (*see* **Note 27**). These neurons are considered classically mature at days in vitro (DIV) 21 and should be fixed around this stage.

3.2.2 Fixation

1. Retrieve the neuron culture dishes from the incubator and transfer them to a certified fume hood for subsequent processing.
2. Remove the conditioning media from the glass-bottom dish, leaving media within the 10 mm micro-well.
3. Add 1 mL 4% PFA to each dish and incubate for 20–30 min at RT (*see* **Note 6**).

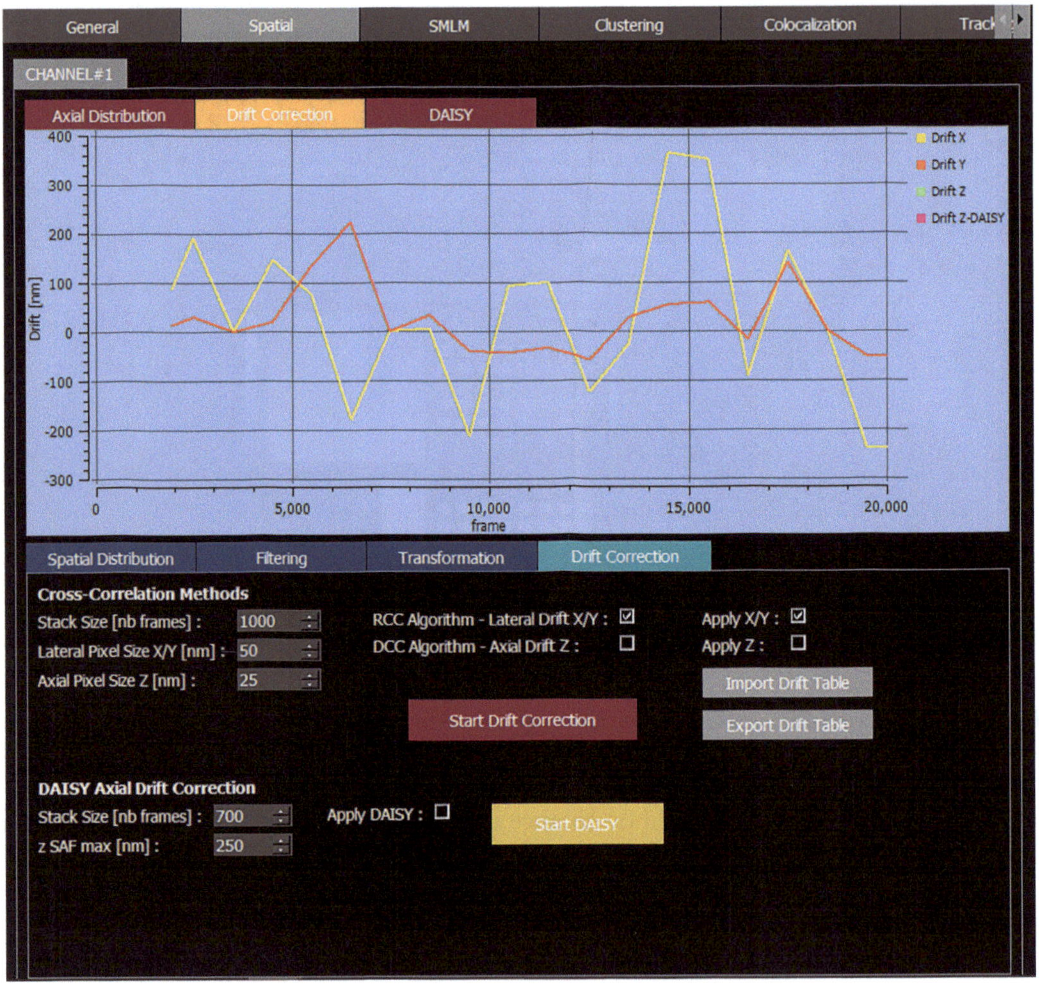

Fig. 14 NEO Analysis Spatial tab showing drift correction via cross-correlation

4. Following fixation, discard the PFA into an appropriate chemical waste container and gently wash each glass-bottom dish with 1 mL PBS. Incubate the cells in PBS for 5 min.
5. Repeat the PBS wash step three times (3 × 5 min).

3.2.3 Immunolabelling

1. Remove the remaining PBS and permeabilize the neurons by adding 1 mL 0.2% Triton X-100 to each dish. Incubate for 5 min at RT (*see* **Note 28**).
2. Discard the Triton X-100 into an appropriate chemical waste container and gently wash each glass-bottom dish with 1 mL PBS. Incubate the cells in PBS for 5 min.
3. Repeat the PBS wash step three times (3 × 5 min).
4. Remove the PBS and add 1 mL Blocking Buffer to each dish. Block for 2 h at RT.

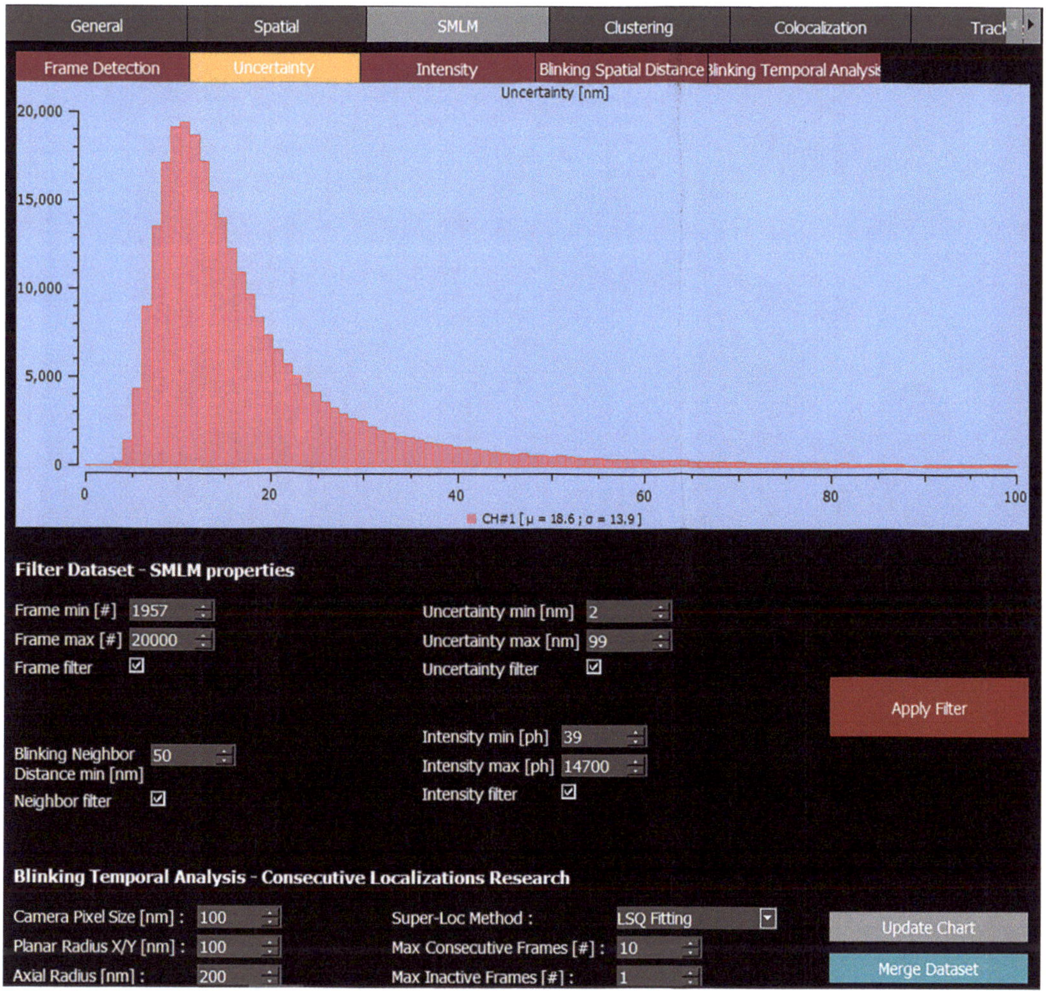

Fig. 15 NEO Analysis SMLM tab, showing the Localization Precision (or Uncertainty) histogram, and the available detection-filter settings

5. Prepare the primary antibodies in Blocking Buffer (*see* **Note 29**).
6. Incubate the neurons with the primary antibody solution overnight (15–20 h) at 4 °C (*see* **Note 30**).
7. Discard the antibody solution and gently wash each glass-bottom dish with 1 mL PBS. Incubate the cells in PBS for 5 min.
8. Repeat the PBS wash step three times (3 × 5 min).
9. Prepare the secondary antibodies in Blocking Buffer (*see* **Notes 29** and **31**).
10. Add the secondary antibody solution to the neurons and incubate for 90 min at RT (protected from light).

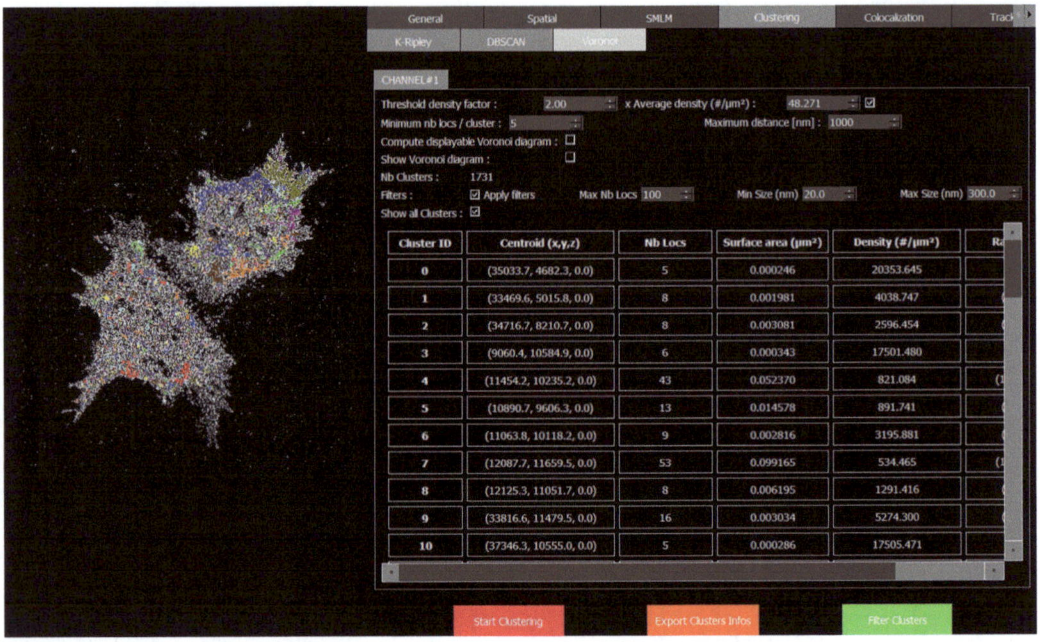

Fig. 16 NEO Analysis Clustering tab, showing cluster analysis performed using Voronoi tessellation

11. Complete a post-staining fixation by incubating cells in 4% PFA buffer for 20 min at RT.
12. Store samples in Storage Buffer at 4 °C (protected from light).

3.2.4 dSTORM Image Acquisition Workflow

As mentioned above, custom-built SMLM setups require significant in-house expertise to build and maintain, and so these experiments are primarily performed on commercial systems. This protocol was implemented using the Zeiss Elyra PS.1 SMLM system with ZEN Black 2012 software and is compatible with most SMLM systems.

1. Prepare the dSTORM Imaging Buffer on ice and take note of the time to ensure image acquisition is within the buffer's lifetime.
2. Replace the Storage Solution within the glass-bottom dish with Imaging Buffer and mount the specimen on the stage using immersion oil on the Zeiss alpha Plan-Apochromat 100×/ 1.46 NA oil-immersion objective.
3. Locate the neurons using your neuronal marker and bring them into focus using the oculars.
4. In the Zeiss ZEN Black software, activate the appropriate lasers in the Laser tab, selecting those corresponding to your secondary antibodies (Fig. 17a). In the Imaging Setup tab, select *Laser WF* (Fig. 17b). This will automatically select the Andor iXon3 897 EMCCD camera (*see* **Note 32**).

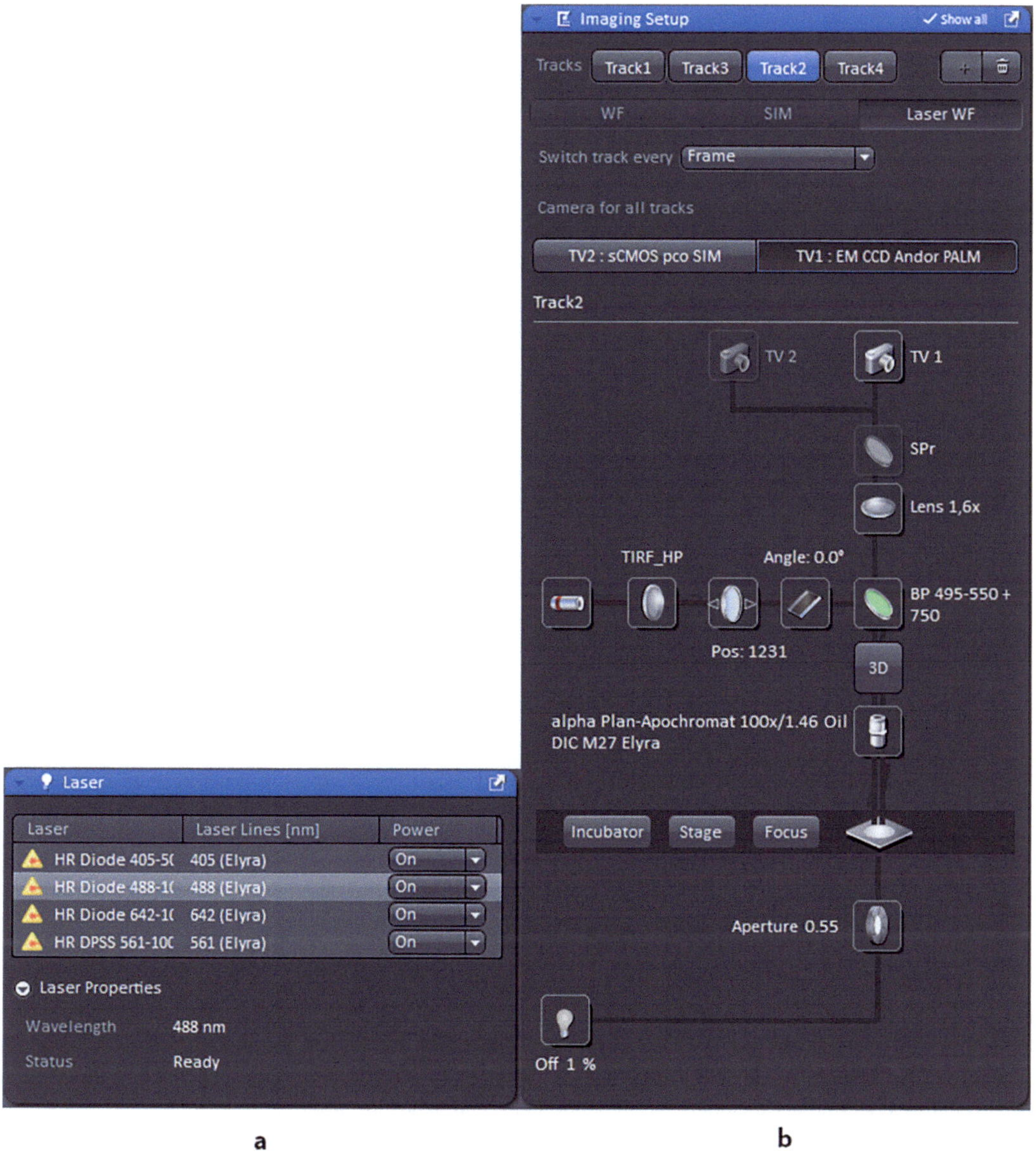

Fig. 17 Laser and Imaging Setup tabs in Zeiss ZEN Black. (**a**) Laser tab showing active 405-, 488-, 561-, and 640-nm lasers. (**b**) Imaging setup illustrating the Laser WF mode, selected EMCCD camera, imaging objective, TIRF illumination configuration, collimator position, TIRF angle adjustment, emission filter, and magnifying lens settings

5. In the Acquisition Mode tab, select the image frame size and set the bit depth to 16-bit (Fig. 18a; *see* **Note 33**).
6. In the Channels tab, enable the track to be imaged (Fig. 18b). Assign the channel color, excitation laser wavelength, laser power, illumination mode, TIRF angle, collimator position, EMCCD gain, and exposure time (*see* **Note 34**).
7. Activate Definite Focus to minimize z-drift.

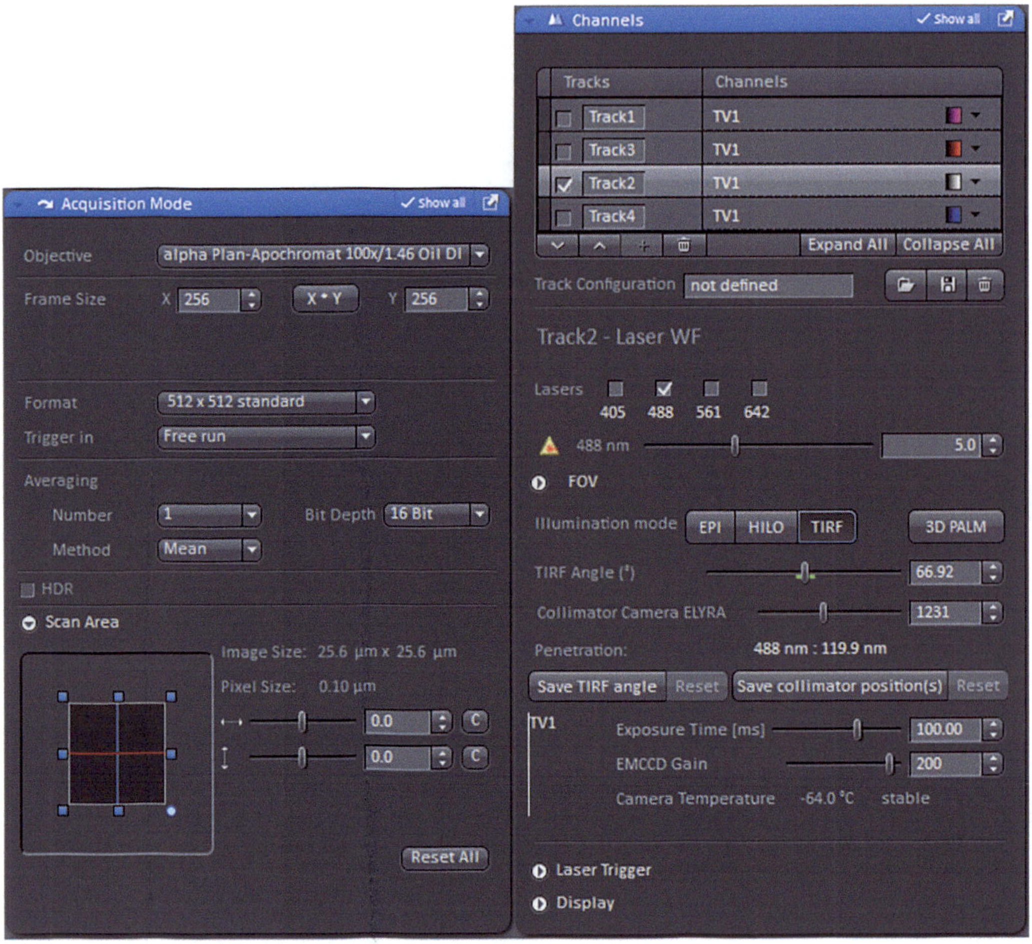

a b

Fig. 18 Zeiss ZEN Black Acquisition Mode and Channels tab. (**a**) Acquisition mode showing imaging objective, frame size, and bit depth. (**b**) Channels tab showing the configured imaging tracks, excitation laser wavelength and power, illumination mode, TIRF angle and corresponding penetration depth, collimator position, and EMCCD camera parameters including temperature, gain, and exposure time

8. Use *Live* mode to visualize the specimen and ensure the structures of interest are correctly focused.
9. To bleach the fluorophores and bring the molecules into the dark state, set up a time series of 20,000 frames with a 0-ms time interval (Fig. 19a). Start the initial time series experiment with a low gain (e.g., 10) and an exposure time that prevents dropped frames (e.g., 30 ms for 512 pixels × 512 pixels; Fig. 18b; *see* **Note 35**). Slowly increase the laser power from 5% to 100%. Once all molecules are in the dark state (with no molecules blinking), slowly increase the EMCCD gain from 10 to 200, avoiding saturation (*see* **Note 36**).

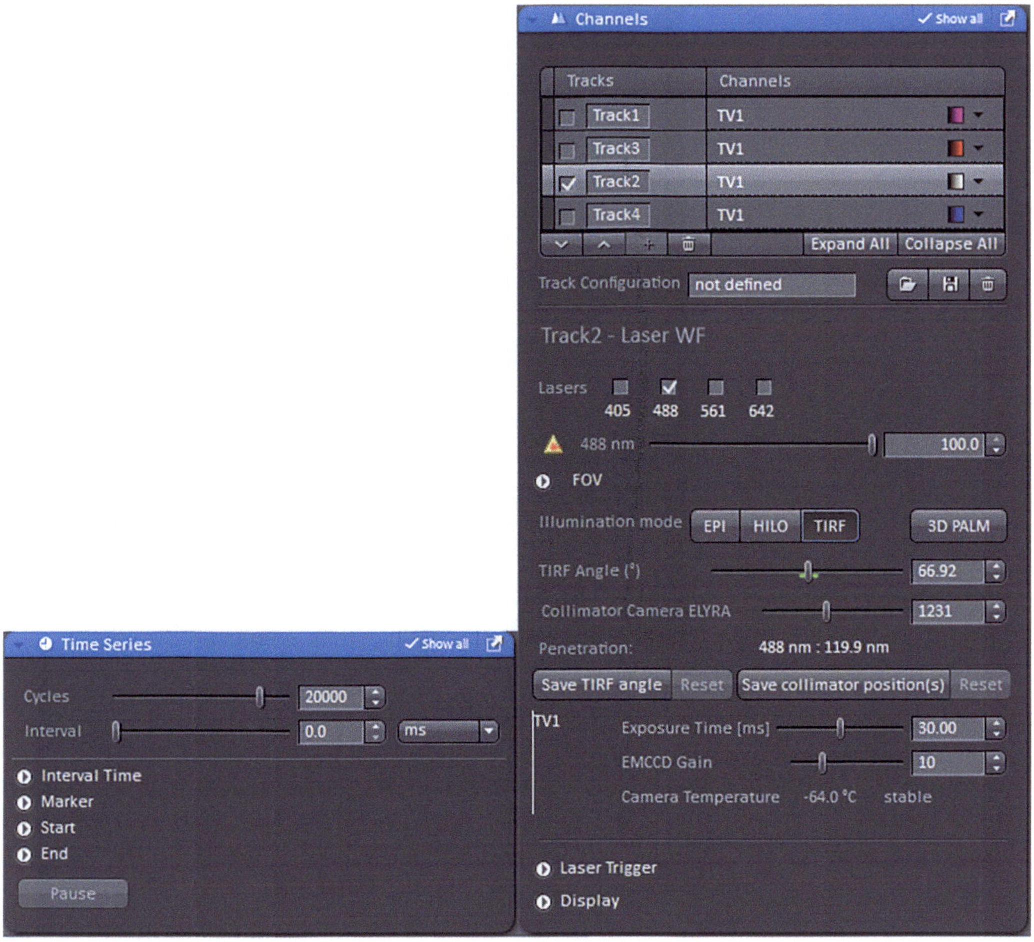

a b

Fig. 19 Zeiss ZEN Black software settings used to induce fluorophore dark state. (**a**) Time series setup of 20,000 frames with zero interval. (**b**) Tail end of the bleaching step, showing 100% laser power and low EMCCD gain

10. After completing the bleaching step and once clear single-molecule blinking is observed, acquire a dSTORM time-series. We recommend 60,000 frames per acquisition (Fig. 20; *see* **Note 37**).

3.2.5 SMLM Data Analysis

We recommend processing the raw data in Zeiss ZEN Black 2012, as it provides full access to instrument-embedded metadata and system-optimized localization tools, including background subtraction, detection and fitting parameters, and drift and distortion correction.

1. To begin determining the PSF value, open the Processing tab, select *PALM*, and then select *PALM* again to launch the peak-finding and localization tool (Fig. 21a). Under Method

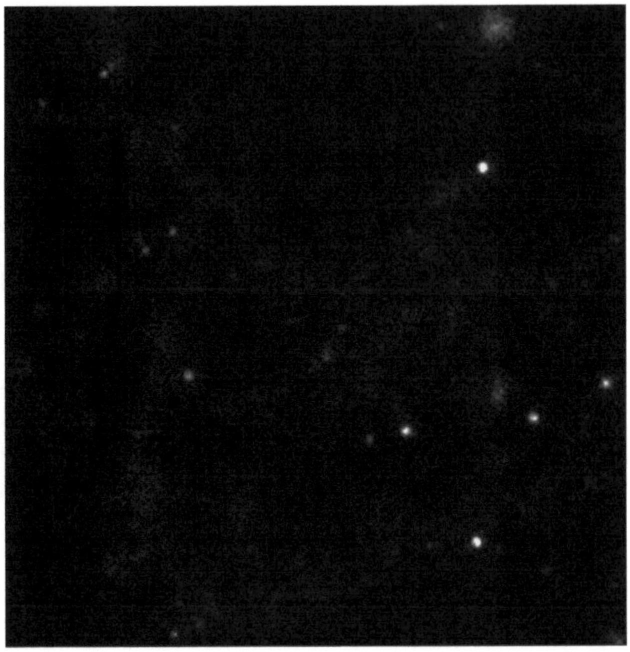

Fig. 20 One frame from a raw SMLM dataset showing individual fluorophores blinking

Parameters click *Select* to process the image stack (Fig. 21b). From the Settings drop-down, choose *Discard overlapping molecules* to restrict the analysis to isolated emitters suitable for determining the PSF (*see* **Note 38**).

2. Next, set the *Peak Mask Size* to define the pixel neighborhood used for molecule detection (*see* **Note 39**), and adjust the *Peak Intensity to Noise* filter threshold to optimize the discrimination of true emitters from background fluctuations (*see* **Note 40**).
3. After peak detection is configured, the Peak Finder window will show the detected molecules based on the parameters set (Fig. 22a). Adjust filters if necessary to ensure that isolated, single-molecule peaks are correctly identified.
4. Under the Localizer panel, choose *x,y 2D Gauss Fit* and click *Apply* to perform the initial Gaussian localization of single emitters (*see* **Note 41**). The software will then display the reconstructed dataset (Fig. 22b).
5. To extract the PSF half-width from this preliminary localization, navigate to the PAL-Statistics tab, choose *Histogram* for *Plot type* and *PSF half width* for *Histogram Source* (Fig. 23a). The resulting histogram (Fig. 23b) typically displays a dominant peak corresponding to the characteristic PSF half-width of the dataset (*see* **Note 42**).

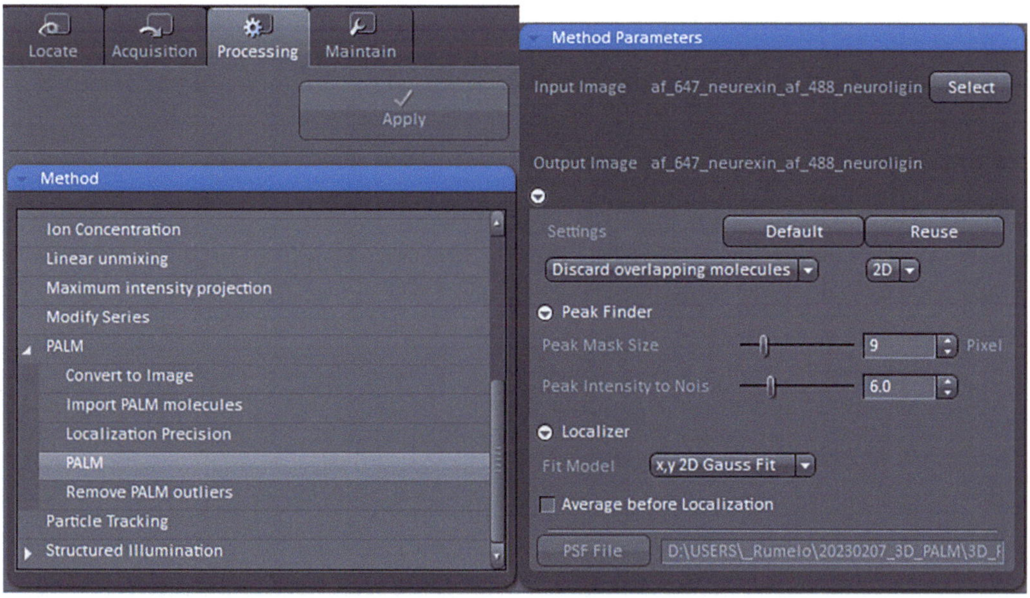

a b

Fig. 21 Launching SMLM analysis in Zeiss ZEN Black. (**a**) Processing tab with PALM analysis selected. (**b**) Method Parameters tab showing peak-finder settings, including the Discard overlapping molecules option enabled

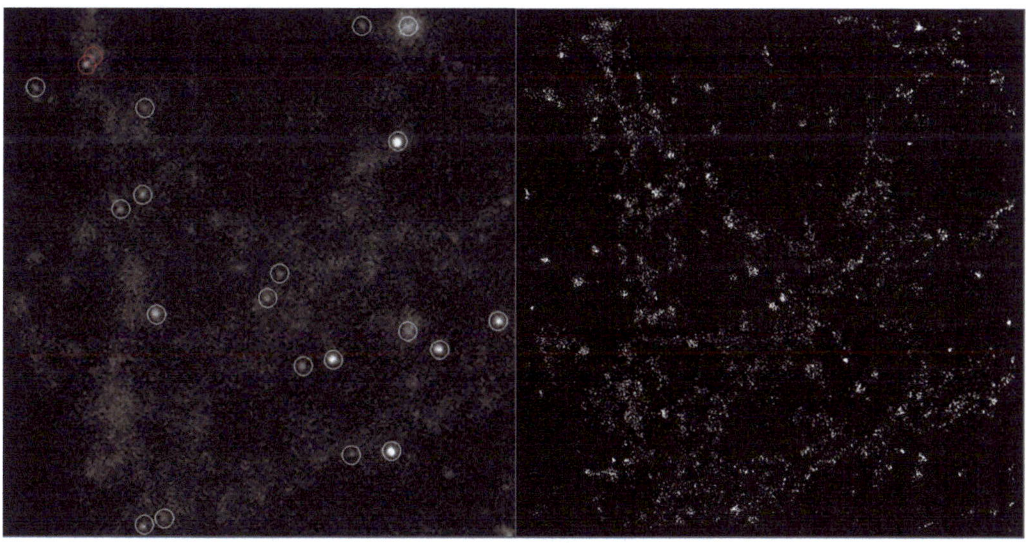

a b

Fig. 22 Initial SMLM processing results. (**a**) Peak Finder window showing a preview of detected molecules. (**b**) Localized dataset generated after applying the initial 2D Gaussian fit

Fig. 23 Determining the PSF half-width. (**a**) PAL-Statistics tab showing the settings for plotting the PSF width histogram and table. (**b**) Resulting PSF width histogram with a peak at ~140 nm

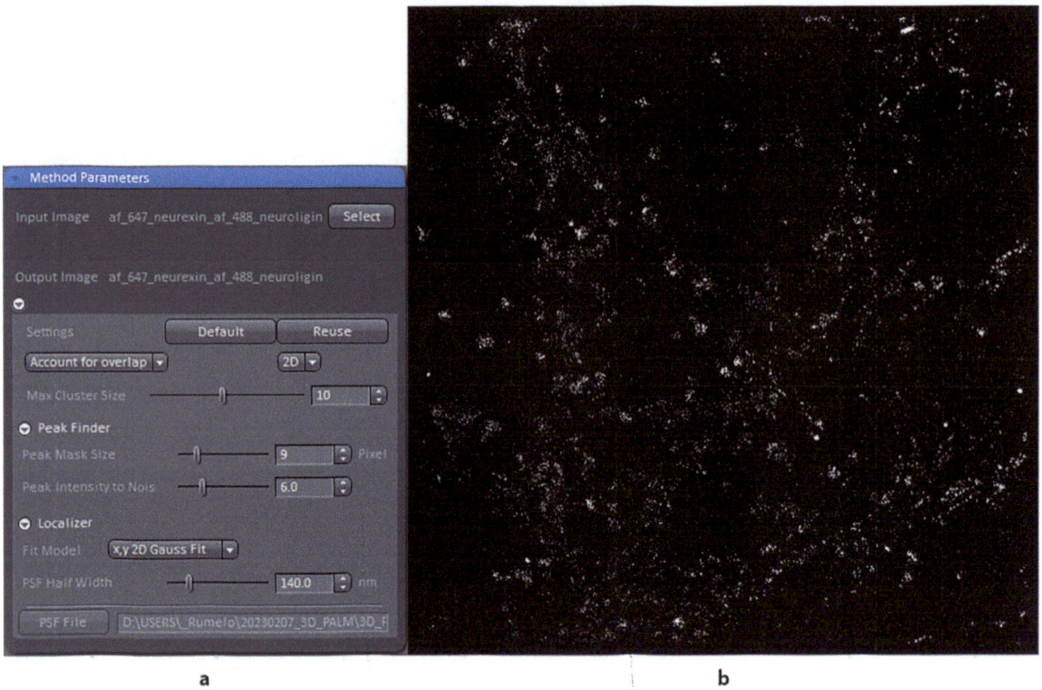

Fig. 24 SMLM processing with multi-emitter fitting. (**a**) Method Parameters tab with Account for overlap selected and the measured PSF half-width used. (**b**) Final localized dataset incorporating multi-emitter events

6. To refine the localization and incorporate information from overlapping emission events, rerun the PALM processing by selecting *Account for overlap* in the Settings drop-down menu (Fig. 24a; *see* **Note 43**).
7. Under the Localizer panel, again select *x,y 2D Gauss Fit* and enter the PSF width measured in the previous steps (e.g., 140 nm). Click *Apply* to compute the final localization dataset (Fig. 24b; *see* **Note 44**).
8. To correct sample drift during acquisition, open the PAL-Drift tab. Select *Model-Based* for *Correction Type* and *Drift Diagram* for *Display*. Select *Automatic* from the *Segments* drop-down menu and specify the maximum number of segments (e.g.,

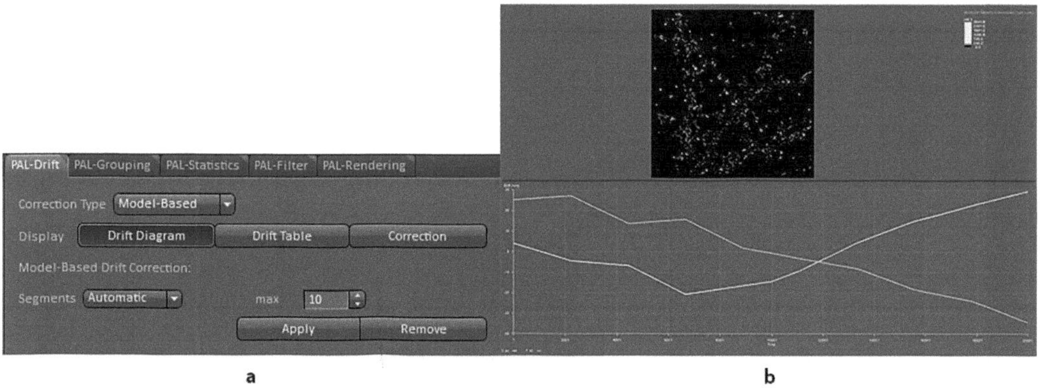

Fig. 25 Drift correction in Zeiss ZEN Black. (**a**) PAL-Drift tab showing parameters used for the correlation-based drift estimation. (**b**) Drift diagram and corresponding drift-corrected dataset

Max 10) to allow the algorithm to determine appropriate temporal subdivisions for drift estimation (Fig. 25a; *see* **Notes 20** and **45**). Click *Apply* to display the drift diagram (Fig. 25b). To save the calculated drift table values, select *Drift Table* in the PAL-Drift tab, right-click, and choose *Save Table.*

9. To filter the data by localization precision, open the PAL-Statistics tab, choose *Statistics* for *Display*, *Histogram* for *Plot type* and *Precision* for *Histogram Source* (Fig. 26a; *see* **Note 46**). Determine the *Full Width at Half Maximum (FWHM)* of the localization precision histogram to set the filter range (e.g., 12 nm to 33 nm; Fig. 26b; *see* **Note 47**). Apply the filter by going to the PAL-Filter tab, ticking *Localization Precision* and entering the FWHM range (Fig. 26c; *see* **Note 48**). ZEN will then update the dataset and render a precision-filtered high-resolution SMLM image (Fig. 26d).
10. To group persistent emitters (*see* **Note 49**), in the PAL-Grouping tab specify the *Capture Radius*, *Max On Time* and *Off Gap* parameters, then click *Group* (Fig. 27a; *see* **Note 50**). From a molecule statistics table of ungrouped data (Fig. 27b) a modified molecule statistics table of grouped data will be displayed (Fig. 27c; *see* **Note 50**).
11. To render the final image, open the PAL-Render tab and set the *Pixel Resolution XY* (*see* **Note 51**). Select the appropriate *Display Mode* (e.g., Gauss, Centroid, Molecule Density, etc.) to visualize the localization data in a manner suited to the experimental question (Fig. 28; *see* **Note 52**). If necessary, adjust the *Expansion factor* to scale the rendering appropriately for presentation without altering localization coordinates (*see* **Note 53**).

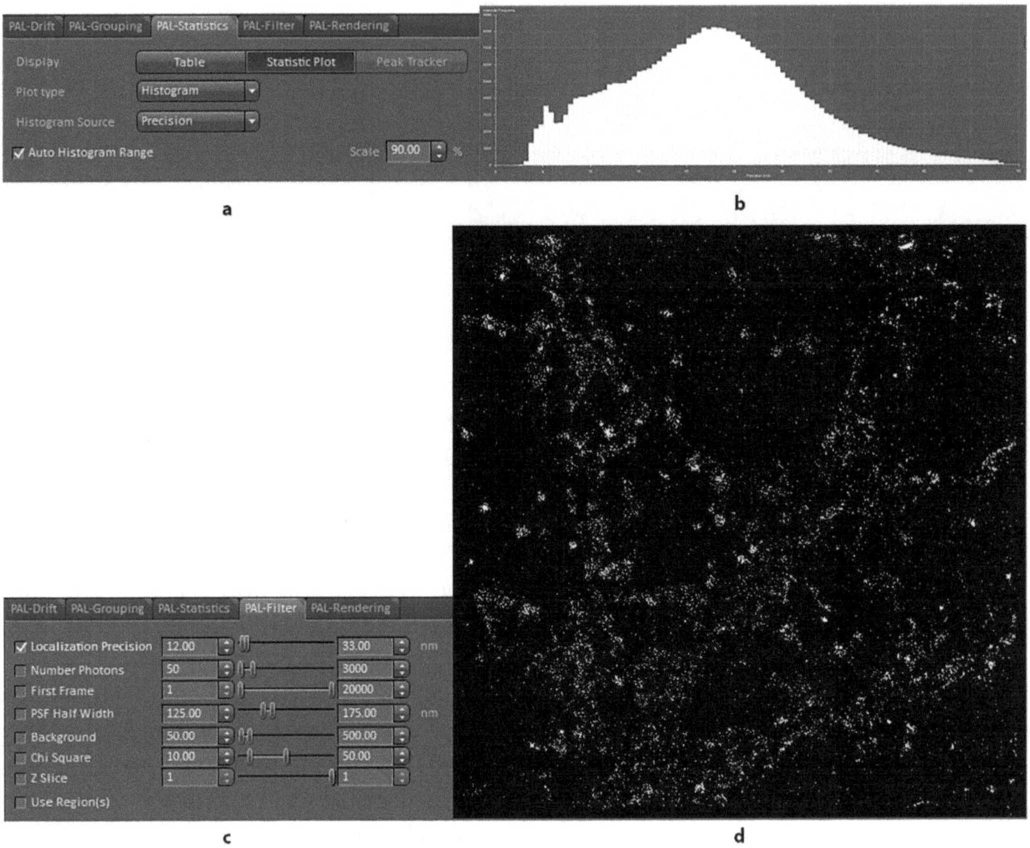

Fig. 26 Filtering localization by precision. (**a**) PAL-Statistics tab set up to display the localization precision histogram. (**b**) Localization precision histogram used to determine the FWHM. (**c**) PAL-Filter tab configured to apply the localization-precision filter. (**d**) Localization precision-filtered SMLM dataset

4 Notes

1. Carefully adjust the pH of the Tris-HCl buffer. Ensure that the pH does not drop below 8.0, as lower pH can reduce switching efficiency and buffer performance for dSTORM imaging.
2. Lipofectamine should ideally be added to the DNA and not the other way around. This allows the cationic lipids to gradually and evenly coat the negatively charged DNA. This slow, controlled mixing forms uniform lipoplexes that efficiently enter cells with low toxicity. If DNA is added directly into concentrated Lipofectamine, it may neutralize the lipids too quickly, causing aggregates that reduce transfection efficiency and increase cell stress.
3. Exposing the neurons to air will kill the culture. When removing the conditioning media, only remove the media outside of the 10 mm well.

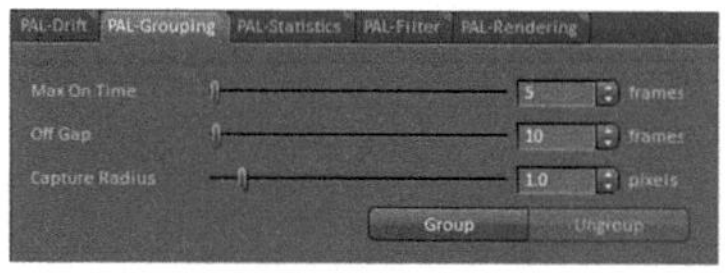

a

Index	First Frame	Number Frames	Frames Missing	Position X [nm]	Position Y [nm]	Precision [nm]	Number Photons	Background variai	Chi square	PSF half width [nm]	Channel	Z Slice
189147	12497	1	0	6131.3	15482.5	19.7	379	20.82	0.48	140.0	1	1
189148	12497	1	0	24797.3	10419.8	27.2	304	28.88	0.46	140.0	1	1
189149	12497	1	0	21817.7	9030.1	17.9	407	18.82	0.37	140.0	1	1
189150	12497	1	0	19721.2	19087.4	20.5	379	23.12	0.61	140.0	1	1
189151	12497	1	0	19452.7	4142.6	27.6	240	17.61	0.33	140.0	1	1
189152	12497	1	0	7907.4	21097.2	25.3	313	25.80	0.50	140.0	1	1
189153	12497	1	0	7638.0	18649.5	23.1	332	23.54	0.45	140.0	1	1
189154	12497	1	0	17454.3	7065.7	16.1	523	25.51	0.45	140.0	1	1
189155	12498	1	0	5925.5	16555.1	22.4	369	27.35	0.51	140.0	1	1
189156	12498	1	0	16564.0	8538.1	21.9	409	32.78	0.54	140.0	1	1
189157	12498	1	0	19412.2	4179.7	28.0	281	26.23	0.43	140.0	1	1
189158	12498	1	0	20393.6	8506.5	23.2	314	20.80	0.41	140.0	1	1
189159	12498	1	0	4765.5	3190.3	23.6	373	32.23	0.56	140.0	1	1
189160	12498	1	0	22905.4	7142.3	24.2	318	23.86	0.40	140.0	1	1
189161	12498	1	0	7911.1	21132.4	22.3	337	22.11	0.50	140.0	1	1
189162	12498	1	0	24842.3	10437.0	27.0	303	28.23	0.50	140.0	1	1
189163	12498	1	0	15906.4	14592.2	22.8	624	91.57	0.75	140.0	1	1
189164	12498	1	0	6600.0	15663.5	21.6	383	27.35	0.51	140.0	1	1
189165	12498	1	0	16081.9	14957.7	26.0	543	91.57	0.75	140.0	1	1
189166	12498	1	0	7639.8	18682.5	29.7	252	23.85	0.39	140.0	1	1
189167	12498	1	0	19739.0	19099.6	17.5	409	17.90	0.43	140.0	1	1
189168	12498	1	0	21811.1	9056.4	18.1	377	16.27	0.39	140.0	1	1
189169	12498	1	0	17457.8	7073.8	13.1	612	20.49	0.39	140.0	1	1
189170	12499	1	0	15152.2	15259.5	27.4	329	35.09	0.75	140.0	1	1
189171	12499	1	0	8164.8	16292.1	24.0	312	22.39	0.45	140.0	1	1
189172	12499	1	0	20357.1	8512.5	17.1	536	32.12	0.62	140.0	1	1
189173	12499	1	0	19392.7	4169.6	26.2	314	28.19	0.47	140.0	1	1
189174	12499	1	0	4194.7	1437.1	16.5	409	15.00	0.39	140.0	1	1
189175	12499	1	0	17449.9	7044.0	25.0	295	22.01	0.48	140.0	1	1
189176	12499	1	0	15294.6	14556.8	23.9	382	35.09	0.75	140.0	1	1
189177	12499	1	0	22909.7	7189.0	28.1	283	26.85	0.39	140.0	1	1
189178	12499	1	0	6605.6	15675.5	26.6	310	28.78	0.48	140.0	1	1
189179	12499	1	0	5476.9	16105.3	32.6	355	61.72	0.77	140.0	1	1
189180	12499	1	0	16529.9	8534.8	22.9	316	20.50	0.46	140.0	1	1
189181	12499	1	0	8892.6	3490.9	14.5	492	16.05	0.50	140.0	1	1

b

Index	First Frame	Number Frames	Frames Missing	Position X [nm]	Position Y [nm]	Precision [nm]	Number Photons	Background variai	Chi square	PSF half width [nm]	Channel	Z Slice
189147	19824	2	0	19774.9	19068.6	13.6	539	21.25	0.48	140.0	1	1
189148	19825	2	2	10023.7	5822.2	12.4	672	28.35	0.57	140.0	1	1
189149	19828	2	0	19773.8	19082.5	12.2	566	17.06	0.49	140.0	1	1
189150	19829	2	5	9990.3	5848.7	14.5	655	38.96	0.55	140.0	1	1
189151	19829	2	4	11586.0	10392.1	14.1	528	22.15	0.55	140.0	1	1
189152	19830	2	3	9990.8	5811.0	12.1	666	25.35	0.58	140.0	1	1
189153	19831	2	0	9985.7	5830.5	14.0	649	35.28	0.61	140.0	1	1
189154	19832	2	2	21837.4	9034.5	12.1	677	25.73	0.51	140.0	1	1
189155	19837	2	0	19765.6	19053.3	14.5	511	21.70	0.49	140.0	1	1
189156	19841	2	8	19764.9	19039.9	13.6	521	20.04	0.42	140.0	1	1
189157	19861	2	0	19743.6	19046.5	13.3	566	22.85	0.48	140.0	1	1
189158	19871	2	1	21848.2	9034.7	12.9	623	24.57	0.51	140.0	1	1
189159	19878	2	3	19782.6	19069.3	13.2	578	23.48	0.52	140.0	1	1
189160	19881	2	3	21837.7	8986.2	13.6	589	26.63	0.51	140.0	1	1
189161	19882	2	0	1786.3	9388.9	12.5	668	26.22	0.67	140.0	1	1
189162	19883	2	10	24006.6	6971.1	12.6	672	26.20	0.58	140.0	1	1
189163	19884	2	3	19755.4	19075.2	12.1	607	20.44	0.38	140.0	1	1
189164	19889	2	2	4076.3	13844.1	12.9	534	17.29	0.39	140.0	1	1
189165	19890	2	4	19768.1	19034.7	13.0	538	19.34	0.44	140.0	1	1
189166	19891	2	0	21821.0	9021.0	12.7	590	22.50	0.56	140.0	1	1
189167	19907	2	6	19787.5	19068.3	12.0	600	19.62	0.56	140.0	1	1
189168	19912	2	0	24652.1	7349.6	12.6	620	23.00	0.50	140.0	1	1
189169	19914	2	10	4707.6	23476.2	12.6	606	23.42	0.73	140.0	1	1
189170	19914	2	1	21817.9	9021.4	13.9	548	23.07	0.43	140.0	1	1
189171	19918	2	4	2356.9	9144.8	13.0	662	31.19	0.63	140.0	1	1
189172	19920	2	7	24847.3	7378.0	14.1	522	21.42	0.45	140.0	1	1
189173	19933	2	2	19757.6	19079.9	12.4	592	20.82	0.45	140.0	1	1
189174	19934	2	7	21815.1	8993.4	12.6	587	21.40	0.37	140.0	1	1
189175	19939	2	0	21849.5	8999.7	12.6	597	21.56	0.44	140.0	1	1
189176	19950	2	4	21805.0	9039.5	12.2	622	21.01	0.43	140.0	1	1
189177	19951	2	9	19727.5	19067.1	13.4	629	28.67	0.48	140.0	1	1
189178	19962	2	0	5327.1	17800.3	12.2	720	31.68	0.52	140.0	1	1
189179	19964	2	0	19743.4	19047.2	12.6	596	21.13	0.49	140.0	1	1
189180	19981	2	0	16565.1	8524.5	13.5	481	16.48	0.45	140.0	1	1
189181	19981	2	3	12302.1	1452.7	12.8	662	27.91	0.60	140.0	1	1

c

Fig. 27 Grouping persistent molecules. (**a**) PAL-Grouping tab, showing settings for the capture radius, maximum on time, and off gap. (**b**) Molecule statistics table of ungrouped data. (**c**) Molecule statistics table of grouped data

4. Washing the neurons with fresh neuronal culture media removes excess lipofectamine/DNA and lowers the toxicity of the transfection. The fresh media must be sterile, warm, and equilibrated before being added to the neurons.
5. Add 1 mL cell suspension to each glass-bottom dish (29 mm diameter). Do not confine the cell suspension to the well but add it to the whole surface of the dish, distributing the cells evenly.

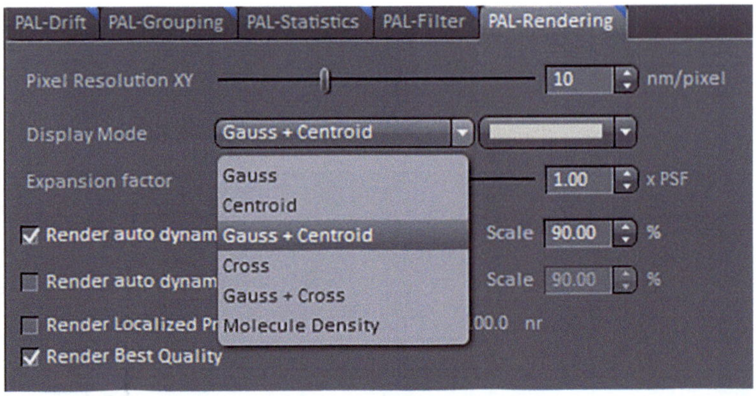

Fig. 28 Display options and rendering parameters available in the PAL-Render tab

6. Paraformaldehyde is toxic and a potential carcinogen and therefore should be prepared in the hood with appropriate PPE (gloves, lab coat, safety glasses, and mask).
7. The hardware configuration defines the optical paths and components used for imaging.
8. Save the project folder on a solid-state drive (SSD), if available, to minimize missed frames when acquiring data at high speeds.
9. The ROI size dictates the frame size. Avoid using binning to preserve resolution. The software displays the resulting acquisition speed in frames per second (FPS). Example settings: 1024 pixels × 1024 pixels sensor size, 611 pixels × 611 pixels ROI size, 50-ms exposure, and 20,000 frames for a stream acquisition.
10. Example: 7% of the 488-nm laser output with the 500 ± 25 nm band-pass filter to collect green emission from un-activated mEos. For single-camera mode, Filter Wheel 1 is used.
11. The Live view allows real-time adjustment of parameters.
12. The illumination should match the ROI size. Select the illumination mode (e.g., TIRF for focusing on focal adhesions) and angle to achieve the desired penetration depth. The depths displayed are estimates.
13. Ticking *ROI* saves the raw data, and ticking *Nano-2D* saves the reconstructed super-resolution image. The software performs single-molecule detections in real-time.
14. Example: Use the 405-nm (low power) for photo-activation, the 561-nm laser for excitation, and the 600 ± 25 nm filter for collecting the activated mEos signal.
15. The software will acquire the number of frames set in the camera tab, performing real-time localization.

16. All data and associated files are saved in the project folder upon completion or stopping of the stream acquisition.
17. It is recommended to process the raw data (ROI TIFF file) in NEO Analysis rather than using the default outputs from NEO Live Imaging, as the initial detection parameters may not be optimal. Tick *Show Overview* to see the processing overview. For more information, please see the 2021 Abbelight User Guide for Data processing with NEO Software.
18. Test different options via the drop-down menu to find the ideal result; refer to the NEO Analysis guide for descriptions. Use the defaults for a first run.
19. Start with a detection threshold of 0.5 for low-signal-to-noise-ratio (SNR) data and 1.0 for high-SNR data.
20. Drift refers to the gradual movement of the sample relative to the imaging system over time. This movement can arise from thermal expansion, mechanical instability, or environmental fluctuations. If left uncorrected, drift broadens structures and reduces effective resolution. This step corrects for drift of the sample during long acquisition streams.
21. The software will load the data, perform localization, apply drift correction, and save the coordinate tables (.csv) and super-resolved images. A confirmation message ("Process Dataset Finished Successfully") will appear upon completion.
22. The 3D Viewer is where you inspect the final super-resolution image and refine the data using various filters.
23. Use *New Scene* to open another dataset and use the arrows to switch between scenes. Use command buttons to take a screenshot, select an ROI, crop an ROI, and save to file.
24. The Drift Correction sub-tab here allows performing drift correction using cross-correlation if not done in the batch process.
25. These tools allow for precise filtering of the final localization data based on quality metrics such as localization precision (or uncertainty, in nm) and photon count.
26. Cluster analysis uses the precise spatial coordinates of SMLM detections to determine whether molecules form non-random nanoscale assemblies. By applying algorithms such as Ripley's K [23, 24], DBSCAN [25] or Voronoi [26], it identifies regions of elevated localization density and quantifies cluster properties including number, size, and density.
27. PC12 cells and other strongly adherent cell lines can also be used for dSTORM. When substituting these for primary neurons, adjust the plating density and fix the cells at the appropriate morphological stage. All subsequent labelling and imaging steps remain largely unchanged.

28. The optimal Triton X-100 permeabilization time depends on the target antigen. Permeabilization can be extended up to 30 min to improve antibody penetration; however, longer incubations may compromise fine cellular morphology. Ideally conditions should be optimized for each marker/cell type.
29. Prepare and handle all primary and secondary antibodies on ice. Apply only enough antibody solution to cover the central well, dispensing it directly onto the center of the dish (typically 100–150 μL for a 10 mm well). For antibody optimization, begin with a 1:500 dilution or the manufacturer's recommended concentration. While extended incubation can improve penetrance in tissue samples, cultured cells generally benefit more from adjusting antibody concentration rather than the incubation time.
30. Wrap the dishes with parafilm to ensure that the primary antibody solution does not evaporate.
31. Alexa Fluor 647 (AF647) is among the most reliable fluorophores for dSTORM, owing to its exceptional brightness (high photon output per switching event) and low duty cycle. As dSTORM image quality is strongly influenced by photon yield, duty cycle, fluorophore stability, and the number of attainable switching cycles, careful selection of secondary antibodies is essential. As a general guideline, fluorophores should be prioritized in the following order: AF647 > AF555 > AF488. We recommend using the 488 channel for a neuronal marker (e.g., MAP2) to easily locate your ROI under the oculars.
32. This selection automatically chooses the Andor iXon3 897 EMCCD camera. The Imaging Setup displays the objective, TIRF mode, collimator position, TIRF angle, emission filter, and magnifying lens. Please see the 2013 Carl Zeiss Microscopy ELYRA Operating Manual: ZEN 2012 (black edition) for more information on the operation of the instrument.
33. The Andor iXon3 897 EMCCD camera has 512 pixels × 512 pixels (full frame) with a 16-μm × 16-μm pixel size. Do not use pixel binning as it lowers the resolution. Use smaller frame sizes like 256 pixels × 256 pixels or 128 pixels × 128 pixels if needed. Use 16-bit depth for greater dynamic range. Do not perform averaging.
34. Recommended starting parameters: 5% laser power, 100-ms exposure, and a gain of 200. ZEN software provides an estimated penetration depth as the TIRF angle changes. Caution: The penetration depth is an estimate and should only serve as a guide.
35. A low gain (e.g., 10) protects the EMCCD sensor from saturation.

36. The EMCCD gain range on the Andor iXon3 897 EMCCD is 0 to 300, but using a gain of 300 is not recommended as it increases camera noise and shortens the lifetime of the sensor. If blinking density remains too high, continue bleaching with the excitation laser and/or adjust illumination conditions; then increase gain as needed to optimise detection without saturation.
37. If imaging multiple fluorophores, sequential dSTORM acquisition of the same region can be performed, starting with the 649-nm channel followed by the 555-nm channel. Following initiation, photo-switching should reach a steady state level. Adequate blinking is fundamental to SMLM; therefore, antibody concentrations, buffer conditions, and laser power should be optimized for experiments.
38. *Discard overlapping molecules* is used only in this first step to obtain a value for the PSF, as it only fits single emitters and discards multi-emitter events.
39. The *Peak Mask Size* defines the radius in pixels used to cut out a detected peak intensity for coordinate system conversion. The default of 9 pixels for 2D works well in our hands.
40. The *Peak Intensity to Noise* threshold defines the threshold for an identified peak's signal strength over the background to be considered "real." The *Peak Mask Size* and *Peak Intensity to Noise* parameters together specify the peak-finding criteria for the preliminary localization pass.
41. The *x,y 2D Gauss Fit* drop-down option fits the intensity distributions of the identified peaks to a full 2D Gaussian function.
42. The purpose of this initial localization is to generate a dataset from which the PSF half-width can be empirically measured. The main peak of the PSF half-width histogram (e.g., ~140 nm) represents the characteristic width of single-molecule emission events in the dataset. This value must be entered in the final localization step, where it serves as a constraint for the multi-emitter fitting algorithm and is critical for accurate reconstruction in regions of high emitter density.
43. *Account for overlap* is the setting that accounts for multi-emitter events when fitting and is essential for processing densely labelled samples.
44. It may be useful to test several PSF width values centered around the measured peak (e.g., 100, 120, 140, 160, 180 nm) to assess the robustness of the reconstruction. In the example shown, a PSF width of 140 nm was used.
45. The Model-Based drift correction algorithm uses cross-correlation between temporal segments of the localization

data to estimate and correct lateral drift. The *Max* parameter specifies the maximum number of independent temporal segments used for fitting. A single segment is usually sufficient for linear drift. If abrupt positional jumps occur (e.g., due to accidental disturbance of the microscope), the data should be segmented so that each interval between jumps is corrected separately and then recombined. If more than ~10 jumps are detected, the acquisition is likely unusable, and the experiment should be repeated.

46. We do not recommend applying *First Frame* filtering. As emitters blink stochastically, genuine localization events occur in every frame. By removing early frames, it discards valid data and can bias the reconstruction. Filtering by localization precision is a more appropriate and data-preserving method for improving dataset quality.
47. The localization precision filter range is determined from the full width at half maximum (FWHM) of the precision histogram. In the example shown, the peak occurs at 23 nm with a maximum count of 8200. Half of this value (4100) defines the points on either side of the peak where the histogram falls to half height, corresponding to precision values of ~12 nm and ~ 33 nm. These values form the recommended filter range.
48. Applying the localization precision filter instructs the software to retain only those localizations falling within the defined FWHM range. The resulting dataset is then used to produce the filtered high-resolution SMLM reconstruction.
49. In SMLM, persistent emitters are molecules that often appear in several consecutive frames. Grouping these events prevents them from being counted multiple times and aggregates their positional and temporal information into a single localization entry. Thus, improving the accuracy of molecule-level measurements.
50. *Capture Radius* sets the maximum pixel radius within which peaks of consecutive frames must lie to be considered the same molecule. *Max On Time* sets the maximum number of frames a peak is allowed to be detected within the capture radius; peaks exceeding this are discarded to remove debris/artifacts that persist for a long sequence. *Off Gap* sets the maximum number of missing frames allowed within the capture radius for peaks to still be grouped as the same molecule. Grouped data often have brighter areas than un-grouped data because localization precision is encoded by brightness. Also note that columns such as Number Frames and Frames Missing may be useful indicators of emitter persistence.

51. *Pixel Resolution XY* is the size of each pixel in nm on the super-resolved image. A good starting point is half of the peak value from the localization precision histogram.
52. The software provides multiple rendering modes for visualizing SMLM data. In *Gauss mode*, each localization is displayed as a Gaussian distribution weighted by its localization uncertainty. *Centroid mode* represents each localization as a single dot, while *Gauss + Centroid* and *Gauss + Cross* combine Gaussian rendering with point or cross overlays. *Molecule Density* mode generates a color-coded heatmap representing molecular density in molecules per nm^2 and is useful for assessing spatial organization.
53. The *Expansion Factor* scales the rendered spot size relative to the PSF. Increasing the factor enlarges spots, reducing apparent resolution but increasing visual confidence. Decreasing it below 1 sharpens the rendering but reduces confidence. The software issues a red warning if the value is set below 0.7, indicating that the image is being rendered with a precision exceeding the limits of standard localization fitting.

Acknowledgments

R.A. led the experimental work, including super-resolution imaging experiments, data analysis, figure generation (Figs. 2, 3, 4, 5, 6, 7, 8, 9, 10, 11, 12, 13, 14, 15, 16, 17, 18, 19, 20, 21, 22, 23, 24, 25, 26, 27, and 28) and contributed to manuscript preparation. A.M. prepared samples for PALM imaging, provided an existing dataset for dSTORM analysis and contributed to the refinement of the materials, data acquisition and analysis descriptions. A.J.M. designed Fig. 1 in consultation with R.A. F.A.M. provided financial support and critically reviewed the manuscript. S.F.L. conceived the research framework and was responsible for drafting, critical revision, and overall organization of the manuscript. National Health and Medical Research Council (NHMRC) Fellowship (grant no. 1155794) awarded to F.A.M.. This word was supported by equipment funding from the Australian Research Council (ARC) (LE130100078), as well as funding from the University of Queensland Strategic Initiatives Fund grant no. DVCR22052A awarded to F.A.M..

Conflict of Interest The authors of this chapter declare no conflict of interest.

References

1. Hückel E (1930) Zur quantentheorie der doppelbindung. Z Phys 60(7):423–456. https://doi.org/10.1007/BF01341254
2. Jabłoński A (1935) Über den mechanismus der photolumineszenz von farbstoffphosphoren. Z Phys 94(1):38–46. https://doi.org/10.1007/BF01330795
3. Stokes GG (1852) On the change of refrangibility of light. Phil Trans R Soc Lond 142:463–562. https://doi.org/10.1098/rstl.1852.0022
4. Born M, Wolf E (1999) Principles of optics: electromagnetic theory of propagation, interference and diffraction of light, 7th edn. Cambridge University Press, Cambridge. https://doi.org/10.1017/CBO9781139644181
5. Abbe E (1873) Beiträge zur theorie des mikroskops und der mikroskopischen wahrnehmung. Arch Mikrosk Anat 9(1):413–468. https://doi.org/10.1007/BF02956173
6. Rayleigh (1879) Investigations in optics, with special reference to the spectroscope. The London, Edinburgh, and Dublin Philosophical Magazine and Journal of Science, 8(49): 261–274. https://doi.org/10.1080/14786447908639684
7. Erickson HP (2009) Size and Shape of Protein Molecules at the Nanometer Level Determined by Sedimentation, Gel Filtration, and Electron Microscopy. Biol Proced Online 11, 32. https://doi.org/10.1007/s12575-009-9008-x
8. Gustafsson MGL (2000) Surpassing the lateral resolution limit by a factor of two using structured illumination microscopy. J Microsc 198(2):82–87. https://doi.org/10.1046/j.1365-2818.2000.00710
9. Chen X, Zhong S, Hou Y et al (2023) Super-resolution structured illumination microscopy reconstruction algorithms: a review. Light Sci Appl 12(1):172. https://doi.org/10.1038/s41377-023-01204-4
10. Hell SW, Wichmann J (1994) Breaking the diffraction resolution limit by stimulated emission: stimulated-emission-depletion fluorescence microscopy. Opt Lett 19(11):780–782. https://doi.org/10.1364/ol.19.000780
11. Vicidomini G, Bianchini P, Diaspro A (2018) Sted super-resolved microscopy. Nat Methods 15(3):173–182. https://doi.org/10.1038/nmeth.4593
12. Eric Betzig et al. (2006) Imaging Intracellular Fluorescent Proteins at Nanometer Resolution. Science 313:1642–1645. https://doi.org/10.1126/science.1127344
13. Hess ST, Girirajan TP, Mason MD (2006) Ultra-high resolution imaging by fluorescence photoactivation localization microscopy. Biophys J 91(11):4258–4272. https://doi.org/10.1529/biophysj.106.091116
14. Rust MJ, Bates M, Zhuang X (2006) Sub-diffraction-limit imaging by stochastic optical reconstruction microscopy (STORM). Nat Methods 3(10):793–796. https://doi.org/10.1038/nmeth929
15. Heilemann M, van de Linde S, Schüttpelz M et al (2008) Subdiffraction-resolution fluorescence imaging with conventional fluorescent probes. Angew Chem Int Ed 47(33): 6172–6176. https://doi.org/10.1002/anie.200802376
16. van de Linde S, Löschberger A, Klein T et al (2011) Direct stochastic optical reconstruction microscopy with standard fluorescent probes. Nat Protoc 6(7):991–1009. https://doi.org/10.1038/nprot.2011.336
17. Rahbek-Clemmensen T, Lycas MD, Erlendsson S et al (2017) Super-resolution microscopy reveals functional organization of dopamine transporters into cholesterol and neuronal activity-dependent nanodomains. Nat Commun 8(1):740. https://doi.org/10.1038/s41467-017-00790-3
18. Butler-Hallissey C, Leterrier C (2024) Super-resolution imaging of the neuronal cytoskeleton. NPJ Imaging 2(1):50. https://doi.org/10.1038/s44303-024-00054-y
19. Siddig S, Aufmkolk S, Doose S et al (2020) Super-resolution imaging reveals the nanoscale organization of metabotropic glutamate receptors at presynaptic active zones. Sci Adv 6(16): eaay7193. https://doi.org/10.1126/sciadv.aay7193
20. Sezgin E (2017) Super-resolution optical microscopy for studying membrane structure and dynamics. J Phys Condens Matter 29(27):273001. https://doi.org/10.1088/1361-648X/aa7185
21. Joensuu M, Martínez-Mármol R, Padmanabhan P et al (2017) Visualizing endocytic recycling and trafficking in live neurons by subdiffractional tracking of internalized molecules. Nat Protoc 12(12):2590–2622. https://doi.org/10.1038/nprot.2017.116
22. Kaech S, Banker G (2006) Culturing hippocampal neurons. Nat Protoc 1(5):2406–2415. https://doi.org/10.1038/nprot.2006.356
23. Ripley BD (1976) The second-order analysis of stationary point processes. J Appl Probab

13(2):255–266. https://doi.org/10.2307/3212829

24. Ripley BD (1977) Modelling spatial patterns. J R Stat Soc Ser B Methodol 39(2):172–192. https://doi.org/10.1111/j.2517-6161.1977.tb01615.x
25. Ester M, Kriegel H-P, Sander J et al (1996) A density-based algorithm for discovering clusters in large spatial databases with noise. In: Proceedings of the second international conference on knowledge discovery and data mining KDD 96, pp 226–231. https://doi.org/10.5555/3001460.3001507
26. Levet F, Hosy E, Kechkar A et al (2015) Sr-tesseler: a method to segment and quantify localization-based super-resolution microscopy data. Nat Methods 12(11):1065–1071. https://doi.org/10.1038/nmeth.3579

Chapter 6

Single Particle Tracking Photo-Activated Localization Microscopy

Alex J. McCann, Christopher Small, and Frédéric A. Meunier

Abstract

Synaptic neurotransmission is an adaptive process in which synapses exhibit short- and long-term plasticity, dynamically strengthening or weakening their output in response to stimulation. Central to this fine-tuning is the reorganization of synaptic machinery and architecture, with nanoscale alterations in receptor clustering, vesicle trafficking and cytoskeletal rearrangements taking place over a short time frame. Traditionally, visualizing the dynamic organization of the synapse through confocal imaging has been challenging, considering the synapse's small size and the resolution limits imposed by light diffraction. The emergence of super-resolution imaging, and especially single-molecule localization microscopy (SMLM), over the last couple of decades, has allowed scientists to achieve nanoscale resolution and start unravelling the molecular underpinnings of synaptic transmission and plasticity in real-time. One such approach is single particle tracking Photo-Activated Localization Microscopy (sptPALM), which takes advantage of total internal reflection fluorescence (TIRF) and sparse, stochastic activation of fluorophores to achieve low spatiotemporal density labelling required to track single synaptic proteins in real time. In this chapter, we provide a conceptual breakdown of sptPALM: its key principles, implementation, and impact on the field of neurotransmission. Further, we detail a systematic protocol for performing sptPALM imaging of synaptic proteins in primary neuronal cultures. This technique allows the experimenter to determine the mobility, cluster patterns, and behavior of the synaptic proteins with nanoscale localization precision.

Key words Synapse, Exocytosis, Endocytosis, Plasma membrane, Photoconversion, Fluorophore, TIRF microscopy, Single-particle tracking, Neurotransmission, Nanoclustering

1 Introduction

Communication between neurons underpins a wealth of different processes in the brain. This communication occurs at the junctions between pre- and post-synapses, which form connections between neurons. Neurons can deliver coordinated bursts of neurotransmitter release with millisecond precision at low energy costs at the pre-synapse [2]. This process involves the docking of synaptic

Alex J. McCann and Christopher Small must be listed as Contributed equally

Frédéric A. Meunier (ed.), *Super-Resolution Microscopy: Methods and Protocols*, Methods in Molecular Biology, vol. 3034,
https://doi.org/10.1007/978-1-0716-5268-8_6,

vesicles (SVs), containing neurotransmitters, at the active zone [3] of the pre-synapse prior to their Ca^{2+}-dependent fusion [4] with the plasma membrane via exocytosis [5], triggering neurotransmitter release and neuron-to-neuron communication. The fusion of SVs is rapid and tightly controlled, with a number of proteins acting in concert with one another to mediate exocytosis in a timely manner. Despite operating in a crowded environment, the synaptic release machinery is dynamically organized, with SVs grouped into different pools based on their release properties [6, 7]. Critically, the pre- and post-synapses are able to rapidly adapt their responses to different transmission frequencies (short-term plasticity, STP) [8–10], a process that is mediated by changes to receptor clustering, cytoskeletal architecture, lipid composition, and changes in the extracellular matrix of synapses [11]. How synapses maintain and fine-tune their nanoscale organization, despite synaptic molecules being subjected to thermal energy-induced Brownian motion [12], is poorly understood.

1.1 Diffraction of Light During Optical Imaging

The emergence of fluorescent proteins (green fluorescent protein (GFP)) [13, 14] revolutionized optical imaging, making it possible to examine the real-time behavior of proteins in their native, cellular environment, with minimal background signal compared to antibody labelling in fixed cells. Much of our understanding of cellular processes, including protein trafficking, receptor clustering, and cytoskeletal rearrangements, stems from confocal imaging of fluorescently tagged proteins. However, this added temporal information came with inherent limitations to spatial resolution when compared to electron microscopy (EM). Traditional confocal microscopy is not optimal for characterizing the organization of synaptic machinery within the tight, densely packed confines of synapses [15]. This is due to the crowdedness of synaptic proteins, combined with light diffraction, which imposes limits on spatial resolution [16]. The passage of light waves through the aperture of a microscope creates a spreading effect, or diffraction, in the emission profile of each fluorophore. This profile takes on the appearance of an Airy disk, roughly 100-fold the size of the molecule itself [17] and the simultaneous emission of many fluorophores precludes precise localization at the synapse. Accordingly, the point at which separate emission profiles overlap and are indistinguishable from each other dictates the spatial resolution limit. This diffraction limit corresponds to ~200 nm across the lateral (x, y) axes. For this reason, structures smaller than the diffraction limit of light cannot be resolved through conventional means. However, modern super-resolution optical imaging techniques have circumvented the limits imposed by the diffraction of light, achieving unprecedented, nanoscale resolution.

1.2 Single Particle Tracking Photo-Activated Localization Microscopy (sptPALM)

The localization of a fluorophore is inferred by isolating the point spread function of its emission profile. However, this cannot be achieved when the emission spectra of fluorophores overlap at high labelling densities (<25 molecules/μm^2) [18]. To avoid this issue, single-molecule localization microscopy (SMLM) implements sparse, low-density labelling. This can be achieved in several ways, but a common approach involves the stochastic photoactivation, or photoswitching, of fluorophores tagged to a molecule of interest. Typically, this process involves switching from a green-emitting to a red-emitting state upon UV-induced excitation [19]. Photo-Activated Localization Microscopy (PALM) takes advantage of specially engineered photoactivatable fluorophores (PA-FPs) to achieve a sparse, spatiotemporal labelling of proteins [20]. In PALM imaging, the majority of fluorophores are held in a non-emissive state. A select portion of fluorophores is stochastically activated or photoconverted to a red-emitting state, before being bleached (Fig. 1a, b). This process is irreversible; theoretically, fluorophores are detected only once [21]. The low-density, stochastic activation of single fluorophores allows for the separation of different emission profiles that would have previously been indistinguishable. The localization of a given fluorophore is determined by fitting the point spread function of each emission profile to a Gaussian curve, thereby approximating the localization of the emitter with nanometer precision (Fig. 1c i) [17, 18, 20, 22, 23]. Used in living cells, PALM can allow for the multiple tracking of single emitters in space and time (single particle tracking Photo-Activated Localization Microscopy, or sptPALM) to determine their nanoscale organization and motion in real-time [24–30] (Fig. 1c ii). By imaging molecules at high frame rate (e.g., 20 frames per second), molecules can be localized across consecutive frames and their trajectories reconstructed (Fig. 1d) [17, 31]. Thus, sptPALM allows the experimenter to resolve the lateral movement of proteins and build structural maps that provide spatiotemporal information of protein localization and mobility, with nanoscale precision [32].

1.2.1 Photoconvertible Fluorophores

As previously mentioned, sptPALM relies on stochastic, low-density activation of proteins tagged with PA-FPs. A range of different PA-FPs can be used for sptPALM. Initially, PA-FPs were limited to photoactivatable green fluorescent protein (PA-GFPs) [33]. Subsequently, PA-GFPs expanded to include Dendra2 [34] and mMaple [35]. Derivatives of the red fluorescent protein (RFPs) have also been engineered [36], including photo-activatable mCherry (PA-mCherry) [37]. One drawback of PA-FPs is their low photon yield (which limits spatial resolution) compared to organic dyes used in other forms of SMLM [17]. However, another subset of probes commonly used are the photoconvertible Eos Fluorescent Proteins (EosFPs). These fluorophores switch from

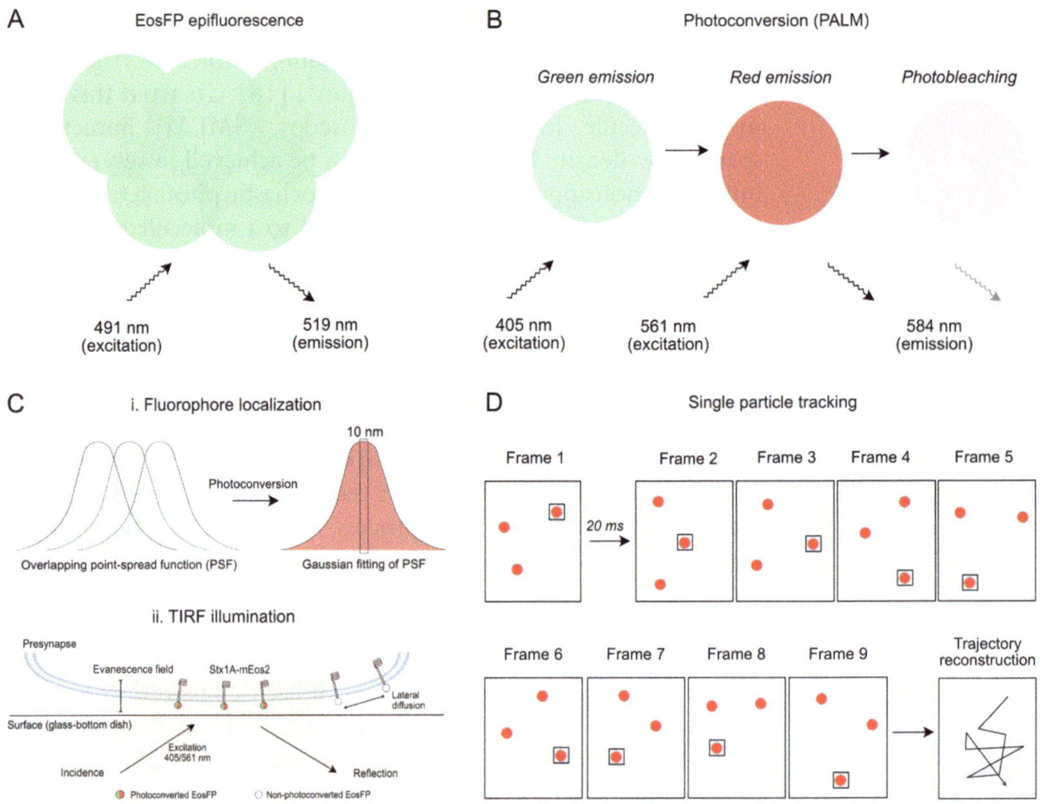

Fig. 1 Photoconversion of EosFP to perform single particle tracking. (**a**) Overlapping emission profiles of non-photoconverted EosFP. (**b**) Photoconversion of a small population of EosFP from green-emitting to red-emitting state for low density labelling. (**c**) **i.** Photoconversion prevents overlapping emission spectra. Localization can be achieved by applying a Gaussian fitting of the point spread function (PSF) of photoconverted-EosFP. **ii.** TIRF illumination triggers photoconversion of Eos-tagged synaptic proteins at the plasma membrane (presynaptic Stx1A-mEos2 shown as an example). (**d**) Tracking single molecules across multiple steps (20 ms time intervals) during stochastic, low-density photoconversion

an emission of 519 nm to 584 nm following excitation at 405 nm and provide an elevated photon output compared to PA-FPs [38]. Until recently, the most commonly used EosFP has been mEos2, due to its high quantal yield affording localization precision (~10–30 nm) and maturation at physiological temperatures [39]. However, mEos2 has the disadvantage of oligomerizing at high concentrations, leading to potential imaging artefacts. To address this issue, monomeric variants of mEosFP have been engineered (mEos3.1, mEos3.2) [40]. At present, the mEos3.1 and mEos3.2 fluorophores have become the gold standard of fluorophores used in sptPALM. Other variants (mEosEM), optimized for correlative light and electron microscopy, are also available [41].

1.2.2 Total Internal Reflection Fluorescence (TIRF)

In order to achieve high signal-to-noise for PALM and sptPALM, total internal reflection fluorescence (TIRF) imaging [42] is commonly used [31]. Unlike conventional epifluorescence, excitation laser beams are projected at an angle, leading to total reflection on the coverslip (Fig. 1c ii). In doing so, a thin optical layer called the evanescent field (~200 nm) forms. This decays exponentially with distance, allowing a high signal-to-noise ratio of the imaged sample. By restricting the evanescence field, TIRF allows for selective excitation of fluorophores in close proximity to the plasma membrane. This illumination is optimal for membrane-associated proteins, including the exocytic machinery (Syntaxin1A, Stx1A and Munc18-1/STXBP-1 [43–46], endocytic adaptor proteins (clathrin, dynamin) [47, 48]. Furthermore, TIRF helps minimize any potential background signal or autofluorescence [20]. A more oblique illumination angle can be used to achieve increased penetration depth (up to ~500 nm). In neurons, this margin (200–500 nm) corresponds to the approximate distance position of the synaptic vesicles from the plasma membrane. Thus, sptPALM can be used to track membrane-bound proteins involved in exo-endocytosis, as well as the sub-synaptic organization of different vesicle pools [49, 50] (*see* Chapter 9).

1.3 Impact of sptPALM on the Field of Neurotransmission

Since its inception, sptPALM imaging has allowed a much deeper investigation of the nanoscale organization of a range of different synaptic proteins, in both a physiological and pathological context. Combined with other SMLM techniques, sptPALM has demonstrated that the organization of synaptic proteins is heterogeneous, oscillating between mobile and immobile pools. When immobile, synaptic proteins are typically arranged into nanodomains, or nanoclusters [51, 52]. Typically, these domains comprise hubs of activity designed to speed up the enzymatic efficiency of biological reactions [50, 51, 53–57], segregate proteins with competing interactions [58], or regulate intracellular protein traffic [15, 48, 59].

sptPALM has shaped our understanding of how presynaptic proteins can self-assemble into nanodomains during synaptic transmission. This is especially true of the exocytic machinery, which forms nanoclusters that contribute to vesicle fusion in response to nerve stimulation. Key SNARE proteins, including syntaxin-1A (Stx1A), are recruited into nanoclusters to drive exocytic events [60–62]. sptPALM has shown that Stx1A molecules are organized within nanodomains through interaction with polyphosphoinositides [62] and Munc18 [46]. These nanodomains comprise a concentrated zone of Stx1A, which promotes vesicle fusion [63]. Altered Stx1A nanoclustering is a hallmark of propofol-induced general anesthesia [60]. Similarly, sptPALM has shaped our understanding of how dynamin isoforms are recruited to the plasma membrane before being trapped in an endocytic nanocluster to initiate endocytosis [47]. In addition to the synaptic machinery, sptPALM can track vesicular proteins, and by extension, SVs. sptPALM of vesicular glutamate transporter-1 (VGlut-1-mEos2) and synaptophysin (Syp-mEos3.2) has been used to track synaptic vesicles, to demonstrate how liquid-phase separation can dynamically anchor vesicles within synaptic, biomolecular condensates, thereby maintaining neurotransmission [25, 28, 64].

At the level of the post-synapse, sptPALM has also allowed us to understand how the dynamic recruitment of ligand-gated ion channel receptors into surface nanodomains fine-tunes neurotransmission. Excitatory AMPA receptors, which form a mobile pool capable of diffusing rapidly between synaptic and extrasynaptic regions, are recruited into surface nanodomains [54]. Interestingly, these postsynaptic, AMPA receptor nanoclusters align spatially and temporally with multiple, clustered presynaptic proteins that comprise the active zone, where exocytosis occurs [54]. Inhibitory GABA-A receptors display similar properties of surface nanoclustering within synaptic and extrasynaptic regions [65]. Membrane-associated tyrosine kinases (Fyn), which promote NMDA receptor activity, form subsynaptic nanoclusters that initiate intracellular signaling cascades that enhance synaptic activity [55, 56], and are also linked to synaptotoxicity during frontotemporal dementia (FTLD) [55].

1.4 sptPALM—The Pros and Cons

sptPALM imaging possesses numerous advantages compared to other super-resolution imaging techniques. First, the technique is optimized for live imaging. Therefore, sptPALM can determine the real time mobility and spatiotemporal organization of synaptic proteins, and how these properties are altered on a short-time scale in response to a change in conditions, such as stimulation. As a form of live imaging, sptPALM avoids the use of fixatives that may alter both the photo-physical properties of fluorophores [66] and cellular organization [67]. The second advantage of sptPALM is that the technique can be used to resolve proteins across different sub-synaptic compartments. Other forms of live SMLM based on fluorescent ligand binding and/or internalization (uPAINT, sdTIM) [68–72] are restricted to imaging a select pool of molecules, such as membrane receptors or endocytic vesicles. In contrast, sptPALM can resolve a wide range of membrane-associated synaptic proteins. This includes vesicular proteins (synaptophysin, VGlut1, SV2A), cytoskeletal proteins (tau, actin), and cytosolic proteins that associate with slow-moving organelles [73, 74] or with the plasma membrane [46]. Further, mEos-tagged nanobodies can be expressed in cells to access the nanoscale organization of endogenous proteins in critically important conformations [50] (*see* Chapter 10). Lastly, whereas other SMLM techniques rely on sparse labelling of proteins using antibodies tagged with organic dyes, sptPALM relies on the expression of fluorescently tagged proteins, ideal for single-step photobleaching, with the 1:1 ratio between protein and fluorophore allowing for accurate estimates of protein stoichiometry [21].

Despite its wealth of advantages, sptPALM is not without drawbacks. For example, due to its reliance on live TIRF imaging, sptPALM is restricted to the tracking of molecules present near the plasma membrane. Trajectory reconstruction in sptPALM is mostly limited to two dimensions, in contrast to other forms of super-resolution imaging, which are capable of 3D reconstruction (3D-SIM, 3D-STORM, fixed PALM). The live imaging nature of sptPALM makes co-localization between different fluorescent markers difficult, as it requires reconciling the mobility of synaptic proteins with the inherent undersampling of photoactivated molecules, while also accounting for acquisition drift introducing spatial misalignment over time. Consequently, apparent proximity between two fluorphores may not reflect a stable association at a given time point. This contrasts with fixed SMLM, where a greater sampling of the probe is possible, and where fixatives immobilize the various cellular components. Furthermore, the use of photoconvertible Eos fluorophores in sptPALM requires low-level exposure to UV light, which can induce phototoxicity [75]. In the majority of cases, sptPALM involves the overexpression of molecules tagged with a PA-FP, which may affect a number of cellular processes. This requires transfection of neurons with transfection agents, such as lipofectamine, which negatively affect the physiology of neurons. To avoid these drawbacks, endogenous proteins can be labelled with fluorophores using TALEN or CRISPR engineering (*see* Chapter 10). Otherwise, care must be taken to ensure that expression of fluorescently tagged proteins is consistent between groups [52, 55]. Finally, due to the rapid photoconversion and bleaching of mEosFP, the trajectories sampled from sptPALM imaging are shorter than those derived from ligands conjugated with organic dyes (such as atto647N) used in other variations of SMLM. Consequently, mEosFP trajectories provide less information regarding the motion of individual proteins over an extended duration. In certain cases, therefore, sptPALM is less ideal for tracking individual molecules over greater time windows, compared with techniques that employ stochastic labelling of proteins using labelled nanobodies, such as uPAINT or sdTIM [68].

In summary, sptPALM can be used to track synaptic proteins tagged with photoconvertible fluorophores, in order to track the nanoscale mobility, clustering, and localization of synaptic proteins (membrane-associated or cytoskeletal) in live neurons. In this book chapter, we provide a step-by-step protocol for performing sptPALM to track mEosFP-labelled proteins at the pre or postsynapse of neurons (Subheadings 3.1, 3.2, and 3.3). The following protocol is optimized for acquiring single particle tracking images in live, hippocampal neurons in vitro, which is described extensively in Joensuu et al., 2017 [70]. Here, we discuss the process for seeding and transfecting hippocampal neurons with an EosFP-tagged protein of interest (e.g., SV2A-mEos2 [55, 57, 58, 76]), and a secondary fluorescent marker (e.g., GFP) to highlight neuronal morphology. Of note, sptPALM can also be applied to secondary cell lines [46, 47] and in vivo, such as in drosophila larvae [62] (*see* Chapter 8). We also provide a step-by-step protocol for analyzing the acquired sptPALM data using a software pipeline described in depth in Gormal et al., 2025 [77] (Subheading 3.4). Software in this pipeline includes TrackMate [78], Super Res Data Wrangler [77], SharpViSu [45], and NASTIC [32, 77]. Although the primary purpose of NASTIC is to perform spatiotemporal nanocluster analysis on spt data, it can also be used to generate panels such as those shown in Fig. 2. For information on how to perform spatiotemporal cluster analysis on spt data using NASTIC, see Chapter 13 [32, 77].

2 Materials

2.1 Cell Culture Preparation

Prepare solutions and media in advance and filter regularly to avoid contamination. Ideally, solutions are made up on the day experiments are carried out. If conserved overnight, solutions are maintained at 4 °C. Prewarm prepared media to 37 °C in a water bath before using.

1. Glass-bottom dish (29 mm diameter, 20 mm well, no 1.5, D29-20-1.5-N, Cellvis, CA, USA) coated with poly-L-lysine (PLL, 1 mg/mL).
2. Filtered Milli-Q water or 1× phosphate-buffered saline (PBS).
3. Plating media: 100 U/mL penicillin-100 μg/mL streptomycin, 1× GlutaMAX supplement, 1× B27, fetal bovine serum (FBS) (5%) in neurobasal media.
4. Culturing media: 100 U/mL penicillin-100 μg/mL streptomycin, 1× GlutaMAX supplement, 1× B27 in neurobasal media (*see* **Note 1**).
5. Ready to seed hippocampal neurons (80000–100,000 per dish), resuspended in plating media, from embryonic day 16.5 (E16.5) mouse embryos.

2.2 Transfection

1. Plasmid encoding an EosFP-tagged protein of interest (e.g., SV2A-mEos2 or Stx1A-mEos2).
2. Plasmid encoding a secondary fluorescent marker (*see* **Note 2**).
3. Transfection reagent (e.g., Lipofectamine-2000) (*see* **Note 3**).
4. Neurobasal media.
5. Inverted light microscope.

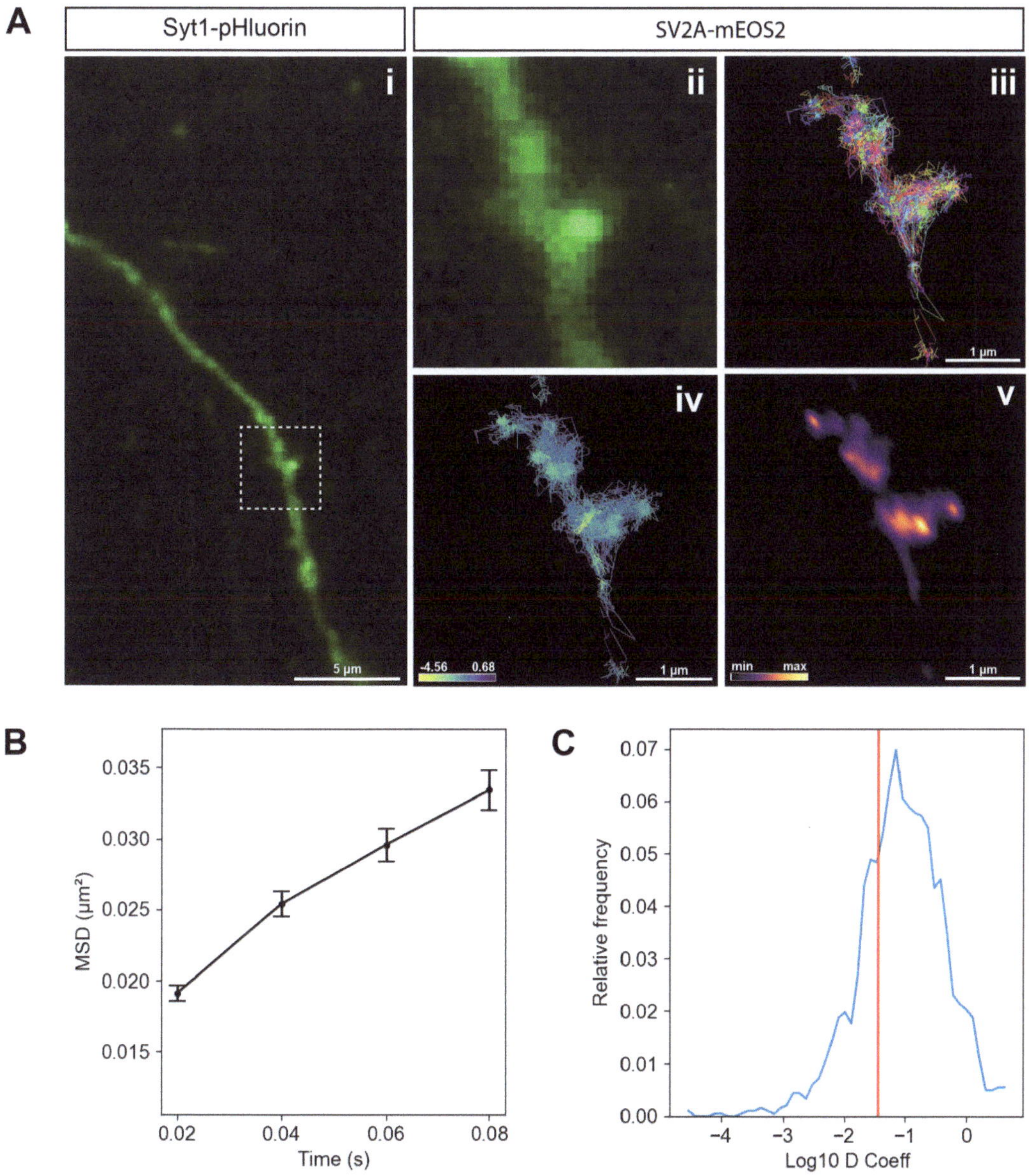

Fig. 2 Example showing sptPALM imaging of SV2A-mEos2 in the presynapse of a hippocampal neuron. (**a**) **i.** GFP epifluorescence (Syt1-pH) highlights axonal morphology. White dashed outline is shown magnified in **ii.** Trajectories colored with **iii** arbitrary colors and **iv** their average $\log_{10}$ diffusion coefficient value (brighter colors = less mobile; darker colors = more mobile). **v.** 2D Kernel Density Estimation plot (darker colors = less detections; brighter colors = more detections). (**b**) Average mean square displacement (MSD) over time (higher MSD = more mobile, lower MSD = less mobile). (**c**) Log_{10} diffusion coefficient frequency distribution histogram (blue) with immobile/mobile threshold shown in red (left of the threshold = immobile population; right of the threshold = mobile population)

2.3 sptPALM Imaging

2.3.1 Buffers

1. Calibration buffer: TetraSpeck™ microspheres (0.1 μm, Invitrogen-Thermo Fisher Scientific) diluted at ~1:1000 phosphate-buffered saline (PBS) (*see* **Note 4**).
2. Low [K^+] buffer/imaging buffer (for imaging at rest): 0.5 mM $MgCl_2$, 2.2 mM $CaCl_2$, 5.6 mM KCl, 145 mM NaCl, 5.6 mM D-glucose, 0.5 mM ascorbic acid, 0.1% BSA, 15 mM HEPES, pH 7.4, 290–310 mOsm.
3. High [K^+] buffer (for stimulation and imaging): 0.5 mM $MgCl_2$, 2.2 mM $CaCl_2$, 56 mM KCl, 95 mM NaCl, 5.6 mM D-glucose, 0.5 mM ascorbic acid, 0.1% BSA, 15 mM HEPES, pH 7.4, 290–310 mOsm.

2.3.2 Microscope Setup

1. Roper Scientific Ring-TIRF microscope.
2. CFI Apo 100×/1.49 N.A. oil-immersion objective (Nikon Instruments).
3. Immersion oil (35 °C).
4. Dual Evolve 512 Delta Electron-Multiplying Charge Coupled Device (EMCCD) cameras (Photometrics, AZ, USA).
5. 405 nm laser (100 mW Vortran Laser Technology) and 561 laser (150 mW Cobolt Jive)
6. TwinCam LS Image Splitter (Cairn Research) (*see* **Note 5**).
7. iLas2 double laser illuminator (Roper Scientific) (*see* **Note 6**),
8. Perfect Focus System (Nikon).
9. TIRF-quality ultra-flat quadruple beam splitter (ZT405/488/561/647rpc; Chroma Technology) (*see* **Note 7**).
10. QUAD emission filter (ZET405/488/561/647 m; Chroma Technology).

2.3.3 Software

1. TrackMate [78] (v7.9.2) (ImageJ/FIJI plugin) (*see* **Note 8**).
2. SharpViSu [45] GitHub, v1) https://github.com/andronovl/SharpViSu (*see* **Note 9**).
3. NASTIC software [32, 77], including NASTIC and Super Res Data Wrangler (GitHub, v1.0.12) (https://github.com/tristanwallis/smlm_clusteringClustering) (*see* **Notes 10** and **11**).
4. ImageJ/FIJI (ImageJ 1.54j) (for running TrackMate).
5. MATLAB (MathWorks, v2017b) (for running SharpViSu).
6. Python (v3.8+) and Python modules (colorama v0.4.6, matplotlib v3.8.4, matplotlib-venn v0.11.7, numpy v1.23.2, pandas v1.4.4, Pillow v9.2.0, FreeSimpleGui v5.2.0.post1, rtree v1.0.0, scikit-learn v1.1.2, scipy v1.13.1, seaborn v0.12.0 and statsmodels v0.13.2) (for running NASTIC).
7. Adobe Illustrator (for making figures).

3 Methods

3.1 Cell Culture Preparation

This protocol covers the plating of hippocampal neurons derived from E16.5 (embryonic day 16.5) mouse embryos into PLL-coated glass-bottom dishes, and maintenance of this neuronal culture until the day of transfection.

1. Seed hippocampal neurons (80,000–100,000) in the glass-bottom of each PLL-coated dish (*see* **Note 12**).
2. Store your hippocampal neurons in an incubator at near-physiological temperatures (37 °C at 5% CO_2) in plating media (5% FBS) for 2 h.
3. Replace the plating media with culturing media (no FBS) and incubate (37 °C with 5% CO_2) until the day of transfection (*see* **Note 13**). Supplement the neurons with culturing media every 5 days (*see* **Note 14**).

3.2 Transfection

Transfections are performed a minimum of 24 h prior to imaging, to allow enough time for the fluorescently-tagged protein of interest to express. Transfections are carried out with plasmids expressing a protein of interest tagged to an EosFP (in this case, SV2A-mEos2) (Fig. 2a i–ii). Although not mandatory, we recommend co-transfecting neurons with a secondary fluorescent marker (*see* **Note 2**).

1. Prepare the transfection mixture by adding the plasmid DNA and lipofectamine reagent to neurobasal media (no added glutaMAX, B27 or FBS) (*see* **Notes 3** and **15**).
2. Incubate the transfection mixture at room temperature (22–25 °C) for 15 min.
3. Collect the conditioning (culturing) media from the periphery of each dish and store at 37 °C with 5% CO_2 (*see* **Note 16**).
4. Add the DNA:lipofectamine mix dropwise to the center of the glass bottom where neurons were previously seeded.
5. Incubate the neuronal cultures (now administered with DNA: lipofectamine solution) for a 2 h duration (37 °C, 5% CO_2).
6. To remove the DNA:lipofectamine mix, perform a series of wash steps with the conditioning media (2 × 300 μL). Subsequently, add the remaining culturing media to each dish (1400 μL per dish).
7. Incubate overnight (minimum 24 h, 37 °C, 5% CO_2).
8. Prior to imaging, verify the health of the neurons and transfection efficiency (level of epifluorescence) on an inverted light microscope. Look for signs of unhealthy neurons such as blebbing or fragmented dendritic branches.

3.3 sptPALM Imaging

Alignment of the EMCCD cameras and calibration of the TIRF illumination angle should be performed before each experiment (*see* **Note 17**). Here, we include a detailed breakdown of camera alignment during dual-color single particle tracking [79].

3.3.1 Camera Alignment

1. Switch on the Roper microscope and iLas^2 illumination system.
2. Apply the calibration buffer to the glass bottom of a 29 mm glass-bottom dish (*see* **Note 4**). Allow the microspheres some time (~5 min) to settle at the bottom of the dish.
3. Mount the dish containing microspheres on the stage Roper microscope.
4. Before starting the calibration, check that the correct dichroic filters are installed, with cameras turned off.
5. Ensure that the Photometrics Evolve 512 Delta EMCCD cameras are mounted on a TwinCam LS Image Splitter (*see* **Note 5**).
6. In the acquisition software of the microscope, open the live view window to visualize the TetraSpeck beads across both camera images.
7. Adjust the Emission Beam Splitter angle by turning the screws marked as "H" and "V" (*see* **Note 18**).

3.3.2 Calibration of TIRF Laser Angle

1. Select point mode in the calibration tab of iLas^2.
2. Calibrate each coordinate so that the TetraSpeck™ microspheres are faintly visible on-screen (*see* **Note 19**).

3.3.3 Image Acquisition

1. Pre-warm the stage of the Roper microscope to near-physiological temperatures (37 °C, 5% CO_2).
2. Select a dish to perform sptPALM imaging. Keep the remaining dishes in reserve in a nearby incubator (37 °C, 5% CO_2).
3. Remove culturing media by performing a series of wash steps (2 × 1000 μL) with low [K^+] imaging buffer (*see* **Note 16**). Subsequently, re-incubate the dish in 1.5 mL of low [K^+] imaging buffer.
4. Place the first dish on the (pre-heated) stage of the Roper microscope.
5. Identify neurons based on the fluorescence of the secondary marker (*see* **Notes 2** and **20**).
6. To take a static image, set the "frames to average" parameter to 20, the "Digitizer" parameter to 10 MHz, and set the exposure time to 100 ms before pressing the "Acquire" button to acquire the image.

7. Prior to acquisition of the timelapse, ensure that the acquisition parameters are set to the following: frames to average (1), gain (1, 1×), EM Gain (111), Digitizer (20 MHz).
8. To trigger the photoconversion of mEosFP, activate the 405 nm (wide-field) laser (2%, surface power density of 3.46×10^{-5} kW/cm^2) (*see* **Note 21**).
9. Simultaneously, activate the 561 nm laser (70%, surface power density of 0.204 kW/cm^2) to image the pool of photoconverted mEosFPs (*see* **Note 22**).
10. Acquire 16,000 frames at 20 ms exposure time (50 Hz) (*see* **Note 23**).

3.4 Image Processing and Data Analysis

This section of the protocol involves the use of an spt analysis software pipeline (described extensively in Gormal et al. [77]) to analyze the acquisition files that were generated in the previous sections (Subheadings 3.1, 3.2, and 3.3), and to generate the panels shown in Fig. 2 (panels A iii–v, B and C). This analysis pipeline involves 5 key steps as follows. (1) TrackMate [78] (ImageJ/Fiji plugin) is used to localize and reconnect detections found in the raw acquisition file in order to generate a trajectory file (*see* **Note 8**). (2) Super Res Data Wrangler [77] (Python-based software; NASTIC family) is used to convert the trajectory file from the TrackMate.csv format to the SharpViSu.ascii format (*see* **Notes 10** and **11**). (3) SharpViSu [35] (MATLAB software) is used to perform drift correction on the converted trajectory files (*see* **Note 9**). (4) Super Res Data Wrangler is used to convert the drift corrected SharpViSu files to the.trxyt format required for subsequent analysis in NASTIC [32, 77] (*see* **Notes 10** and **11**). (5) NASTIC (Python-based software) is used to obtain the panels shown in Fig. 2a ii–v, b and c (*see* **Note 10**). Of note, the panels in Fig. 2a–c were obtained using NASTIC without performing spatiotemporal clustering analysis. For in-depth information on performing spatiotemporal clustering analysis using NASTIC, see Chapter 13.

3.4.1 Localization and Tracking—TrackMate

1. Open the ImageJ/Fiji software (*see* **Note 8**).
2. Load the acquisition file by either dragging and dropping the acquisition file onto the ImageJ/Fiji toolbar, or by clicking "File" > "Open..." in the ImageJ/Fiji toolbar to browse for and select the acquisition file.
3. (Optional) if the brightness of the loaded acquisition file is too low, click on "Image" > "Adjust" > "Brightness/Contrast" followed by "Auto" to automatically adjust the brightness of the image.
4. Set the spatial and temporal resolution of the acquisition file by clicking "Image" > "Properties..." to open a popup window

and changing the value of "Frames (t):" to 16,000, "Pixel width:" and "Pixel height:" to 0.106 μm, and "Frame interval:" to 0.02 s (*see* **Note 24**).

5. (Optional) draw a region of interest (ROI): click on "Image" > "Stacks" > "Z Project..." in the ImageJ/Fiji toolbar to open a "ZProjection" window. Select "Max intensity" as the "Projection type" and click "OK." A new window displaying the maximum intensity image of the timelapse acquisition will appear. Click on the maximum intensity image to select it and follow **step 3** if the brightness is too low. Click on either the "Rectangle" or "Free selection" buttons in the ImageJ/Fiji toolbar and draw directly on the acquisition, then click "Analyze" > "Tools" > "ROI Manager..." followed by "Add[t]" in the ROI Manager. Save the ROI (for use in NASTIC in later steps) by clicking on "Properties..." in the ROI manager to bring up a Properties window, ticking the "List coordinates" checkbox followed by "OK" in the Properties window to open an XY_Coordinates window, and clicking "File" > "Save as..." in the XY_Coordinates window to save the ROI file, making sure that the name of the saved ROI file follows the *XY_Coordinates*.csv naming format (* = any character) so that it can be recognized by NASTIC. (*see* **Note 25**).
6. Open TrackMate by clicking "Plugins" > "Tracking" > "Trackmate" in the ImageJ/FIJI toolbar. Check that the calibration settings are correct before clicking the "Next" button.
7. Click on the dropdown box to select "LoG detector," and then click the "Next" button (*see* **Note 26**).
8. Define the detection size using the "Estimated object diameter:" parameter. We recommend starting with a diameter of 0.5 μm. Press "Preview" to view detections that have been identified in the current frame using the selected detection size (identified detections are outlined by colored circles). Navigate to various timeframes in the acquisition and press "Preview" to view the identified detections and adjust the value of the "Estimated object diameter:" parameter accordingly (*see* **Note 27**).
9. (Optional) Define the detection threshold using the "Quality threshold" parameter. We recommend starting with the default value. This parameter can be adjusted by either replacing the current numerical value in the input box with a new one or by clicking and dragging the vertical line on the quality histogram (located below thresholding parameters) towards the left (lower threshold = more detections) or right (higher threshold = less detections). As with **step 8**, press "Preview" to view detections that have been identified in various timeframes, and adjust the parameter as required (*see* **Note 28**).

10. Run the localization using the selected parameters by clicking "Next."
11. Select the initial thresholding value by either pressing "Auto" to automatically select the detections that will be used for tracking, or click and drag the threshold to the desired thresholding value. Click and drag the window all the way to the left to include all identified detections in the tracking step. Press "Next" to apply the filter.
12. Press "Next" to get to the "Select a tracker" window of TrackMate. In this window, click on the dropdown box and select "Simple LAP tracker," before clicking "Next" (*see* **Note 29**).
13. Define the parameters of the tracker before clicking "Next" to perform tracking. If using the Simple LAP tracker, we recommend starting with a "Linking max distance" of 0.5 μm, and without gap closing by setting "Gap-closing max frame gap" to 0. The "Gap-closing max distance" parameter can be left at 0.5 μm. (*see* **Notes 30** and **31**).
14. After tracking is complete, click on the "Next" button to get to the "Set filters on tracks" screen. Define the minimum trajectory length by clicking on the green plus symbol to add a filter option (which will appear at the top of the "Set filters on tracks" window), clicking on the dropdown box and selecting the "Number of spots in track" filter, dragging the sliding window to the minimum trajectory length, and clicking on the "Above" radio option to include all trajectories with lengths above the selected sliding window. We recommend using a minimum trajectory length of 8 (*see* **Note 32**).
15. Define the maximum trajectory length by clicking on the green plus symbol to add another filter option, select "Number of spots in track," drag the sliding window to the maximum trajectory length, and click the "Below" radio option to include all trajectories with lengths below the selected sliding window value. We recommend using a maximum trajectory length of 1000 (*see* **Note 32**). Click "Next" to apply the minimum and maximum trajectory length filters.
16. Export the trajectory data for further analysis by clicking "Tracks" > "Export to csv" (*see* **Notes 11 and 33**).

3.4.2 Download and Install Python, NASTIC, and Python Modules Required for NASTIC

1. Go to the Python downloads page (https://www.python.org/downloads/). Select and download a version of Python (version 3.8 or above) that is appropriate for your operating system.
2. Run the Python Installer. Make sure that the "Add to PATH" option is ticked. Complete the installation.

3. Open an instance of the command line. Type **python -m pip install colorama==0.4.6 matplotlib==3.8.4 matplotlib-venn==0.11.7 numpy==1.23.2 pandas==1.4.4 Pillow==9.2.0 FreeSimpleGui==5.2.0.post1 rtree==1.0.0 scikit-learn==1.1.2 scipy==1.13.1 seaborn==0.12.0 statsmodels==0.13.2** followed by the return key.
4. Go to the NASTIC GitHub (https://github.com/tristanwallis/smlm_clusteringClustering). On the right-hand side of the screen under the "Releases" heading, click on the green "Latest" button to the right of "NASTIC." Click on the NASTIC zip folder (e.g., NASTICv1_0_12.zip) to download the latest version of NASTIC.
5. Unzip the downloaded NASTIC folder to a location of your choosing.
6. Test that the download has been successful by double-clicking on the **nastic_gui.py** script and the **super_res_data_wrangler_gui.py** script. The graphical user interfaces (GUIs) should appear after a short period of time.

3.4.3 File Conversion #1—Super Res Data Wrangler

The layout of trajectory information within the exported TrackMate.csv file depends on the version of TrackMate used (e.g., columns may be in a different order and may have different headings). Based on this, conversion of TrackMate.csv files to the SharpViSu.ascii format using the Super Res Data Wrangler requires the user to inspect the TrackMate.csv file in Excel, and then use this information to select the column number that contains (1) trajectory number, (2) x-coordinate, (3) y-coordinate, and (4) time information, as well as the number of headers at the top of the file, the units of the x- and y-coordinate columns, and the units of the time column. Here we will describe the parameters that match the output of TrackMate version 7.9.2; however these can be modified as needed for the version of TrackMate used. (*see* **Note 34**).

1. Double-click on the **super_res_data_wrangler_gui.py** script. Alternatively, open an instance of the command line, navigate to the location of the script (in the unzipped NASTIC folder), and type **python super_res_data_wrangler_gui.py** followed by the return key. The GUI should appear after a short period of time, with the "Select filetypes" tab displayed.
2. Change the "Input filetype" on the left to "csv(TrackMate)," by clicking on the downwards facing arrow on the right-hand side of the dropdown box, and selecting the "csv(TrackMate)" option.
3. Change the "Output filetype" field on the right to "ascii."

4. Press the "CONFIRM FILETYPES" button at the bottom of the tab to progress to the "Load files" tab.
5. Press the "Browse folder (directory)" radio option at the top of the "Load files" tab.
6. Press the "Browse folder" button to browse for and select a directory containing multiple TrackMate.csv trajectory files to convert.
7. Press the "Find files" button to populate the table below with all files in the selected directory that have the ".csv" extension.
8. (Optional) Narrow the list of found files by typing in a phrase in the "(Optional) Filenames contain" field, before clicking on the "Find files" button to update the list with the new results.
9. (Optional) exclude files from analysis by clicking on the checkbox of the file to highlight it, then click on the checkbox again to untick the checkbox.
10. Press the "LOAD DATA" button at the bottom of the tab to load the selected file(s) and change to the "Set parameters" tab.
11. Open one of the TrackMate.csv files in Excel to check the format of the data. Identify the number of headers, the column numbers of the columns that contain the trajectory numbers, x-coordinate, y-coordinate, and time information, as well as the units of the x- and y-coordinates and time columns. (*see* **Note 34**). Use this information as a reference when setting the parameters in the Super Res Data Wrangler GUI.
12. In the "Set parameters" tab of the Super Res Data Wrangler GUI, set "X,Y units" to the spatial units used in the x- and y-coordinate columns of the Excel file (um) (*see* **Note 34**).
13. Make sure that "Pixel size (um/px)" is set to 0.106 (*see* **Note 23**).
14. Set "Time units" to the temporal units used in the time column of the Excel file (s) (*see* **Note 34**).
15. Make sure that "Acquisition frequency (Hz)" is set to 50.0 (*see* **Note 23**).
16. Set "Number of headers" to the number of headers in the Excel file (4) (*see* **Note 34**).
17. Set "Trajectory col" to the column number that contains the trajectory information in the Excel file (3) (*see* **Note 34**).
18. "Set parameters" tab: Set "X-position col" to the column number that contains the x-coordinate information in the Excel file (5) (*see* **Note 34**).
19. "Set parameters" tab: Set "Y-position col" to the column number that contains the y-coordinate information in the Excel file (6) (*see* **Note 34**).

20. Set "Time col" to the column number that contains the time information in the Excel file (8) (*see* **Note 34**).
21. Press the "CONVERT" button at the bottom of the tab to convert the selected trajectory file(s) to the ".ascii" file format.

3.4.4 Drift Correction—SharpViSu

Drift correction works best on high-density data sets, as higher numbers of detections/trajectories enable greater estimation of drift even with larger subdivisions of acquisition frames (more subdivisions ensure smoother drift correction) (see **step 14**). We recommend performing drift correction on each cell, and then comparing the drift corrected and uncorrected images to assess whether to use the drift corrected ascii or the uncorrected (original) ascii for subsequent analyses (see **step 20**). It is important to note that uncorrected ascii files are in pixels, whereas drift-corrected ascii files are in nanometers. SharpViSu [35] assumes that uncorrected ascii files have a pixel size of 0.100 μm/px. This is a hard-coded value that will need to be changed if working with a different pixel size (e.g., 0.106 μm/px) (see **step 1**). Alternatively, if the pixel size of your acquisition file is 0.100 μm/px, skip **step 1** and go straight to **step 2**.

1. (Optional) If using a pixel size other than 0.100 μm/px (e.g., 0.106 μm): in the downloaded SharpViSu folder, open the adjustformat.m script. Change the "100" value in Line 15 to "106": "A(:,4:5) = A(:,4:5) * 106; % x, y in nm." Change the "100" value in Line 16 to "106": "A(:,8:9) = A(:,8:9) * 106; % sigmas in nm."
2. Create a new folder that contains the ascii file that will be drift-corrected and the corresponding id file. When drift correcting multiple files, make sure that each ascii/id file is placed into a separate folder in order to prevent the previously drift-corrected file from being overwritten by a subsequent drift-corrected file (output files are generated using the same name).
3. Open MATLAB (Version 2017b). In the "HOME" tab at the top of MATLAB, press "New Script" to display the Command Window under the Editor.
4. In the "HOME" tab, under "ENVIRONMENT" click "Set Path" to open the "Set Path" popup window. Click the "Add with Subfolders..." to open a folder browser. In the folder browser, navigate to the location of the folder containing SharpViSu, click on the folder and press the "Select Folder" button to confirm. In the "Set Path" window, click "Close," then click "No" in the popup window that appears.
5. Run SharpViSu by typing SharpViSu in the MATLAB Command Window, followed by the return key. A window displaying the SharpViSu graphical user interface will appear.

6. "Initial eventlists" section (top left): make sure that the "Format" parameter is set to "Leica LAS AF" (*see* **Note 35**).
7. "Initial eventlists" section: press the "Browse..." button under the "Red channel (e.g., 642 nm) eventlist" heading to browse for and select the.ascii file that will be used for drift correction. The "Red" channel will represent the drift corrected version of the.ascii file.
8. "Initial eventlists" section: press the "Browse..." button under the "Green channel eventlist" heading to browse for and select the same.ascii file that was selected for the previous step. The "Green" channel will represent the uncorrected (original) version of the.ascii file.
9. "Z-stack" section (bottom left): make sure that the "Mounting n" parameter is set to "1.51 (TDE)."
10. "Drift estimation" section (middle left): make sure that the "Method:" parameter is set to "histogram."
11. "Drift estimation" section: set the "pixel size, nm" parameter for the "Red" channel and "Green" channel to 100 (*see* **Note 36**).
12. "Drift estimation" section: set the initial values for the "Subsets (> = 2)" parameter to 4 for both the "Red" and "Green" channels (*see* **Note 37**).
13. "Drift estimation" section: Press the "Calculate" button. The estimated level of drift will be displayed as red and green arrows (see **Note 38**).
14. "Drift estimation" section: adjust the "Subsets > = 2" parameter for the "Red" channel as needed, and then press "Calculate" to replot the estimated drift with the new value. The selected value must be able to divide the total number of frames into even groups (e.g., 4, 8, 16, 32 for a 16,000 frames acquisition). Select the highest subset number that produces a good result. Dividing the acquisition into too many subsets will reduce the datapoints in each subset and can result in a correction error. The red channel arrows should follow the overall direction of the green channel arrows; however, the line generated by the red channel arrows should appear smoother due to having more subdivisions (the green arrows are subdivided into several red arrows). Try to keep the values for the "Subsets > = 2" parameter similar across cells belonging to the same dataset.
15. "Eventlist correction" section (top right): set the "Iterations of drift correction" parameter to 4 for the "Red" channel and "0" for the "Green" channel (*see* **Note 39**).

16. "Eventlist correction" section: make sure that the "Consecutive channels" and "Chromatic aberrations" checkboxes are unticked.
17. "Auto save in current folder" section (bottom right): make sure that the "Corr. Eventlists" checkbox is ticked.
18. "Eventlist correction" section: press the "Apply" button. Open a file browser and navigate to the location of the.ascii file that was selected for drift correction in **steps** 7 and **8**. Two files will be generated in this location (el_grl_corrected.ascii—uncorrected green channel; and el_red_corrected.ascii—drift corrected red channel). Wait until both files are created and are approximately the same file size, indicating that drift correction has finished.
19. "2D Reconstruction" section (middle right): press the "XY..." button to view the results. If the display is too dim, increase the "multiplier" values.
20. "Eventlist" section (top right): click on the "Corrected" and "Original" radio buttons to switch between the corrected and uncorrected images, respectively, to assess whether drift correction was successful. If a drifted data set has been corrected appropriately for axial drift, ideally the image should look sharper/less diffuse and points should no longer be elongated/skewed in the same direction, and the morphology of the cell should look more like the static image taken of the cell.
21. Open the uncorrected (el_grl_corrected.ascii) and drift corrected (el_red_corrected.ascii) files in Excel for comparison. The x,y coordinates of the first line should be the same in both files, and the distance between subsequent frames (level of x,y drift correction) should be reasonable (and align with the visible estimation of drift observed in the original acquisition).
22. Rename the el_red_corrected.ascii file in order to differentiate it from other drift corrected.ascii files (e.g., date_sample_dc.ascii). Rename the.id file to match the new name of the drift corrected ascii file (e.g., date_sample_dc.id) so that it will be recognized by the Super Res Data Wrangler during file conversion. Including "dc" in the filename can help differentiate drift corrected ascii files from uncorrected ascii files during file conversion.

3.4.5 File Conversion #2—Super Res Data Wrangler

1. Double-click on the **super_res_data_wrangler_gui.py** script. Alternatively, open an instance of the command line, navigate to the location of the script (in the unzipped NASTIC folder), and type **python super_res_data_wrangler_gui.py** followed by the return key. The GUI should appear after a short period of time, with the "Select filetypes" tab displayed.

2. "Select filetypes" tab: change the "Input filetype" field to "ascii (drift corrected)" (*see* **Note 40**).
3. "Select filetypes" tab: change the "Output filetype" field to "trxyt.""Select filetypes" tab: Press the "CONFIRM FILETYPES" button to progress to the next tab.
4. "Load files" tab: Press the "Browse folder (directory)" radio option followed by the "Browse folder" button to browse for and select a directory containing multiple drift corrected.ascii trajectory files to convert.
5. "Load files" tab: Press the "Find files" button to populate a list with all files in the selected directory that have the ".ascii" extension. (*see* **Note 41**).
6. (Optional) narrow the list of found files by typing in a phrase in the "(Optional) Filenames contain" field, before clicking on the "Find files" button to update the list with the new results. If drift corrected files were saved with the filename format date_-sample_dc.ascii, then typing the phrase _dc in the "(Optional) Filenames contain" field will only show drift corrected ascii files.
7. "Load files" tab: Exclude files from analysis by clicking on the checkbox of the file to highlight it, then clicking on the checkbox again to untick the checkbox.
8. "Load files" tab: Press the "LOAD DATA" button to load the file(s) selected for conversion and change to the next tab.
9. "Set parameters" tab: Make sure that "Acquisition frequency (Hz)" is set to 50.0 (*see* **Note 23**).
10. "Set parameters" tab: press the "CONVERT" button to convert the selected trajectory file(s) to the ".trxyt" file format.

3.4.6 Data Analysis—NASTIC

This protocol only covers the generation of images and metrics shown in Fig. 2. For detailed information on how to perform spatiotemporal nanoclustering analysis using NASTIC [32, 77], see Chapter 13.

1. Double-click on the **nastic_gui.py** script. Alternatively, open an instance of the command line, navigate to the location of the script (in the unzipped NASTIC folder), and type **python nastic_gui.py** followed by the return key. The GUI should appear after a short period of time, with the "File" tab displayed.
2. "File" tab: Click "Browse" to browse for and select a trxyt file to load.
3. "File" tab: Set the "Minimum trajectory length" parameter to 8 (*see* **Note 32**).

4. “File” tab: Set the “Maximum trajectory length” parameter to 1000 (*see* **Note 32**).
5. “File” tab: Press “PLOT RAW DETECTIONS” to load and view the detections in the trxyt file and switch to the next tab.
6. “ROI” tab: Press “Load” to browse for and select the ROI file that was used for TrackMate analysis (Subheading 3.4.1, **step 5**). The TrackMate ROI file should be in the *XY_Coordinates*.csv file format (* = any character) in order to be recognized by NASTIC. Make sure that the value for the “Pixel (um)” parameter is correct (e.g., 0.106 for Roper) (*see* **Note 24**). (Optional): If the “Pixel(um)” value has been adjusted, press the “Replot ROIs” button to replot the ROI using the new pixel size.
7. “ROI” tab: Press “SELECT DATA IN ROIS” to progress to the next tab.
8. “Clustering” tab: Set the “Acquisition time (s)” to 320 (*see* **Note 42**).
9. “Clustering” tab: Set the “Frame time (s)” to 0.02 (*see* **Note 24**).
10. “Clustering” tab: tick the “Diffusion coefficient screen” (log10) checkbox, set the “Min” parameter to −5 and the “Max” parameter to 1 (*see* **Note 33**).
11. “Clustering” tab: The remaining parameters can be left as default values as this chapter does not focus on clustering (*see* Chapter 13 for clustering in NASTIC).
12. “Clustering” tab: Press “CLUSTER SELECTED DATA” to progress to the next tab (*see* **Note 43**).
13. “Display” tab: At the top of the tab, make sure that “Traj.”, “Scalebar” and “Colorbar” is ticked, and that “Centr.”, “Clust.” and “Hotsp.” is unticked.
14. “Display” tab: In the middle of the “Trajectory” subtab, click on the “Random” checkbox to tick it. Press the “PLOT CLUSTERED DATA” button at the bottom of the screen to replot the graph (*see* **Notes 44** and **45**). Do not close the plot before saving. Figure 2a iii shows an example plot with arbitrarily colored trajectories.
15. (Optional): “Display” tab: Adjust the aesthetics of the scalebar by navigating to the “Scalebar” subtab, adjusting the parameters as required, and replotting the graph by pressing the “PLOT CLUSTERED DATA” button. Do not close the plot.
16. (Optional) “Display” tab: navigate to the “Export” subtab and adjust the format of the plot as desired. Do not close the plot.
17. Click on the “Metrics” tab at the top of the screen to change to the “Metrics” subtab.

18. "Metrics" tab: make sure that the "Plot All MSD" checkbox is ticked and that the "Plot legend" button is unticked, before pressing the MSD button to plot the MSD of all trajectories (MSD plot with black line; regardless of clustering status). Do not close the All MSD plot. Figure 2b shows an example MSD plot (*see* **Note 46**).
19. "Metrics" tab: Set the "Res" parameter next to the "KDE" button to 0.9. Then press the "KDE" button to plot the 2D Kernel Density Estimation plot (*see* **Notes 47** and **48**). Do not close the plot before saving. Figure 2a v shows an example KDE plot.
20. "Metrics" tab: make sure that the "Pixel(um):" parameter is set to the correct pixel size (e.g., 0.106) before pressing the "Diffusion coefficient" button to plot trajectories colored by their average $\log_{10}$ diffusion coefficient value, and to plot the $\log_{10}$ diffusion coefficient frequency histogram with immobile/mobile threshold shown. Do not close the plots before saving. Figure 2a iv shows an example plot with trajectories colored by their average $\log_{10}$ diffusion coefficient values. Figure 2c shows an example $\log_{10}$ diffusion coefficient frequency distribution histogram plot (blue) with the immobile/mobile threshold shown (red) (*see* **Notes 49** and **50**).
21. "Metrics" tab: Press the "SAVE ANALYSES" button to save the open plots, metrics, and ROIs used.
22. Navigate to the location of the trxyt file that was loaded into NASTIC. The saved plots will be located in a date-stamped folder in the same directory as the trxyt file. The plots can be used to make figure panels using software such as Adobe Illustrator (*see* Fig. 2).

4 Notes

1. We recommend using phenol-free neurobasal media to avoid any potential background in the imaging process. However, this is not mandatory, so long as the culturing media is removed when washing with low [K^+] buffer prior to imaging.
2. Examples of secondary fluorescent markers include empty-vector green fluorescent protein (GFP) or mCardinal. Co-transfection with GFP or mCardinal allows the experimenter to identify neurons, discern their morphology, and select an appropriate region of interest (ROI). To pinpoint active presynaptic boutons, neurons can be transfected with vesicular proteins tagged with pHluorin, a pH-sensitive GFP variant that fluoresces upon transitioning from the acidic environment of the vesicular lumen to the neutral extracellular pH

following vesicle fusion [80]. Commonly used vesicular markers include VAMP2- [69] SV2A- [48] or Syt1-pHluorin [15, 48]. Be aware of possible interactions between mEosFP-tagged and pHluorin (pH)-tagged proteins. Interactions between proteins may disrupt their mobility and nanoscale organization. For example, examining the presynaptic mobility of mEosFP-tagged dynamin, while using Syt1-pH to label active nerve terminals, is problematic considering the interaction between both molecules [81]. Figure 2a i and ii show an example of using a secondary marker (Syt1-pH) to highlight axonal morphology and assist with ROI selection (presynapse).

3. Options other than lipofectamine are available (e.g., calcium phosphate treatment or electroporation).
4. The calibration buffer contains microspheres that fluoresce in a range of different spectra (blue, green, red, and far-red) and are therefore ideal for aligning cameras and adjusting the TIRF angle of illumination. This buffer can also be applied to the dish when performing sptPALM acquisitions, as these microspheres are fixed to the bottom of the dish, they provide a reference point for subsequent drift correction.
5. Allows precise xy-alignment of transmitted and reflected ports.
6. Enables 360° TIRF illumination.
7. Required for distortion-free reflection of lasers.
8. Numerous algorithms are available for single particle tracking [82, 83]. For this protocol, we recommend the use of TrackMate [78] a freely accessible tracking tool that can run within ImageJ/Fiji.
9. Spatiotemporal intensity, mobility, and trajectory maps are subject to drift, which typically occurs throughout the duration of an acquisition. For drift correction prior to cluster analysis, we recommend using SharpViSu [45], which runs in MATLAB.
10. Various spt analysis software can be used to generate trajectory, diffusion coefficient, and density maps, as well as mean square displacement (MSD) and diffusion coefficient frequency distribution plots. Here, we used a custom Python toolkit called NASTIC [32, 77] to generate the panels shown in Fig. 2 (panels A iii–v, B and C), which were obtained without performing spatiotemporal cluster analysis. The NASTIC family of software also includes the Super Res Data Wrangler, which is used in this protocol to convert between spt trajectory file types. The main purpose of NASTIC is to quantify the spatiotemporal dimensions of nanoclusters (area, radius, apparent lifetime, and density), which is covered extensively in Chapter 13.

11. Different analysis software has different input file type requirements. Output files from spt analysis software may need conversion before subsequent analysis in a different software. We recommend using the Super Res Data Wrangler [77] (custom Python script with graphical user interface (GUI) and command line (cli) versions available) for conversion between various trajectory file types.
12. If coating dishes: Prior to dissection, coat the glass-bottom dishes with PLL and incubate for a minimum of 2 h (37 °C with 5% CO_2). Only coat the glass bottom well of each dish (where neurons are seeded and imaged) with PLL—the periphery of the dish should remain dry. Wash the glass bottom of each dish (3 × 5 min) with filtered Milli-Q water or phosphate-buffered saline (PBS).
13. Depending on the nature of the sptPALM imaging, transfection is performed at different time points. We recommend transfection at 15–16 days in vitro (DIV15-16) to ensure maturity of presynaptic nerve terminals prior to imaging (DIV18).
14. When carrying out transfections, it is important to ensure in advance that a sufficient volume of conditioning media is present in each dish. Conditioning media refers to the culturing media collected from the dish at the time of transfection. This media contains growth factors that promote neuronal survival. Conditioning media is used to wash away the DNA-lipofectamine solution applied to neuronal cultures during transfection.
15. The amount of DNA, lipofectamine and pure neurobasal media volume added is dependent on the number of dishes being transfected. Here, we recommend using a 1:1 ratio of DNA:lipofectamine (per dish: 2 μg of DNA, 2 μL of lipofectamine and 50–100 μL neurobasal media).
16. It is important to leave a small volume of liquid covering the glass bottom of each dish at all times to prevent the neuronal culture from drying out.
17. Camera alignment is essential when combining sptPALM with another SMLM technique (e.g., uPAINT, *see* Chapter 9). Camera alignment must be performed prior to adjusting the TIRF angle (i.e., each realignment necessitates a new TIRF angle calibration).
18. This allows you to align the beads between each channel. It is also possible to rotate one channel in reference to the other. Here, it is important to note that the beads located at the periphery of your field of view will have a greater tendency to misalign. Thus, when combining sptPALM with another mode

of SMLM, it is important to restrict your field of view to a region (ideally the center) where pixel-to-pixel alignment is achieved.

19. In the iLas2 illumination system, the galvanometer-scanning mirrors control TIRF alignment. Four coordinates (north/south/east/west) define the annular illumination pattern.

20. Use 491 nm excitation if GFP or pHluorin was co-transfected as the secondary marker. Use 647 nm excitation if mCardinal was co-transfected as the secondary marker.

21. Excitation at 405 nm triggers photoconversion of the mEosFP. The power of the 405 nm laser is minimized for two reasons: first, to avoid UV-induced phototoxicity of neurons, and second, to achieve continuous stochastic photoactivation and conversion of mEosFP from green to red. Photoactivation must be slow, at frequencies that avoid the overlap of fluorescent emitters.

22. Due to the low intensity 405 nm excitation, only a small portion of mEosFP is in an emissive, photoconverted state. Therefore, the intensity of 561 excitation can be much higher than that of the 405 nm laser. Excitation of photoconverted mEosFP using the 561 nm laser intensity dictates the number of molecules imaged, and by extension, the mobility, longevity, and number of trajectories reconstructed from sptPALM imaging. For this reason, it is highly recommended to keep the intensity of the 561 nm laser constant between experimental groups to derive equivalent sample sizes and trajectories between groups.

23. Note that higher imaging rates are optimal for faster-moving molecules. Conventionally, 16,000 frames are acquired, although this may vary depending on the experiment in question.

24. The spatial resolution (pixel size) refers to the size of a single pixel in microns and differs between microscopes. For the Roper (used in this protocol), this value is 0.106 μm. The temporal resolution (acquisition frequency) is the rate at which each individual frame is acquired. This value is 20 ms (or 50 Hz) if one frame is acquired every 20 ms. If using TrackMate: although TrackMate will detect the metadata of most raw files, it is essential to check that the correct spatial and temporal values have been set.

25. This will ensure that only detections that fall within the defined region are used for generating trajectories. ROIs are useful for comparing the mobilities of proteins of interest across different cellular compartments, e.g., in the axon as compared to the presynapse.

26. We recommend using the Laplacian of Gaussian (LoG) detector as it is better suited for detecting single fluorescent particles (due to having a Gaussian-like shape) against a noisy background compared to other detection algorithms that are better used for deriving the shapes of cellular structures [84].
27. The detection size is used to exclude areas of localised intensity that are too large (i.e., likely to be caused by background/artefacts).
28. The detection threshold is used to determine whether an area of localised intensity (that falls within the defined detection size) is identified as a detection (used for trajectory generation) or as background (omitted from trajectory generation). If the threshold is too high (everything under this value is included), false positives may be included, and the high density of the data may lead to increased linkage errors. If the threshold is too low (everything over this value is excluded), the data may be too sparse, leading to a low number of trajectories. Selected threshold values are subject to user variability; therefore, we recommend that the thresholding values for a particular dataset be determined by the same user, using the same thresholding criteria.
29. TrackMate [78] offers several tracking options, but we recommend the Simple LAP tracker for fast reconstruction of high-density trajectory data [85].
30. The "Linking max distance" parameter is the maximum distance that can occur between detections (of neighboring frames) for these detections to still be considered part of the same trajectory. Setting this parameter will minimize potential mistracking issues.
31. The "Gap-closing max distance" and "Gap-closing max frame gap" parameters are the maximum distance and time (respectively) that can occur between two trajectory segments and still be merged into a single trajectory. Setting the "Gap-closing max frame gap" parameter to 0 means that no gap closing will be performed, making the "Gap-closing max distance" parameter irrelevant.
32. Trajectories that are very short may not provide sufficient information to reliably infer mobility behaviors. Mean square displacement (MSD) calculations require at least 4 time points, and therefore only trajectories with at least 5 frames should be included. Trajectories that are very long may also be a result of artefacts/background and are more likely to be affected by drift. They may exhibit behaviors that are not representative of the rest of the population. We recommend only including trajectories that are between 8 and 1000 frames in length.

33. TrackMate does not filter trajectories based on their diffusion coefficients. Trajectories with very low diffusion coefficients may be a result of artefacts/background or trapped molecules that exhibit a confined behavior that is not representative of the entire population. Trajectories with very high diffusion coefficients (higher than expected) may also be due to mistracking. We recommend excluding trajectories with $\log_{10}$ [D] of less than −5 and greater than 1. This step can be performed manually (after localization and tracking) by filtering the exported xyt trajectory file in Excel, or in NASTIC (*see* Subheading 3.4.6).
34. The layout of TrackMate.csv files may differ depending on the version of TrackMate used. In the version used in this protocol (v7.9.2), there are 4 header rows: the first 3 rows contain column headings (e.g., LABEL, ID, TRACK_ID, QUALITY, POSITION_X, POSITION_Y, POSITION_Z, POSITION_T, FRAME, ...), and the fourth row contains the units of each of the columns (e.g., (blank), (blank), (blank), (quality), (μm), (μm), (μm), (sec), (blank), ...). In this version of TrackMate (v7.9.2), the column that contains the trajectory number information is column 3 (TRACK_ID), the column that contains the x-coordinate information is column 5 (POSITION_X) which is in microns (μm), the column that contains the y-coordinate information in column 6 (POSITION_Y) which is in microns (μm), and the column that we use in this protocol for time information in column 8 (POSITION_T) which is in seconds (sec).
35. The "Leica LAS AF" file format refers to the SharpViSu.ascii format. It is important to note that the uncorrected ascii file format is in pixels (px), whereas the drift corrected file format is in nanometers (nm).
36. This value refers to the pixel size of the histogram, not the pixel size of the acquisition file, and therefore will not change the aspect ratio of ascii file following drift correction.
37. SharpViSu divides the total number of frames into subsets (each containing the same number of frames) in order to estimate drift. For a 16,000 frame acquisition, setting the "Subsets" parameter to 4 creates 4 groups, each consisting of 4000 frames. The Red "Subsets" parameter can be adjusted (it must be a number that can divide the total number of frames into even groups), whereas the Green "Subsets" parameter should remain at 4.
38. The arrows represent the initial estimation of overall drift observed in each channel, with the individual arrows representing the drift between each subset (calculated from cross-correlation of the frame subsets), and are used to infer the level of drift correction that is required.

39. The "Iterations of drift correction" parameter is the number of iterations that are used to correct a particular channel (by subtracting the observed drift). Make this value 0 to leave a particular channel uncorrected.
40. Alternatively, if converting uncorrected (original).ascii files to the.trxyt format, select "ascii" as the input filetype if using the uncorrected (original).ascii file. It is important to note that the x and y coordinates of uncorrected ascii files are in pixels, whereas drift-corrected ascii files are in nanometers.
41. Both uncorrected and drift corrected.ascii files will appear in the list as they have the same file extension. Make sure that you are converting the correct files using the correct input file type. Saving drift corrected files with "dc" in the filename can help differentiate the two filetypes.
42. This is the total amount of time (in seconds) taken to acquire the image stack (e.g., 16,000 frames each taken at a frame rate of 0.02 s (50 Hz) = 320 s).
43. If there is an error message popup (e.g., "Alert: No unique spatiotemporal clusters containing trajectories found in the selected ROI. Please try adjusting the ROI or clustering parameters"): Although this protocol does not involve spatiotemporal cluster analysis (only using NASTIC to generate panels in Fig. 2), NASTIC requires that at least 1 cluster is identified before proceeding to the "Display" tab. To satisfy this requirement, try increasing the "Time threshold (s)" parameter to 320 (entire length of acquisition in seconds), and try to increase the "Radius factor" and "Maximum cluster radius" parameters to the size of the acquisition (entire length of acquisition in microns). Also try decreasing the "Minimum cluster radius" parameter to the smallest possible value. These changes should result in all trajectories throughout the acquisition time series being encompassed into a single giant pseudo "cluster," which enables the user to move to the "Display" tab. Importantly, these clustering parameters do not impact the output of Fig. 2, which are obtained independently of clustering (MSD, VAR, and $\log_{10}$ diffusion coefficient plots and analyses are performed on all trajectories regardless of their clustering status).
44. The "Random" checkbox in the "Trajectory" subtab arbitrarily colors trajectories using 8 bright colors. The aesthetic of the trajectory lines can be adjusted using the "Width" (line width) and "Opacity" (line transparency) options. Pressing the "PLOT CLUSTERED DATA" button again will randomly color the trajectories again.
45. Trajectory maps show the spatial movement of molecules and can be used to observe individual behaviors (e.g., confined vs

diffusing) across different areas of the sample, which can be compared to the corresponding areas in the 2D Kernel Density Estimation and $\log_{10}$ diffusion coefficient maps.

46. The mean square displacement (MSD) ($\mu m^2 s^{-1}$) is a measurement of the mobility of a molecule over a certain period of time. Molecules with larger MSDs exhibit higher mobilities compared to molecules with lower MSDs. MSD can also be calculated manually by applying the following equation: MSD $(t) = a + 4Dt$, where t = time, a = y intercept and D = diffusion coefficient ($\mu m^2 s^{-1}$). The area under the MSD curve is calculated in order to draw statistical comparisons between the MSDs of different groups.
47. The KDE plot is very slow; please wait until the plot has appeared before doing anything else.
48. 2D Kernel Density Estimation plots show the spatial density of detections, which can be useful for identifying whether molecules are uniformly distributed or form nanoclusters.
49. Diffusion coefficient plots show the spatial mobilities of molecules, which can be useful for identifying regions of high and low mobilities.
50. The frequency distribution histogram of $\log_{10}$ [D] enables the visualization of different mobility subpopulations (e.g., a population of molecules with lower mobilities on the left of the immobile/mobile threshold, and a population of molecules with higher mobilities on the right of the immobile/mobile threshold). From this thresholded histogram, the mobile-to-immobile ratio can be calculated, which is useful for comparing different groups.

Acknowledgments

sptPALM was carried out at the Queensland Brain Institute (QBI) Advance Microscopy Facility. C.S. performed and wrote the steps for transfection and data acquisition, generated Fig. 1, and acquired the single-particle tracking recordings used for the generation of Fig. 2. A.J.M. performed and wrote the steps for data analysis, and generated Fig. 2. C.S., A.J.M. and F.A.M. edited the manuscript. The Protein Expression Facility (PEF) at UQ (Brisbane, Australia) generated the SV2A-mEos2 plasmid, which was previously described in Small et al., 2024 [38]. We thank the Clem Jones Centre for Ageing Dementia Research (CJCADR) for its support. We also thank Dr. Parnayan Syed for their input and edits on TrackMate, and Dr. Rachel Gormal for their input and edits on SharpViSu. National Medical and Research Council (NHMRC) Fellowship (grant no. 1155794) awarded to F.A.M.. This work was supported by equipment funding from the Australian Research Council (ARC)(LE130100078), as well as funding from the University of Queensland Strategic Initiatives Fund (grant no. DVCR22052A) awarded to F.A.M..

Conflict of Interest The authors of this chapter declare no conflict of interest.

References

1. Moerner WE (2007) New directions in single-molecule imaging and analysis. Proc Natl Acad Sci USA 104(31):12596–12602
2. François-Martin C, Rothman JE, Pincet F (2017) Low energy cost for optimal speed and control of membrane fusion. Proc Natl Acad Sci USA 114(6):1238–1241
3. Couteaux R, Pecot-Dechavassine M (1970) Synaptic vesicles and pouches at the level of "active zones" of the neuromuscular junction. C R Acad Hebd Seances Acad Sci D 271(25): 2346–2349
4. Schiavo G, Gu Q, Söllner T et al (1996) Calcium-dependent switching of the specificity of phosphoinositide binding to synaptotagmin. Proc Natl Acad Sci USA 93(23):13327–13332
5. Südhof TC (2012) The presynaptic active zone. Neuron 75(1):11–25
6. Rizzoli SO, Betz WJ (2005) Synaptic vesicle pools. Nat Rev Neurosci 6(1):57–69
7. Denker A, Rizzoli SO (2010) Synaptic vesicle pools: an update. Front Synaptic Neurosci 2: 135
8. Doussau F, Schmidt H, Dorgans K et al (2017) Frequency-dependent mobilization of heterogeneous pools of synaptic vesicles shapes presynaptic plasticity. elife 6:e28935
9. Regehr WG (2012) Short-term presynaptic plasticity. Cold Spring Harb Perspect Biol 4(7):a005702
10. Pan B, Zucker RS (2009) A general model of synaptic transmission and short-term plasticity. Neuron 62(4):539–554
11. Choquet D, Triller A (2013) The dynamic synapse. Neuron 80(3):691–703
12. Kusumi A, Sako Y, Yamamoto M (1993) Confined lateral diffusion of membrane receptors as studied by single particle tracking (nanovid microscopy). Effects of calcium-induced differentiation in cultured epithelial cells. Biophys J 65(5):2021–2040
13. Heim R, Cubitt AB, Tsien RY (1995) Improved green fluorescence. Nature 373(6516):663–664
14. Tsien RY (1998) The green fluorescent protein. Annu Rev Biochem 67(1):509–544
15. Joensuu M, Syed P, Saber SH et al (2023) Presynaptic targeting of botulinum neurotoxin type a requires a tripartite PSG-Syt1-SV2 plasma membrane nanocluster for synaptic vesicle entry. EMBO J 42(13):e112095
16. Heilemann M (2010) Fluorescence microscopy beyond the diffraction limit. J Biotechnol 149(4):243–251
17. Manley S, Gillette JM, Lippincott-Schwartz J (2010) Single-particle tracking photoactivated localization microscopy for mapping single-molecule dynamics. Methods Enzymol 475: 109–120
18. Zhong H (2010) Photoactivated localization microscopy (PALM): an optical technique for achieving~ 10-nm resolution. Cold Spring Harb Protoc 2010(12):pdb.top91
19. Ando R, Hama H, Yamamoto-Hino M (2002) An optical marker based on the UV-induced green-to-red photoconversion of a fluorescent protein. Proc Natl Acad Sci USA 99(20): 12651–12656
20. Betzig E, Patterson GH, Sourgrat R et al (2006) Imaging intracellular fluorescent proteins at nanometer resolution. Science 313(5793):1642–1645
21. Đurisić N, Laparra-Cuervo L, Sandoval-Álvarez Á et al (2014) Single-molecule evaluation of fluorescent protein photoactivation efficiency using an in vivo nanotemplate. Nat Methods 11(2):156–162
22. Cheezum MK, Walker WF, Guilford WH (2001) Quantitative comparison of algorithms for tracking single fluorescent particles. Biophys J 81(4):2378–2388
23. Thompson RE, Larson DR, Webb WW (2002) Precise nanometer localization analysis for individual fluorescent probes. Biophys J 82(5): 2775–2783
24. Gormal RS, Meunier FA (2022) Nanoscale organization of the pre-synapse: tracking the neurotransmitter release machinery. Curr Opin Neurobiol 75:102576

25. Longfield SF, Gormal RS, Feller M et al (2024) Synapsin 2a tetramerisation selectively controls the presynaptic nanoscale organisation of reserve synaptic vesicles. Nat Commun 15(1): 2217
26. Joensuu M, Martínez-Mármol R, Mollazade M, Padmanabhan P et al (2020) Single-molecule imaging of recycling synaptic vesicles in live neurons. In: Yamamoto N, Okada Y (eds) Single molecule microscopy in neurobiology. Neuromethods. Humana, New York, p 154
27. Small C, Martínez-Mármol R, Amor R et al (2021) Combining single molecule super-resolution imaging techniques to unravel the nanoscale organization of the presynapse. In: Niedergang F, Vitale N, Gasman S (eds) Exocytosis and endocytosis. Methods in molecular biology. Humana, New York, p 2233
28. Longfield SF, Mollazade M, Wallis TP et al (2023) Tau forms synaptic nano-biomolecular condensates controlling the dynamic clustering of recycling synaptic vesicles. Nat Commun 14(1):7277
29. Chhabra Y, Wong HY, Nikolajsen LF et al (2018) A growth hormone receptor SNP promotes lung cancer by impairment of SOCS2-mediated degradation. Oncogene 37(4): 489–501
30. Zhou Y, Ariotti N, Rae J et al (2021) Caveolin-1 and cavin1 act synergistically to generate a unique lipid environment in caveolae. J Cell Biol 220(3):e202005138
31. Manley S, Gillette JM, Patterson GH et al (2008) High-density mapping of single-molecule trajectories with photoactivated localization microscopy. Nat Methods 5(2): 155–157
32. Wallis TP, Jiang A, Young K et al (2023) Super-resolved trajectory-derived nanoclustering analysis using spatiotemporal indexing. Nat Commun 14(1):3353
33. Patterson GH, Lippincott-Schwartz J (2002) A photoactivatable GFP for selective photolabeling of proteins and cells. Science 297(5588): 1873–1877
34. Gurskaya NG, Verkhusha VV, Shcheglov AS et al (2006) Engineering of a monomeric green-to-red photoactivatable fluorescent protein induced by blue light. Nat Biotechnol 24(4):461–465
35. McEvoy AL, Hoi H, Bates M et al (2012) mMaple: a photoconvertible fluorescent protein for use in multiple imaging modalities. PLoS One 7(12):e51314
36. Kremers GJ, Hazelwood KL, Murphy CS et al (2009) Photoconversion in orange and red fluorescent proteins. Nat Methods 6(5): 355–358
37. Subach FV, Patterson G, Manley S et al (2000) Photoactivatable mCherry for high-resolution two-color fluorescence microscopy. Nat Methods 6(2):153–159
38. Wiedenmann J, Ivanchenko S, Oswald F et al (2004) EosFP, a fluorescent marker protein with UV-inducible green-to-red fluorescence conversion. Proc Natl Acad Sci USA 101(45): 15905–15910
39. McKinney SA, Murphy C, Hazelwood K et al (2009) A bright and photostable photoconvertible fluorescent protein. Nat Methods 6(2): 131–133
40. Zhang M, Chang H, Zhang Y et al (2012) Rational design of true monomeric and bright photoactivatable fluorescent proteins. Nat Methods 9(7):727–729
41. Fu Z, Peng D, Zhang M et al (2020) mEosEM withstands osmium staining and Epon embedding for super-resolution CLEM. Nat Methods 17(1):55–58
42. Axelrod D (1989) Total internal reflection fluorescence microscopy. Methods Cell Biol 30:245–270
43. Lanoue V, Chai YJ, Brouillet JZ et al (2019) STXBP1 encephalopathy connecting neurodevelopmental disorders with α-synucleinopathies? Neurology 93(3): 114–123
44. Malintan NT, Nguyen TH, Han L et al (2009) Abrogating Munc18-1-SNARE complex interaction has limited impact on exocytosis in PC12 cells. J Biol Chem 284(32): 21637–21646
45. Andronov L, Lutz Y, Vonesch JL et al (2016) SharpViSu: integrated analysis and segmentation of super-resolution microscopy data. Bioinformatics 32:2239–2241
46. Kasula R, Chai YJ, Bademosi AT et al (2016) The Munc18-1 domain 3a hinge-loop controls syntaxin-1A nanodomain assembly and engagement with the SNARE complex during secretory vesicle priming. J Cell Biol 214(7): 847–858
47. Jiang A, Kudo K, Gormal RS et al (2024) Dynamin1 long-and short-tail isoforms exploit distinct recruitment and spatial patterns to form endocytic nanoclusters. Nat Commun 15(1):4060
48. Small C, Harper C, Jiang A et al (2024) SV2A controls the surface nanoclustering and endocytic recruitment of Syt1 during synaptic vesicle recycling. J Neurochem 168(9):3188–3208
49. Gormal RS, Nguyen TH, Martin S et al (2015) An acto-myosin II constricting ring initiates

the fission of activity-dependent bulk endosomes in neurosecretory cells. J Neurosci 35(4):1380–1389
50. Gormal RS, Padmanabhan P, Kasula R et al (2020) Modular transient nanoclustering of activated β2-adrenergic receptors revealed by single-molecule tracking of conformation-specific nanobodies. Proc Natl Acad Sci USA 117(48):30476–30487
51. Owen DM, Rentero C, Rossy J et al (2010) PALM imaging and cluster analysis of protein heterogeneity at the cell surface. J Biophotonics 3(7):446–454
52. Padmanabhan P, Martínez-Mármol R, Xia D et al (2019) Frontotemporal dementia mutant tau promotes aberrant Fyn nanoclustering in hippocampal dendritic spines. Elife 8:e45040
53. Davis S, Gale NW, Aldrich TH et al (1994) Ligands for EPH-related receptor tyrosine kinases that require membrane attachment or clustering for activity. Science 266(5186): 816–819
54. Nair D, Hosy E, Petersen JD et al (2013) Super-resolution imaging reveals that AMPA receptors inside synapses are dynamically organized in nanodomains regulated by PSD95. J Neurosci 33(32):13204–13224
55. Martínez-Mármol R, Small C, Jiang A et al (2023) Fyn nanoclustering requires switching to an open conformation and is enhanced by FTLD-tau biomolecular condensates. Mol Psychiatry 28(2):946–962
56. Hilzenrat G, Pandžić E, Yang Z et al (2020) Conformational states control Lck switching between free and confined diffusion modes in T cells. Biophys J 118(6):1489–1501
57. Gormal RS, Martinez-Marmol R, Brooks AJ et al (2024) Location, location, location: protein kinase nanoclustering for optimised signalling output. elife 13:e93902
58. Chhabra Y, Seiffert P, Gormal RS et al (2023) Tyrosine kinases compete for growth hormone receptor binding and regulate receptor mobility and degradation. Cell Rep 42(5):112490
59. Gimber N, Tadeus G, Martizen T et al (2015) Diffusional spread and confinement of newly exocytosed synaptic vesicle proteins. Nat Commun 6(1):8392
60. Bademosi AT, Steeves J, Karunanithi S et al (2018) Trapping of syntaxin1a in presynaptic nanoclusters by a clinically relevant general anesthetic. Cell Rep 22(2):427–440
61. Ribrault C, Reingruber J, Petković M et al (2011) Syntaxin1A lateral diffusion reveals transient and local SNARE interactions. J Neurosci 31(48):17590–17602
62. Bademosi AT, Lauwers E, Padmanabhan P et al (2016) In vivo single-molecule imaging of syntaxin1A reveals polyphosphoinositide-and activity-dependent trapping in presynaptic nanoclusters. Nature Commun 7(1):13660
63. Bademosi AT, Meunier FA (2022) Unveiling the nanoscale dynamics of the exocytic machinery in chromaffin cells with single-molecule imaging. In: Chromaffin cells: methods and protocols. Springer US, New York, pp 311–327
64. Hoffmann C, Rentsch J, Tsunoyama TA et al (2023) Synapsin condensation controls synaptic vesicle sequestering and dynamics. Nat Commun 14(1):6730
65. Dixon CL, Sah P, Keramidas A et al (2017) γ1-containing GABA-A receptors cluster at synapses where they mediate slower synaptic currents than γ2-containing GABA-A receptors. Front Mol Neurosci 10:178
66. Sun M, Hu K, Bewersdorf J et al (2021) Sample preparation and imaging conditions affect mEos3. 2 photophysics in fission yeast cells. Biophys J 120(1):21–34
67. Li Y, Almassalha LM, Chandler JE et al (2017) The effects of chemical fixation on the cellular nanostructure. Exp Cell Res 358(2):253–259
68. Giannone G, Hosy E, Levet F et al (2010) Dynamic superresolution imaging of endogenous proteins on living cells at ultra-high density. Biophys J 99(4):1303–1310
69. Joensuu M, Padmanabhan P, Durisic N et al (2016) Subdiffractional tracking of internalized molecules reveals heterogeneous motion states of synaptic vesicles. J Cell Biol 215(2): 277–292
70. Joensuu M, Martínez-Mármol, Padmanabhan P et al (2017) Visualizing endocytic recycling and trafficking in live neurons by subdiffractional tracking of internalized molecules. Nat Protoc 12(12):2590–2622
71. Wang T, Martin S, Papadopulos A et al (2015) Control of autophagosome axonal retrograde flux by presynaptic activity unveiled using botulinum neurotoxin type a. J Neurosci 35(15): 6179–6194
72. Wang T, Martin S, Nguyen T et al (2016) Flux of signalling endosomes undergoing axonal retrograde transport is encoded by presynaptic activity and TrkB. Nat Commun 7(1):12976
73. Bademosi AT, Decet M, Kuenen S et al (2023) EndophilinA-dependent coupling between activity-induced calcium influx and synaptic autophagy is disrupted by a Parkinson-risk mutation. Neuron 111(9):1402–1422

74. Vanhauwaert R, Kuenen S, Masius R et al (2017) The SAC 1 domain in synaptojanin is required for autophagosome maturation at presynaptic terminals. EMBO J 36(10): 1392–1411
75. Wäldchen S, Lehmann J, Klein T et al (2015) Light-induced cell damage in live-cell super-resolution microscopy. Sci Rep 5(1):15348
76. Harper CB, Small C, Davenport EC et al (2020) An epilepsy-associated SV2A mutation disrupts Synaptotagmin-1 expression and activity-dependent trafficking. J Neurosci 40(23):4586–4595
77. Gormal RS, Wallis TP, McCann AJ et al (2025) Nanoscale spatiotemporal cluster analysis of expressed and endogenous proteins. Nat Protoc. https://doi.org/10.1038/s41596-025-01209-w
78. Tinevez JY, Perry N, Schindelin J et al (2017) TrackMate: an open and extensible platform for single-particle tracking. Methods 115:80–90
79. Small C, Martínez-Mármol R, Amor R et al (2021) Combining single molecule super-resolution imaging techniques to unravel the nanoscale organization of the presynapse. Methods Mol Biol 2233:265–286
80. Miesenböck G, De Angelis DA, Rothman JE (1998) Visualizing secretion and synaptic transmission with pH-sensitive green fluorescent proteins. Nature 394(6689):192–195
81. McAdam RL, Varga KT, Jiang Z et al (2015) The juxtamembrane region of synaptotagmin 1 interacts with dynamin 1 and regulates vesicle fission during compensatory endocytosis in endocrine cells. J Cell Sci 128(12):2229–2235
82. Chenouard N, Smal I, de Cahumont F et al (2014) Objective comparison of particle tracking methods. Nat Methods 11(3):281–289
83. Kalaidzidis Y (2009) Multiple objects tracking in fluorescence microscopy. J Math Biol 58(1–2):57–80
84. Ershov D, Phan MS, Pylvänäinen JW et al (2022) TrackMate 7: integrating state-of-the-art segmentation algorithms into tracking pipelines. Nat Methods 19(7):829–832
85. Jaqaman K, Loerke D, Mettlen M et al (2008) Robust single-particle tracking in live-cell time-lapse sequences. Nat Methods 5(8):695–702

Chapter 7

Single-Molecule Localization Microscopy in the Nucleus with Self-Labelling Tags: Halo, SNAP, and CLIP Tags

Alex J. McCann and Frédéric A. Meunier

Abstract

In comparison to single-molecule localization microscopy (SMLM) of proteins located on the cell surface, SMLM of proteins within the nucleus requires fluorescent tags that are generally brighter, longer-lasting and (in the case of fluorescent ligands) membrane-permeable. Although fluorescent ligands typically have superior optical qualities compared to those of fluorescent protein (FP) tags, making them ideal candidates for nuclear SMLM, unlike FPs, they are unable to be expressed in cells as a fusion protein with the protein of interest. Self-labelling tags, on the other hand (e.g., HaloTag, SNAP-tag, and CLIP-tag), combine the ability to express a protein tag (e.g., HaloTag protein) as a fusion protein, with the enhanced optical qualities of externally added fluorescent dyes (e.g., dye-conjugated HaloTag ligand which covalently binds to the HaloTag protein). Excitation of only a subset of fluorescently labelled molecules is essential for achieving single-molecule resolution, and therefore SMLM of nuclear proteins requires a microscope setup capable of penetrating the nucleus with a thin beam that is powerful enough to produce an appropriate signal-to-noise ratio. Highly Inclined and Laminated Optical sheet (HILO) microscopy uses an oblique angle (just below that of Total Internal Reflection Fluorescence (TIRF)) to excite fluorescently labelled molecules within a thin, highly inclined plane, the signal of which can be amplified using a high-power TIRF filter.

This chapter describes the following protocols: (i) seeding of cells into glass-bottom imaging dishes, (ii) transfection of cells with a HaloTag-tagged nuclear protein of interest, (iii) incubation of transfected cells with a fluorescent nuclear marker and a fluorescent HaloTag ligand, and (iv) single-molecule imaging of the fluorescently labelled HaloTag-tagged protein of interest in the nucleus of live cells using HILO microscopy with a high-powered TIRF filter.

Key words HaloTag, SNAP-tag, CLIP-tag, Self-labelling protein tags, Single-molecule localisation microscopy, Nucleus, HILO

1 Introduction

Single-molecule localisation microscopy (SMLM) encompasses a variety of super-resolution imaging techniques such as single particle tracking photoactivation localisation microscopy (sptPALM; *see* Chapter 6), stochastic optical reconstruction microscopy (STORM; *see* Chapter 5) and universal point accumulation in

Frédéric A. Meunier (ed.), *Super-Resolution Microscopy: Methods and Protocols*, Methods in Molecular Biology, vol. 3034, https://doi.org/10.1007/978-1-0716-5268-8_7,

nanoscale topography (uPAINT, *see* Chapter 9) [1–7]. SMLM techniques are often used in conjunction with other imaging techniques such as total internal reflection fluorescence (TIRF) microscopy [8] and highly inclined and laminated optical sheet (HILO) microscopy [9]. The usage of these techniques depends upon the cellular location of the protein of interest, among other factors. For example, TIRF microscopy has been used to track the mobilities of various proteins on the cell surface [10–13], as it generates an evanescent wave at the glass/sample interface (~100 nm into the sample) that limits the background from out-of-focus molecules located deeper in the cell [8] (Fig. 1a). Although TIRF is well-suited for imaging individual proteins on the 2D surface of the cell (*see* Chapter 6), alternative techniques are needed to image individual proteins located deeper within the 3D environment of the cell (such as the nucleus) [9, 14]. One such technique, used in this protocol, is HILO microscopy [9].

1.1 Highly Inclined and Laminated Optical Sheet (HILO) Microscopy

Unlike TIRF, which uses an excitation laser at an angle greater than that of the critical angle, HILO microscopy uses an incidence angle below the critical angle, which generates a thin beam that penetrates the sample at an oblique highly inclined angle [9] (Fig. 1b). In this protocol, we use HILO at an incidence angle of 60.18° with a high power TIRF filter to further increase the signal-to-noise ratio (amplifies the HILO beam ~four fold). The depth (z-axis) of the HILO beam can be adjusted so that different sections (z-planes) of the cell and/or tissue can be imaged. An optional step in this protocol is to incorporate a nuclear marker to aid identification of the center (z-plane) of the nucleus (Fig. 1c and d).

Due to its ability to penetrate samples with a relatively high signal-to-noise ratio, HILO has been used to investigate the mobilities of various transcription factors in the nucleus of live cells [15–21]. Of note, a requirement of single particle tracking is that the fluorescently tagged protein of interest remains within the field of view long enough to capture the minimum number of frames needed to accurately generate trajectories and extract diffusion coefficient values. Based on this, freely diffusing molecules are unable to be reliably captured using single particle tracking as their mobilities are inherently higher than that of the acquisition capability of the microscope, and therefore only a subset of molecules with lower mobilities and confined behaviors are likely to be captured. Molecules may exhibit lower mobilities and confined behaviors for a variety of reasons, such as being in the presence of biomolecular condensates (BMCs) [22, 23] or upon interacting with chromatin in the case of transcription factors (TFs) [15, 16, 19, 20, 24].

By performing HILO with different acquisition speeds (e.g., "fast" and "slow" tracking), different mobility behaviors within the nucleus can be extracted [15, 17, 21]. For example, faster

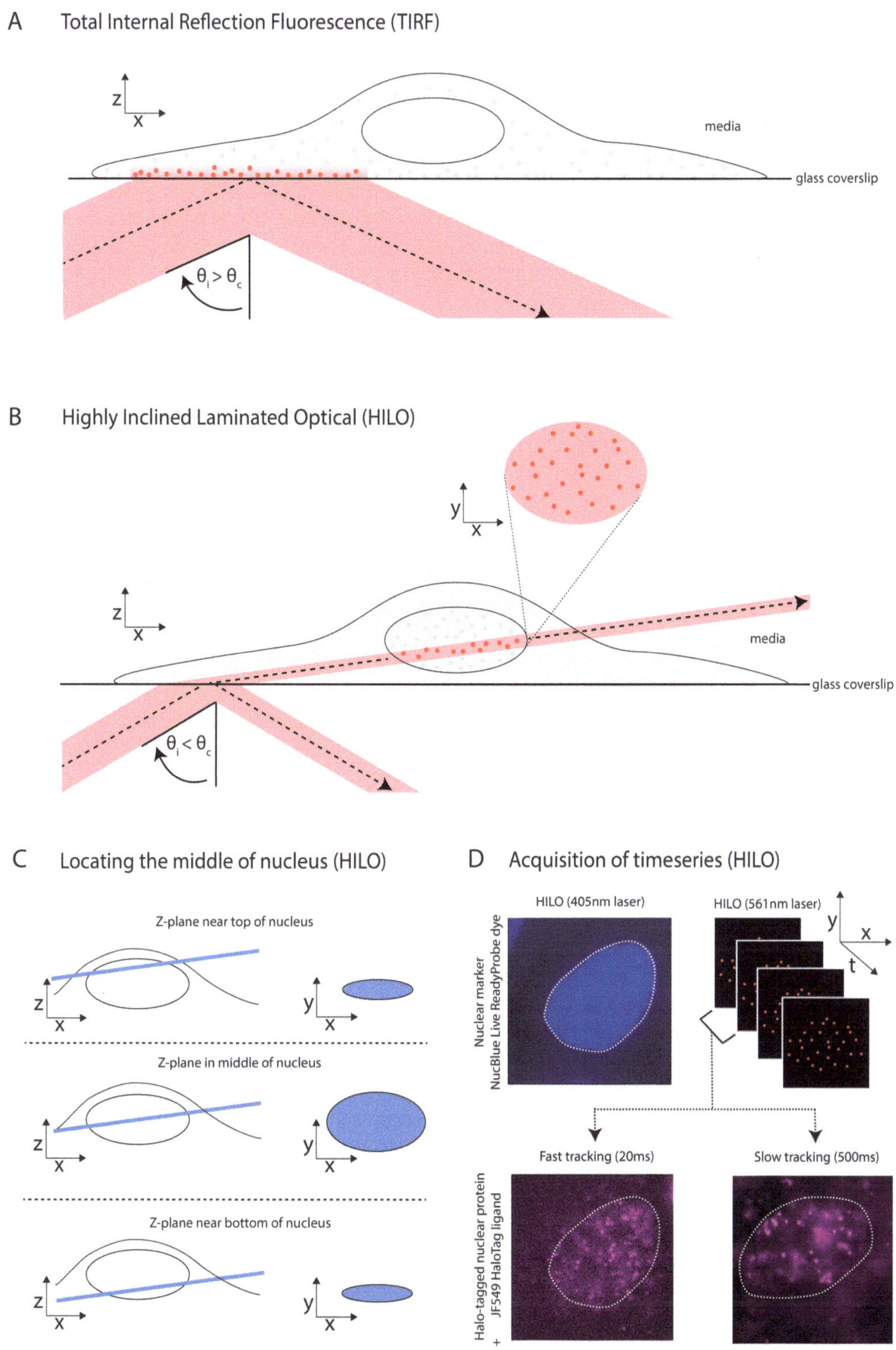

Fig. 1 (**a**) Total internal reflection fluorescence (TIRF) microscopy is used to image individual fluorescently labelled molecules of interest on/near the surface of live cells. The incidence angle (θ_i) of the excitation laser is greater than that of the critical angle (θ_c). (**b**) Highly inclined and laminated optical sheet (HILO) microscopy is

acquisition speeds capture a wide spectrum of mobilities (from slowly diffusing/immobile through to fast diffusing), which can be used to generate trajectories from which analyses such as the mean square displacement (MSD), $\log_{10}$ diffusion coefficient frequency distribution, and mobile-to-immobile ratio can be inferred [7, 10–13, 15, 22] (*see* Chapters 6 and 13). On the other hand, slower acquisition speeds increase the signal-to-noise ratio of the slowly diffusing/immobilized population of molecules by blurring the faster molecules into the background, which can be used to measure the time spent immobilized (lifetime) [15, 17, 21, 24]. In this protocol we use a combination of "fast" (20 ms) and "slow" (500 ms) acquisition speeds to capture different mobility behaviors (Fig. 1d).

1.2 Fluorescent Labelling Strategies

In order to visualize and track the movement of individual molecules, the molecule of interest must be labelled with a fluorescent tag using a method that enables spatial separation of the labelled population of molecules (e.g., sparse labelling) [2, 6]. Photoactivatable and photoconvertible fluorescent proteins (PA-FPs and PC-FPs respectively), are commonly used in SMLM as their labelling density (level of photoactivated or photoconverted fusion protein) can be controlled by adjusting the power of the photoactivation/photoconversion laser [1, 2, 5] (Fig. 2a). An alternative to PA-FPs and PC-FPs are self-labelling protein tags (e.g., HaloTag, SNAP-tag, and CLIP-tag) [25–34], which have grown in popularity due to the development of an array of cell-permeable dye-conjugated ligands with enhanced optical properties that

Fig. 1 (continued) used to image individual fluorescently labelled molecules of interest (such as transcription factors) in the nucleus of live cells. The critical angle (θ_c) is greater than that of the incidence angle (θ_i) of the excitation laser. (**c**) A live cell nuclear dye can be used to locate the center of the nucleus more easily during HILO microscopy with a 405 nm laser. The shape of the nucleus (as indicated by the nuclear dye) changes as the user adjusts the z-axis (top, middle, and bottom panels). The center of the nucleus has been located when the height (y-axis) and width (x-axis) of the nucleus are highest (middle panel). (**d**) Top left panel: Once the center of the nucleus has been located, a snapshot is taken, which can be used as a reference when drawing regions of interest (ROIs) in subsequent data acquisition steps. An example snapshot of a nucleus stained with NucBlue Live Ready Probe dye, captured using HILO microscopy with a 405 nm excitation laser, is shown. Top right panel: a time series is then taken of the nucleus in order to capture individual fluorescently labelled proteins of interest, using an excitation laser that excites the fluorescent tag of the protein of interest, in HILO mode. In this example a 561 nm excitation laser is used to excite the JF549 HaloTag ligand that is bound to a HaloTag-tagged nuclear protein of interest. Each frame captures individual fluorescently labelled proteins of interest, which appear as bright spatially distinct points (red) against a darker background. Different acquisition speeds (time taken to acquire each individual frame), such as "fast" tracking (e.g., 20 ms) and "slow" tracking (e.g., 500 ms), can be used to capture different populations of molecules based on their mobilities. Bottom panels: example (left) "fast" tracking (20 ms) frame and (right) "slow" tracking (500 ms) frame taken of a JF549 HaloTag ligand/HaloTag-tagged nuclear protein of interest, captured using HILO microscopy with a 561 nm excitation laser

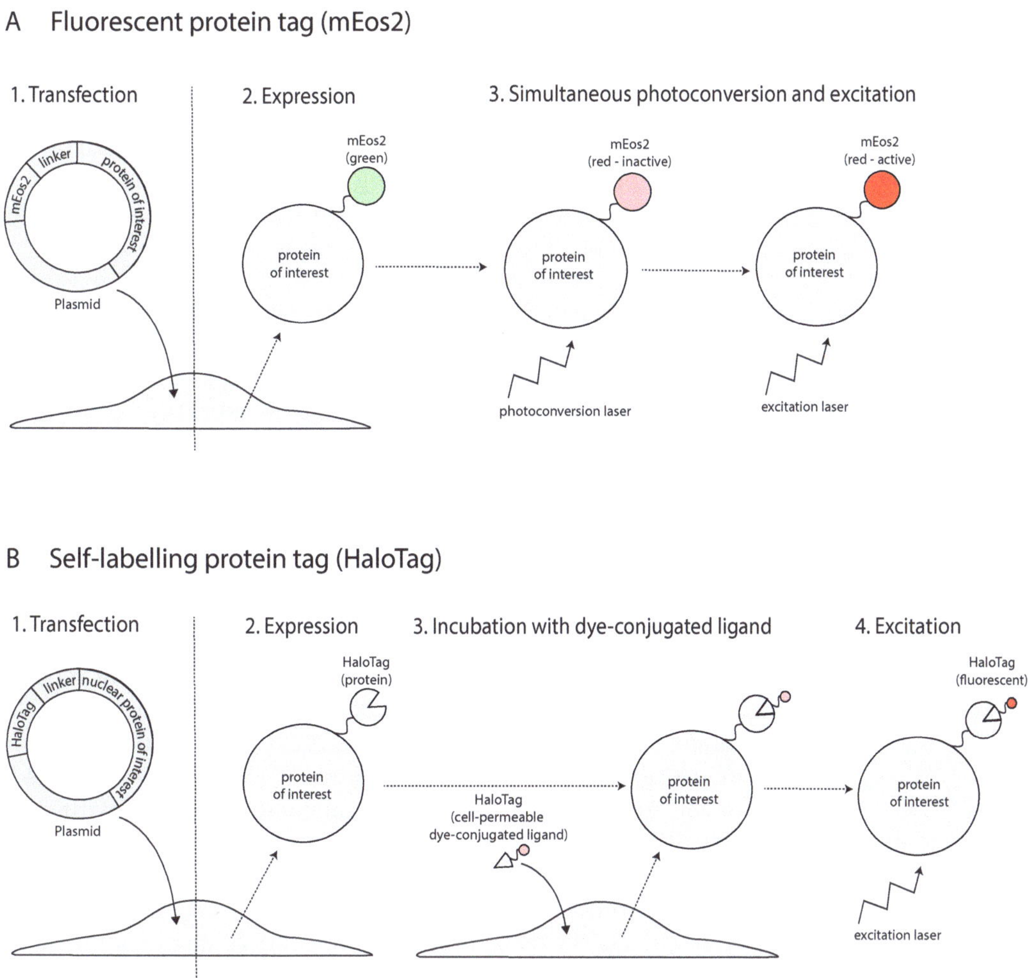

Fig. 2 (**a**) The principle behind using fluorescent protein tags in single-molecule imaging experiments, using photoconvertible fluorescent protein mEos2 as an example. (1) Transfection: cells are transfected with a plasmid encoding a protein of interest tagged with mEos2 (as a fusion protein). (2) Expression: after a period of time, cells begin expressing the protein of interest tagged with mEos2, which will appear as green fluorescent blobs (unable to distinguish individual molecules) under a 488 nm excitation laser. (3) Simultaneous photoconversion and excitation: a photoconversion laser (405 nm) is used at a low laser power to photoconvert a subset of mEos2-tagged molecules. Simultaneously, an excitation laser (561 nm) is used to excite the population of converted mEos2-tagged molecules, which will appear as red spatially distinct points. (**b**) The principle behind using self-labelling protein tags in single-molecule imaging experiments, using the HaloTag as an example. (1) Transfection: cells are transfected with a plasmid encoding a protein of interest tagged with the HaloTag (as a fusion protein). (2) Expression: after a period of time, cells begin expressing the protein of interest tagged with the HaloTag. (3) Incubation with dye-conjugated ligand: immediately before imaging, cells are incubated with a low concentration of fluorescent dye-conjugated HaloTag ligand, which covalently binds to the HaloTag of the HaloTag-tagged protein of interest. (4) Excitation: an excitation laser (561 nm) is used to excite the population of HaloTagged protein of interest that is bound by the dye-conjugated HaloTag ligand, which will appear as red spatially distinct points

covalently-bind to the respective self-labelling protein tag [35–40]. The labelling density of self-labelling protein tags is controlled by the concentration of fluorescent dye-conjugated ligand, which is added externally following the expression of the self-labelling protein tag fusion protein (Fig. 2b), and optionally by an initial photobleaching period prior to imaging [41]. The labelling density of self-labelling protein tags can be further controlled for by using a photoactivatable fluorescent dye-conjugated ligand [36].

1.3 Self-Labelling Protein Tags

Self-labelling protein tags are enzymes that have been modified so that they remain covalently bound to their ligands instead of cleaving them—functioning instead as a versatile two-component protein labelling system. The protein tag is expressed as a fusion protein with the protein of interest, and the ligand to which it binds is conjugated to a functional group such as a fluorescent dye (Fig. 2b), affinity tag or surface (e.g., bead or resin). Self-labelling protein tags provide a number of advantages over traditional tags [26, 28, 36, 41]. For example, the interchangeability of ligands for the same self-labelling protein tag replaces the need to generate multiple tags for the same protein of interest [26, 28]. As the ligands are not interchangeable between the different types of self-labelling tags (e.g., HaloTag ligands only bind to HaloTag proteins), two or more self-labelling tags can be used at the same time to label different proteins of interest [34, 37, 39, 42]. Additionally, self-labelling protein tags can be used together with other types of tags to perform single-molecule tracking in conjunction with other imaging modalities. For example, the SNAP-tag has been used in combination with the STREAMING-tag to simultaneously image the dwell times (chromatin occupancy) of the SOX2 transcription factor (SOX2-SNAP-tag) specifically at the Nanog gene locus (STREAMING-tag; TetO/mTetR-GFP) along with the transcriptional activity of the Nanog gene locus (STREAMING-tag; MS2/MCP-RFP) [43]. Furthermore, the strength of the covalent bond formed between the fusion protein tag and ligand can be exploited to obtain better outcomes in protein yield and purity compared to other affinity tags in protein purification experiments [28]. In this protocol we use self-labelling protein tags, as the optical properties of the fluorescent dyes that are conjugated to self-labelling protein tag ligands typically outperform those of FPs [36, 41]. For example, fluorescent dyes TMR and JF549 have brightness values approximately two-fold and ten-fold greater than that of photoactivatable FP PA-mCherry, respectively, resulting in lower localisation errors and longer trajectories [41]. Similarly, photoactivatable fluorescent dyes PA-JF549 and PA-JF646 were shown to be significantly brighter than the photoconvertible FP mEos3.2, again resulting in lower localisation errors and longer trajectories [36].

Table 1
HaloTag, SNAP-tag, and CLIP-tag

Tag	Origin	Size	Ligand	Refs.
HaloTag	Haloalkane dehalogenase (*Rhodococcus rhodochrous*)	297aa (33.6 kDa)	Chloroalkane (CA)	[25–29]
SNAP-tag	O^6-alkylguanine-DNA-alkyltransferase (human)	182aa (19.4 kDa)	O^6-benzylguanine (BG) O^6-benzylchloropyrimidine (CP)	[30–33, 44–46]
CLIP-tag	O^6-alkylguanine-DNA-alkyltransferase (human)	182aa (19.4 kDa)	O^2-benzylcytosine (BC)	[33, 34]

Amongst the most widely used self-labelling protein tags are the HaloTag [25–29], SNAP-tag [30–32], and CLIP-tag [33, 34] (Table 1).

1.4 HaloTag, SNAP-Tag, and CLIP-Tag

The HaloTag is a 297 amino acid (33.6 kDa) protein derived from the Haloalkane dehalogenase enzyme found in the *Rhodococcus rhodochrous* bacterial species that was modified to covalently bind chloroalkane (CA) derivatives [25–29] (Table 1). The original HaloTag contained a single-point mutation that inactivated its catalytic activity, resulting instead in covalent ligand binding [25–27]. A number of additional mutations have since been introduced into the HaloTag to improve its expression, solubility, and ligand-binding kinetics, giving rise to the latest commercially available HaloTag variant known as HaloTag7 (also referred to as just HaloTag) [28, 29] (Table 2).

The SNAP- and CLIP-tags are smaller 182 amino acid (19.4 kDa) proteins that were originally derived from the Human O^6-alkylguanine-DNA-alkyltransferase (hAGT) enzyme, with the SNAP-tag engineered to covalently bind O^6-benzylguanine (BG) and O^6-benzylchloropyrimidine (CP) derivatives [30–33, 44–46], and the CLIP-tag engineered to covalently bind O^2-benzylcytosine (BC) derivatives [33, 34] (Table 1). The first SNAP-tag consisted of a single point mutation that increased its affinity for BG derivatives compared to the wild-type hAGT protein [30]. Subsequently, a series of point mutations as well as a C-terminal truncation were introduced into the SNAP-tag protein in order to increase its stability and ligand binding kinetics and decrease its DNA-binding affinity and size [31, 32, 34, 45, 46]. The CLIP-tag was generated by mutating the SNAP-tag in order to produce a similar self-labelling protein tag with different ligand binding specificity [34]. Additional mutations were subsequently added to further improve upon the ligand-binding kinetics of the SNAP-tag and CLIP-tag, resulting in the SNAPf-tag and the CLIPf-tag, respectively [33] (Table 2).

Table 2
Commercially available HaloTag, SNAP-tag, and CLIP-tag variants

Tag	Mutations	Outcome	Refs.
HaloTag			
HaloTag7 (HaloTag)	L47V, S58T, D78G, Y87F, L88M, C128F, A155T, E160K, A167V, A172T, K175M, C176G, K195N, A224E, N227D, E257K, T264A, H272N, Y273L, P291S, A292T, 294E, 295I, 296S, 297G	Improved expression, solubility, and ligand-binding kinetics compared to original HaloTag.	[28, 29], *P*
SNAP-tag			
SNAP-tag	C62A, K125A, A127T, R128A, G131K, G132T, M134L, R135S, C150S, N157G, S159E, del183-207	Decreased size, decreased DNA-binding affinity and increased ligand-binding kinetics compared to wild-type hAGT protein.	[31, 32], *N*
$SNAP_f$-tag	E30R, K32I, L33F, C62A, Q115S, Q116H, K125A, A127T, R128A, G131K, G132T, M134L, R135S, C150Q, S151G, S152D, G153L, A154D, N157G, S159E, del183-207	Improved ligand-binding kinetics and decreased DNA-binding compared to original SNAP-tag.	[32, 33, 46], *N*
CLIP-tag			
$CLIP_f$-tag	E30R, M60I, C62A, Y114E, A121V, K125A, R128A, K131N, G132T, M134L, S135D, C150Q, L153S, G157P, E159L, del183-207	Binds different ligand to SNAP-tag, improved ligand-binding kinetics compared to original CLIP-tag.	[33, 34], *N*

Mutations compared to the wild-type protein. [N] = information obtained from NEB website (https://www.neb.com); [P] = information obtained from Promega website (https://www.promega.com)

1.4.1 HaloTag, SNAP-Tag, and CLIP-Tag Vectors

As mentioned above, the first part of the self-labelling protein tag system involves the expression of a fusion protein (protein of interest tagged with a self-labelling protein tag) in the cell line of choice (Fig. 2b).

There are a variety of different vectors available for the expression of HaloTag, SNAP-tag, and CLIP-tag fusion proteins, including those that require insertion of the protein of interest (empty vector) and those that already contain the protein of interest (pre-made vector) (Table 3).

The main features of these vectors are (i) a promoter for expression of the fusion protein (e.g., CMV for mammalian expression and T7 for bacterial expression), (ii) an N-terminal or C-terminal self-labelling protein tag (e.g., HaloTag, SNAP-tag, or CLIP-tag), (iii) the protein of interest (pre-made vector) or multiple cloning site (MCS) for insertion of the protein of interest (empty vector), (iv) a protein linker (between the tag and the

Commercially available HaloTag, SNAP-tag, and CLIP-tag vectors

Tag	Supplier	Tag location	Promoter	Antibiotic resistance	Selection marker	Comment
HaloTag vectors						
HaloTag (HaloTag7)	Promega	N-terminus C-terminus	CMV CMV1d, CMV2d, CMV3d, T7	Ampicillin Kanamycin	Neomycin (neo vectors) Barnase (flexi vectors)	Requires cloning Contains a single MCS at either the N-or C-terminus (depending on the vector) for inserting a protein of interest Stable expression in mammalian cells (neo vectors)
	Kazusa genome technologies (Promega collaboration)	N-terminus	CMV T7	Ampicillin	Barnase	Pre-made. Large number of ORFs (~10,000 human) inserted into the Promega HaloTag flexi vector
	Genecopoeia (Promega collaboration)	N-terminus C-terminus	CMV T7	Ampicillin	Neomycin Puromycin Hygromycin	Pre-made. Large number of ORFs (~20,000 human and ~ 15,000 mouse) inserted at either the N- or C-terminus of the HaloTag. Stable transfection in mammalian cells (neomycin, Puromycin and Hygromycin vectors)
SNAP-tag vectors						
SNAP26b	NEB	N-terminus C-terminus	T7	Ampicillin	N/A	Requires cloning. Contains 2 MCS for inserting a protein of interest into either the N- or C-terminus of the SNAP26b protein. Codon optimized for bacterial expression
$SNAP_f$	NEB	N-terminus C-terminus	CMV	Ampicillin	Neomycin	Requires cloning. Contains 2 MCSs for inserting a protein of interest into either the N- or C-terminus of the $SNAP_f$ protein. Stable transfection in mammalian cells
CLIP-tag vectors						
$CLIP_f$	NEB	N-terminus C-terminus	CMV	Ampicillin	Neomycin	Requires cloning. Contains 2 MCS for inserting a protein of interest into either the N- or C-terminus of the $CLIP_f$ protein. Stable transfection in mammalian cells

Information on each vector was found on the website of the indicated supplier. Promega: https://www.promega.com, NEB: https://www.neb.com, Kazusa Genome Technologies: http://www.kazusa.or.jp/e/index.html, Genecopoeia: https://www.genecopoeia.com

protein of interest/MCS), v) a termination sequence (e.g., SV40 polyA tail), (vi) an antibiotic resistance gene (e.g., ampicillin or kanamycin), and if applicable, vii) a selection marker (e.g., neomycin, barnase, puromycin or hygromycin) (Table 3).

1.4.2 HaloTag, SNAP-Tag, and CLIP-Tag Fluorescent Ligands

Following expression of the fusion protein in the cell line of choice, a small concentration of self-labelling protein tag ligand (e.g., fluorescent dye-conjugated ligand) is added externally to the cells as the second part of the self-labelling protein tag system (Fig. 2b).

A variety of fluorescent-dye conjugated ligands, dyes with reactive groups for conjugating to ligands, and dye-conjugation kits are available for use with the HaloTag, SNAP-tag, and CLIP-tag (Table 4).

Importantly, the selected self-labelling protein tag ligand should match that of the expressed self-labelling protein tag (e.g., HaloTag vector and JF549 HaloTag ligand).

1.4.3 HaloTag, SNAP-Tag, and CLIP-Tag Comparisons

Of note, differences in performance such as labelling kinetics [39, 47], labelling brightness [39–41, 47], photostability [17, 39], and membrane permeability [39] have been observed between different self-labelling protein tags (using the same fluorescent dye) [17, 40, 47] and between different fluorescent dyes (using the same self-labelling protein tag) [39–41, 47]. Furthermore, the protein of interest to which the self-labelling protein tag is fused [40], and the type of sample used (e.g., cell type and fixed vs live cells) [39, 40], has also been shown to have an impact on performance.

Examples of differences in HaloTag, SNAP-tag, and CLIP-tag performances are shown in Table 5.

One such example is fluorescent dye JF549, which was found to be ~2 times more photostable when used as a ligand with the HaloTag as compared to the SNAPf-tag and CLIPf-tag [17]. Another example is near far-red dyes JF711 and JF722 whereby JF711 produced higher brightness values than JF722 in fixed cells, whereas JF722 exhibited faster ligand binding kinetics and a higher photostability than JF711 in live cells [39].

2 Materials

The following three-day protocol details the imaging of a HaloTag-tagged nuclear protein of interest in the nucleus of live HeLa cells, starting from the seeding of cells into imaging dishes. Prepare all media (and Gelatin) in advance and store at 4 °C until use. All work (prior to imaging on Day 3) should be performed in a cell culture hood to prevent contamination.

Table 4
Commercially available cell-permeable dyes, dye-conjugated ligands, and dye conjugation kits for use with self-labelling protein tags

Dye	Ex.	Em.	ε	Φ	Bright.	Ref.	Commercial availability
Coumarin	362	450	–	–	–	P	HaloTag ligand (Promega)
430	421	444–484	a) 56	0.5	28	N	SNAP-tag ligand (NEB)
Oregon green	490^{N} 492^{P}	514^{N} 518^{P}	b) 87^{N}	0.9^{N}	78.3	N, P	SNAP-tag ligand (NEB) HaloTag ligand (Promega)
diAcFAM	492	521	–	–	–	P	HaloTag ligand (Promega)
R110Direct	498	528	–	–	–	P	HaloTag ligand (Promega)
JF503	503	529	c) 83 d) 95	0.87	72.2 82.7	[37]	HaloTag ligand (Promega)
505	504	532	e) 70	0.8	56	N	CLIP-tag ligand (NEB)
505-star	504	532	e) 70	0.8	56	N	SNAP-tag ligand (NEB)
JF525	525	549	c) 94 f) 122	0.91	85.5 111	[37, 39, 42]	HaloTag ligand (Promega) Dye with reactive group for conjugation (Tocris/ HelloBio) Conjugation kit (HelloBio)
JF526	526	550	c) 19 f) 118	0.87	16.5 102.7	[39, 42]	Dye with reactive group for conjugation (Tocris)
JF549	549	571	c) 101 f) 134	0.88	88.9 117.9	[35–39, 42]	HaloTag ligand (Promega) Dye with reactive group for conjugation (Tocris/ HelloBio) Conjugation kit (HelloBio)
TMR	548	572	c) 78	0.41	32	[35]	HaloTag ligand (Promega)
TMRDirect	552	578	–	–	–	P	HaloTag ligand (Promega)
TMR-star	554	580	e) 89	0.28	24.92	N	SNAP-tag ligand (NEB) CLIP-tag ligand (NEB)
PA-JF549	(405) 551 553	570 573	85.9 98.9	0.78 0.85	67 84.1	[36]	Dye with reactive group for conjugation (Tocris)
JFX554	554	576	c) 104	0.8	83.2	[35]	HaloTag ligand (Promega)
JF585	585	609	c) 1.5 f) 156	0.78	1.2 121.7	[37, 39, 42]	HaloTag ligand (Promega) Dye with reactive group for conjugation (Tocris)
JF635	635	652	c) 0.4 g) 167	0.56	0.2 93.5	[37, 42]	HaloTag ligand (Promega) Dye with reactive group for conjugation (Tocris)

(continued)

Table 4 (continued)

Dye	Ex.	Em.	ε	Φ	Bright.	Ref.	Commercial availability
JF646	646	664	c) 5–5.6 g) 148–152	0.52 0.54	2.6–2.9 79.9–82.1	[35–39, 42]	HaloTag ligand (Promega) Dye with reactive group for conjugation (Tocris)
PA-JF646	(405) 649 651	663 665	c) 118–120 g) 154–162	0.47 0.53	55.5–56.4 81.6–85.9	[36]	Dye with reactive group for conjugation (Tocris)
647-SiR	645	661	h) 100	0.39	39	N	SNAP-tag (NEB)
JFX650	650	667	f) 17.6	0.53	9.3	[35]	HaloTag ligand (Promega) Dye with reactive group for conjugation (Tocris)
JF669	669	682	c) 112 g) 116	0.37	41.1 42.9	[38, 39]	Dye with reactive group for conjugation (Tocris)

Ex. = Excitation spectra (nm); Em. = Emission spectra (nm); ε = Extinction coefficient (mM^{-1} cm^{-1}); Φ = quantum yield; Bright. = brightness (ε x Φ); Ref. = references for numerical values. N = values obtained from NEB website (https://www.neb.com); P = values obtained from Promega website (https://www.promega.com). Tocris website: https://www.tocris.com, HelloBio website: https://hellobio.com. Extinction coefficient measurements in (a) methanol, (b) pH 9.0 buffer, (c) 10 mM HEPES pH 7.3, (d) 0.1% v/v Et_3N in TFE, (e) pH 7.4 buffer, (f) 0.1% v/v TFA in TFE, (g) 0.1% v/v TFA in ethanol or (h) aqueous solution

2.1 Day 1—Seeding of Cells into Imaging Dishes

1. ~155,000 ready to seed HeLa cells (60–75% confluency) (*see* **Notes 1** and **2**).
2. 29 mm glass bottom dish with 20 mm micro-well #1.5 cover glass (*see* **Note 3**).
3. 0.5–1% Gelatin (*see* **Note 4**).
4. Phosphate-buffered saline (PBS).
5. "High serum culture media": 10% Fetal Bovine Serum, 1% GlutaMAX, 1% MEM non-essential amino acids in Dulbecco's Modified Eagle Media (*see* **Note 5**).
6. Cell incubator (37 °C, 5% CO_2).
7. Cell culture hood.

2.2 Day 2—Transfection of Cells with HaloTag-tagged Nuclear Protein of Interest

1. Glass-bottom dish containing HeLa cells from Day 1 (*see* Subheading 2.1).
2. X-tremeGENE™ 9 Transfection Reagent kit (Roche) (*see* **Note 6**).
3. Purified plasmid encoding HaloTag-tagged nuclear protein of interest (*see* **Note 7**).
4. "Low serum transfection media": 1% GlutaMAX in FluoroBrite™ DMEM (Gibco) (*see* **Note 8**).

Table 5
Example comparisons of HaloTag, SNAP-tag and CLIP-tag self-labelling protein tags and fluorescent dye-conjugated ligands

Tag 1	Ligand 1	Tag 2	Ligand 2	Findings	Refs.
HaloTag	TMR (CA)	HaloTag	JF549 (CA)	Higher labelling brightness for JF549 at lower laser intensities Lower localisation error for JF549 Higher average track duration for JF549	[41]
	JF549 (CA)		JF646 (CA)	Higher labelling brightness for JF549 Lower localisation error for JF549	
HaloTag	JF549	SNAPf-tag	JF549cp	Higher photostability for HaloTag	[17]
HaloTag	JF549	CLIPf-tag	JF549	Higher photostability for HaloTag	[17]
HaloTag	SiR (CA) TMR (CA) JF549 (CA) JF646 (CA)	SNAPf-tag	SiR (BG) TMR (BG) JF549 (BG) JF646 (BG)	Higher labelling brightness for HaloTag compared to SNAPf-tag (particularly far-red dyes JF646 and SiR) Labelling brightness depends on the protein of interest being tagged	[40]
HaloTag	JF549	HaloTag	JF525 JF552	Higher cell permeability for JF525 and JF552	[39]
			JF571	Higher photostability for JF571	
	JF570		JF593	Higher photostability for JF593	
	JF646		JF669	Higher photostability for JF669	
	JF722		JF711	Higher labelling brightness in fixed cells for JF711 Faster labelling kinetics and higher photostability in live cells for JF722	
SNAP-tag	JF549 (CP)	SNAP-tag	JF552 (CP)	Similar labelling brightness and photostability Lower non-specific staining for JF552	[39]
HaloTag	JF549	SNAP-tag	JF552 (CP)	Similar non-specific staining	[39]
HaloTag	TMR (CA) CPY (CA) Alexa488 (CA)	SNAP-tag	TMR (BG) TMR (CP) CPY (BG) CPY (CP) Alexa488 (BG) Alexa488 (CP)	Faster labelling kinetics for HaloTag compared to SNAP-tag (all ligands) HaloTag ligand with the fastest labelling kinetics was CPY SNAP-tag ligand with the fastest labelling kinetics was TMR	[47]
HaloTag	TMR (CA) CPY (CA) Alexa488 (CA)	CLIP-tag	TMR (BC) CPY (BC) Alexa488 (BC)	Faster labelling kinetics for HaloTag compared to CLIP-tag (all ligands) HaloTag ligand with the fastest labelling kinetics was CPY CLIP-tag ligand with the fastest labelling kinetics was CPY	[47]

(continued)

Table 5 (continued)

Tag 1	Ligand 1	Tag 2	Ligand 2	Findings	Refs.
SNAP-tag	TMR (BG) TMR (CP) CPY (BG) CPY (CP) Alexa488 (BG) Alexa488 (CP) Fluorescein (BG) Fluorescein (CP)	SNAPf-tag	TMR (BG) TMR (CP) CPY (BG) CPY (CP) Alexa488 (BG) Alexa488 (CP) Fluorescein (BG) Fluorescein (CP)	Similar labelling kinetics for TMR (BG) and CPY (BG) Faster labelling kinetics for SNAPf-tag compared to SNAP-tag for all other ligands SNAP-tag ligand with the fastest labelling kinetics was TMR (BG) SNAPf-tag ligand with the fastest labelling kinetics was TMR (BG) Faster labelling kinetics for BG ligands compared to their CP counterparts	[47]
CLIP-tag	TMR (BC) CPY (BC) Alexa488 (BC) Fluorescein (BC)	CLIPf-tag	TMR (BC) CPY (BC) Alexa488 (BC) Fluorescein (BC)	Faster labelling kinetics for CLIPf-tag compared to CLIP-tag (all ligands) CLIP-tag ligand with the fastest labelling kinetics was TMR CLIPf-tag ligand with the fastest labelling kinetics was TMR	[47]
SNAPf-tag	TMR (BG) TMR (CP) CPY (BG) CPY (CP) Alexa488 (BG) Alexa488 (CP) Fluorescein (BG) Fluorescein (CP)	CLIPf-tag	TMR (BC) CPY (BC) Alexa488 (BC) Fluorescein (BC)	Faster labelling kinetics for SNAPf-tag compared to CLIPf-tag (all ligands) SNAPf-tag ligand with the fastest labelling kinetics was TMR CLIPf-tag ligand with the fastest labelling kinetics was TMR	[47]

CA chloroalkane, *BG* O6-benzylguanine, *CP* O6-benzylchloropyrimidine, *BC* O6-benzylcytosine

5. "High serum culture media": 10% Fetal Bovine Serum, 1% GlutaMAX, 1% MEM non-essential amino acids in Dulbecco's Modified Eagle Media (*see* **Note 5**).
6. Cell incubator (37 °C, 5% CO_2).
7. Cell culture hood.

2.3 Day 3—Preparation of Microscope for Imaging

1. Elyra PALM/STORM total internal reflection fluorescence (TIRF) microscope.
2. Andor 897 EMCCD camera.
3. SR Cube 05 RL—BP 420-480/BP570-640/LP 740 filter set.
4. 100× oil 1.46 NA Alpha Plan-Apochromat TIRF objective.

5. High-power TIRF filter (TIRF_HP) (*see* **Note 9**).
6. 405 nm and 561 nm lasers (*see* **Note 10**).
7. Immersion oil.
8. eXcite lamp,
9. Environment chamber with temperature and CO_2 control.
10. ZEISS ZEN blue software (*see* **Note 11**).
11. Stage insert for holding glass-bottom dishes (*see* **Note 3**).

2.4 Day 3—Incubation of Cells with Nuclear Dye and HaloTag Dye

1. Glass-bottom dish containing transfected HeLa cells from Day 2 (*see* Subheading 2.2).
2. "Imaging media": 20 mM HEPES in FluoroBrite™ DMEM (Gibco) (*see* **Note 12**).
3. NucBlue Live ReadyProbes Reagent (Hoechst 33342) (*see* **Note 13**).
4. Cell incubator (37 °C, 5% CO_2).
5. JF549 HaloTag dye ligand (*see* **Note 14**).
6. Alfoil (*see* **Notes 12** and **14**).

2.5 Day 3—Imaging of Cells Using HILO

1. Glass-bottom dish containing HeLa cells incubated with NucBlue Live ReadyProbes Reagent and JF549 Halo-tag ligand (*see* Subheading 2.4).
2. Elyra PALM/STORM total internal reflection fluorescence (TIRF) microscope.
3. Andor 897 EMCCD camera.
4. SR Cube 05 RL – BP 420-480/BP570-640/LP 740 filter set.
5. 100× oil 1.46 NA Alpha Plan-Apochromat TIRF objective.
6. high-power TIRF filter (TIRF_HP).
7. 405 nm and 561 nm lasers.
8. Immersion oil.
9. eXcite lamp.
10. Environmental chamber with temperature and CO_2 control.
11. ZEISS ZEN blue software (*see* **Note 11**).
12. Stage insert for holding glass-bottom dishes (*see* **Note 3**).

3 Methods

3.1 Day 1—Seed the HeLa Cells into Gelatin-Coated Glass-Bottom Imaging Dishes

This protocol is performed when cells are ready for seeding (~60–75% confluency) (*see* top panel of Fig. 3 for overview of Day 1 protocol). Prepare high serum media in advance and store at 4 °C. Prewarm the required volume of high serum media to 37 °C immediately before proceeding with the steps below. All steps should be performed in a cell culture hood to prevent contamination.

1. Coat the glass coverslips of the 29 mm imaging dishes with 0.5–1% gelatin and incubate the dishes at 37 °C for 1 h (*see* **Note 15**).
2. Following incubation, gently aspirate the extra fluid from the dish (*see* **Note 16**).
3. Wash the dishes three times with PBS (using 2 mL each time), aspirate the remaining fluid, then set aside for seeding (*see* **Notes 17** and **18**).
4. Seed ~155,000 HeLa cells (in 2 mL of pre-warmed high serum media) into the prepared imaging dishes (*see* **Notes 2** and **4**).
5. Incubate the dishes containing the cells overnight at 37 °C with 5% CO_2.

3.2 Day 2—Transfect the Cells with the HaloTag-tagged Nuclear Protein of Interest

This protocol is performed 24 h after the cells have been seeded into imaging dishes (see middle panel of Fig. 3 for an overview of the Day 2 protocol). All steps should be performed in a cell culture hood to prevent contamination. Prepare high serum media and low serum media in advance and store at 4 °C. Prewarm the required volume of high serum media to 37 °C immediately before proceeding with the steps below.

1. 24 h after seeding, aspirate the media from the dishes containing the cells and replace with fresh prewarmed high serum media (*see* **Note 17**).
2. Allow XtremeGENE9 reagent, plasmid DNA and low serum media to come to room temperature.
3. In a sterile Eppendorf tube, add 100 μL of low serum media, 3 μL of XtremeGENE9 transfection reagent and 1 μg of purified plasmid DNA. Gently pipette up and down to mix. Incubate the transfection mixture for 15 minutes at room temperature.
4. Add 100 μL of the prepared transfection mixture in a dropwise manner to the center of each glass-bottom dish.
5. Incubate the dishes containing the transfected cells overnight at 37 °C with 5% CO_2.

3.3 Day 3—Prepare the Microscope for Imaging

This step is performed the day after transfection, and immediately before cells are incubated with the fluorescent nuclear marker and HaloTag ligand (see bottom panel of Fig. 3 for overview of Day 3 protocol). (*see* **Note 19**).

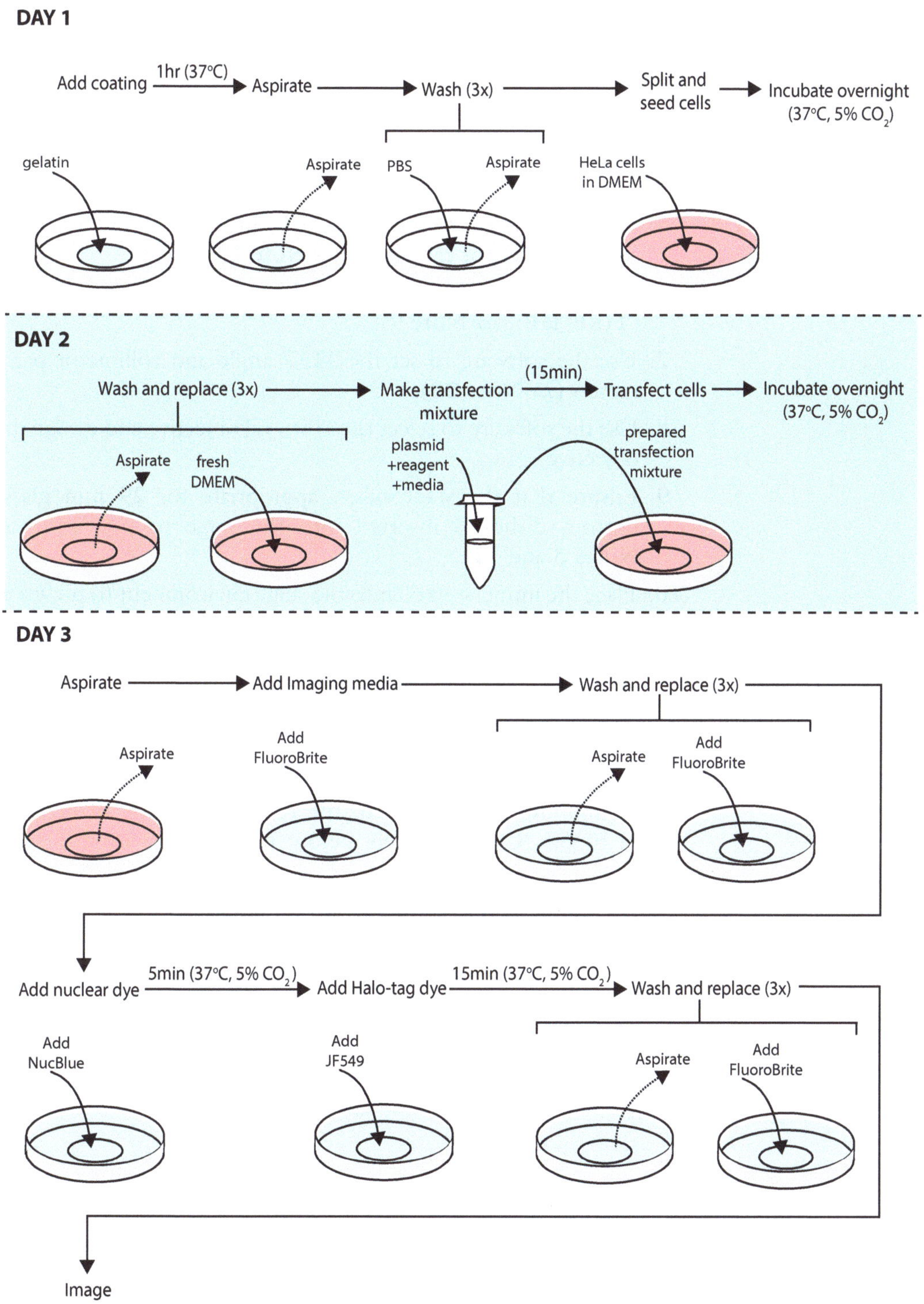

Fig. 3 Overview of the protocol, split into 3 days. Day 1 (Subheading 3.1): Dishes are coated in gelatin, and then cells in high serum media are seeded into the prepared dishes. Day 2 (Subheading 3.2): High serum media is replaced with fresh high serum media, and cells are transfected with a plasmid encoding a HaloTag-

1. Ensure that the stage environment is set to 37 °C with 5% CO_2 (*see* **Note 18**).
2. Turn on eXcite lamp (used for locating cells).
3. Open the Zeiss ZEN BLUE software (*see* **Note 11**).
4. Use the software to check that the correct filter turret is in place.
5. Use the software to select the 561 nm and 405 nm lasers and set the power for each laser (*see* **Note 10**).
6. Use the software to select the high-power TIRF filter (TIRF_HP) (*see* **Note 9**).
7. Use the software to set the TIRF angle and collimator parameters (*see* **Note 20**).
8. Use the software to select the 100× oil objective and escape the objective.
9. Ensure that the stage insert appropriate for 29 mm glass-bottom dishes is inserted into the stage to prewarm (*see* **Notes 3** and **21**).
10. Place the immersion oil into the stage environment to prewarm (*see* **Note 21**).
11. Proceed immediately with the next section.

3.4 Day 3—Incubate the Cells with a Nuclear Dye and HaloTag Dye

This protocol is performed immediately before imaging (see bottom panel of Fig. 3 for an overview of the Day 3 protocol). Prepare imaging media in advance and store at 4 °C (protected from light with aluminum foil). Prewarm the required volume of imaging media to 37 °C immediately before proceeding with the steps below.

1. After preparing the microscope, aspirate media from the cells, wash 2 times with prewarmed imaging media (2 mL each time), and replace with 2 mL of prewarmed imaging media (*see* **Note 17** and **Note 18**).
2. Add 2 drops/mL of NucBlue Live ReadyProbes Reagent (Hoechst 33342) directly to the media and incubate for 5 min (37 °C with 5% CO_2).

Fig. 3 (continued) tagged protein of interest. Day 3 (Subheadings 3.3, 3.4, and 3.5): The microscope, microscope stage, and immersion oil are prewarmed; high serum media from the previous day are replaced with phenol-red-free imaging media, cells are first incubated with a live cell nuclear dye, and then with a dye-conjugated HaloTag ligand, and then cells are imaged on the prewarmed microscope

3. Following incubation, add 2 nM of JF549 HaloTag dye directly to the media and incubate for a further 15 min (37 °C with 5% CO_2).
4. Following the second incubation, aspirate media from the cells, wash 2 times with prewarmed imaging media (2 mL each time) and replace with 2 mL of prewarmed imaging media (*see* **Note 18**).
5. Proceed immediately with the next section.

3.5 Day 3—Image the Cells

This step is performed immediately after incubating the cells with nuclear and HaloTag dyes (see bottom panel of Fig. 3 for an overview of the Day 3 protocol).

1. With the 100× oil objective escaped, immediately transfer the dish containing the cells to the prewarmed microscope stage insert, and use the clips of the microscope insert to secure the dish in place.
2. Place a drop of the prewarmed immersion oil onto the escaped 100× oil objective.
3. Use the coarse focus knob of the microscope to bring the 100× oil objective closer to the sample, until the immersion oil visibly displaces on the surface of the glass bottom dish.
4. Use the eXcite lamp at a low power level (e.g., 12%) and the fine focus knob of the microscope to find the samples and bring them into focus. Turn the power of the lamp to 0% once the cells have been located and focused.
5. Use the software to set the work position and turn on definite focus.
6. Allow ~5 min for the dish to equilibrate to the temperature of the stage (*see* **Note 21**).
7. Look for individual cells with a uniformly shaped nucleus that are not overlapping with neighboring cells, that appear healthy and have a medium density of fluorescently labelled HaloTag-tagged molecules (*see* **Note 22**).
8. Once an ideal cell has been identified, use the 405 nm laser to locate the middle of the nucleus by increasing and decreasing the z-plane and locating the z-plane at which the nucleus is widest in diameter (*see* Fig. 1c), and take a snapshot for reference (*see* **Note 23**).
9. Optional: if required, use the 561 nm laser to bleach the target and surrounding cells for ~1–3 min to reduce the density of the labelled molecules prior to imaging (*see* **Note 24**).
10. Use the 561 nm to capture a time-lapse acquisition of the HaloTag-tagged molecules (*see* **Note 25**).

4 Notes

1. Various cell types can be used; however, the selected cell type should form an adhesive monolayer, as this limits cell movement (drift) and the degree to which cells overlap with neighboring cells (background). Ideally the selected cell type should have a round, flat, and easy-to-identify nucleus that is located close to the cell surface (coverslip), as these qualities improve the signal-to-noise ratio (imaging) and reduce the likelihood of mistracked trajectories (analysis). For these reasons, together with being easy to culture and transfect, HeLa and HEK cells provide good candidates, especially when starting out. Of note, it can also be advantageous to select a cell type that does not normally express the proteins of interest, as this allows their mobilities to be compared without interference from their endogenous counterparts and/or their protein partners.
2. Adjust the seeding density as required to suit your selected cell type, imaging dish, and transfection method. Ideally at the time of imaging, individual transfected cells should be clearly identifiable with as little overlap with neighboring cells as possible (to reduce background) without impacting the viability of the cells. Previously, we found that seeding HeLa cells at a density of ~155,000 cells in 2 mL of high serum media at 60–75% confluency into 29 mm glass-bottom dishes (Day 1) enables sufficient growth for transfection the following day (Day 2), without cells overlapping too much when imaging the day after that (Day 3).
3. Select a No. 1.5 glass-bottom imaging dish with a microwell diameter of 20 mm, and a total diameter that corresponds to that of your microscope stage insert. The dish should ideally fit as snuggly as possible to prevent drift from occurring whilst imaging. Previously we have used Cellvis Cat. No. D29-20-1.5 N for microscope stage inserts that suit 29 mm glass-bottom dishes, and Matek Cat. No. P35-1.5-20C for microscope stage inserts that suit 35 mm glass-bottom dishes.
4. Use a coating that best suits your selected cell type. Previously we have obtained good results in terms of cell adhesion, growth, and density using 0.5–1% gelatin for HeLa cells.
5. Here in this protocol, we use high serum media (containing 10% Fetal Bovine Serum) in order to promote sufficient growth prior to transfection and subsequent imaging. Of note, we have not included antibiotics in this protocol as this can interfere with the transfection efficacy. However, this requires careful handling and vigilant screening of cell culture reagents,

equipment, and cells to ensure that contamination does not occur. Antibiotics can be included in the high serum media as per the user's discretion.

6. Use a transfection method/kit that is suitable for your selected cell type. Previously we have obtained good results with the X-tremeGENE 9 Transfection Reagent kit (Roche), using 1–2 μg of purified plasmid DNA.
7. Use a plasmid that encodes the nuclear protein of interest, separated by a flexible linker with either an N-terminal or C-terminal HaloTag or SNAP-tag. The plasmid should also have a promoter upstream, and a poly-A tail downstream, of the HaloTag-tagged protein of interest, which are appropriate for your selected cell type (e.g., a CMV promoter and bHG polyA tail for expression in mammalian cells). Make sure that the type of self-labelling protein tag that was selected matches the type of dye-conjugated ligand that will be used (e.g., use a HaloTag-tagged protein with a HaloTag dye, and a SNAP-tagged protein with a SNAP-tag dye). As with other protein tags, it is important to assess whether placement of the tag (N-terminal vs C-terminal) interferes with the function of your protein of interest. Techniques such as confocal microscopy (cellular localisation), western blot (protein partner binding), RNA-seq (transcription), and ChIP-seq (chromatin-binding) can be used to compare differences between the Halo/SNAP-tagged protein of interest and its untagged/alternatively tagged counterpart. Of note, various self-labelling protein vectors (either custom-made with the protein of interest already inserted at the N- or C-terminus of the self-labelling protein tag, or with multiple cloning sites for future insertion of the protein of interest at the N- or C-terminus of the self-labelling protein tag) are available to purchase from suppliers (Table 3).
8. Here we use a low serum medium during preparation of the transfection mixture as serum can interfere with the formation of the transfection complex and therefore transfection efficiency.
9. This high-power TIRF filter concentrates the laser (~fourfold), enabling penetration of the laser into the nucleus at a good signal-to-noise ratio.
10. Use the 405 nm laser at a very low power level (e.g., 1%) so that it is just high enough to identify the shape of the nucleus. Use the 561 nm laser at a low power level (e.g., 10%) so that it is just high enough to get a sufficient signal (individual molecules should be bright with minimal background and clearly distinguished from neighboring molecules) without photobleaching the sample too quickly (single molecules should be present

through the duration of the acquisition). Previously, we used the 561 nm laser (oblique illumination with high-power TIRF filter) at a power of 11.6 μW and a power density of 0.728 W/ cm^2.

11. Imaging software often comes with the microscope and therefore varies from microscope to microscope.
12. Use a phenol-red-free imaging media. Previously we have obtained good results using FluoroBrite DMEM, which greatly reduces the level of background fluorescence compared to standard phenol-red-free DMEM. Store the imaging media at 4 °C and protect from light (e.g., by wrapping in alfoil).
13. This step is optional; however, we found that incorporating a nuclear dye makes it much easier to identify individual healthy cells (large, round, and uniformly shaped nucleus) and to locate the middle of the nucleus prior to image acquisition, and to draw regions of interest (ROIs) around the nucleus during the subsequent analysis steps. Previously, we have obtained good results using NucBlue Live ReadyProbes Reagent (Hoechst 33342). Other fluorescent markers can also be used (e.g., by performing the incubation step with an alternative fluorescent dye on Day 3, or co-transfecting the Halo-/SNAP-/CLIP-tagged protein of interest with a spectrally distinct fluorescently labelled nuclear protein on Day 2). Depending on the identity of the protein of interest, an additional nuclear marker may not be needed (e.g., for exclusively nuclear proteins).
14. Aliquots should be made in advance using fresh DMSO and stored protected from light (e.g., with alfoil) at −20 °C, at a small enough concentration to avoid freeze-thaw cycles. Sparse labelling of the self-labelling protein with the externally added fluorescent ligand is essential for achieving the spatial resolution required for SMLM. Over-labelling is prevented by using a small concentration of the fluorescent ligand and incubating for a short period of time. Previously we have obtained good results by adding 2 nM of the JF549 HaloTag ligand in 2 mL of prewarmed FluoroBrite DMEM to HeLa cells and incubating for 15 min. Adjust the concentration and incubation period of the fluorescent ligand as needed.
15. Add just enough coating material to cover the glass coverslip in the microwell of the glass-bottom dishes (leaving the surrounding plastic dry).
16. To aspirate the coating (e.g., gelatin) following incubation, hold the dish at a slight angle and gently aspirate the fluid from the edge of the microwell.
17. To aspirate media or PBS, hold the dish at a slight angle, and gently aspirate the excess fluid from the inside edge of the dish.

18. All wash steps should be performed gently to prevent detachment of the coating and/or cells from the coating. To perform a wash, press the pipette against the inner wall at the edge of the glass-bottom dish, slowly add the fluid, then gently rock the dish for a brief period before aspirating.
19. The microscope is set up in advance to reduce the time spent between preparing cells for imaging (addition of imaging media and incubation with fluorescent ligand) and imaging, and to decrease the amount of drift that occurs whilst imaging. Imaging sessions should be kept short in order to capture cells at their healthiest. Drift occurs due to differences in temperature between the glass-bottom dish and the objective, which results in the mislocalization of single-molecule detections. Although drift correction software exists, drift should be reduced where possible.
20. Adjust as needed. For our system we used a TIRF angle of 60.18° and a collimation of 1891.
21. Ensuring that the stage environment and immersion oil reach 37 °C before imaging reduces the amount of drift that is likely to occur due to differences in temperature with the glass-bottom dish.
22. Overlapping cells generate a background that makes identifying regions of interest (ROIs) such as the nucleus or cell border of the cell being imaged, particularly difficult and increases the likelihood of mistracked trajectories during later analysis steps.
23. Taking a snapshot of the nucleus helps with defining regions of interest (ROIs) during later analysis steps.
24. This step is optional as it depends on the level of fluorescent ligand labelling. We have found that cells that already have the density of labelling required for imaging bleach rapidly throughout the acquisition, whereas cells that have a high level of labelling prior to photobleaching maintain sufficient signal throughout the acquisition.
25. We recommend using 20 ms for 6000 frames for "fast tracking," and 500 ms for 500 frames for "slow tracking." Fast tracking captures molecules with faster mobilities (20 ms or slower), which enables the generation of trajectories from which analyses such as the average mean square displacement (MSD), diffusion coefficient and mobile-to-immobile ratio can be derived. Slow tracking captures molecules with slower mobilities (500 ms or slower). By reducing the level of background generated by faster-moving molecules, the dwell times of molecules (time spent immobilized) can be calculated. This is particularly useful for determining the chromatin occupancy of transcription factors (TFs).

Acknowledgments

HeLa cells were a gift from Professor Geoffrey Faulkner (Queensland Brain Institute/Translational Research Institute, Brisbane, Queensland, Australia). JF549 dye was a gift from Dr. Luke Lavis (Janelia Research Campus, Howard Hughes Medical Institute, Ashburn, Virginia, United States). We would like to thank Dr. Rachel Gormal and Dr. Rumelo Amor for their feedback on Fig. 1. National Medical and Research Council (NHMRC) Fellowship (grant no. 1155794) awarded to F.A.M.. This word was supported by equipment funding from the Australian Research Council (ARC) (LE130100078), as well as funding from the University of Queensland Strategic Initiatives Fund grant no. DVCR22052A awarded to F.A.M..

Conflict of Interest The authors of this chapter declare no conflict of interest.

References

1. Betzig E, Patterson GH, Sougrat R et al (2006) Imaging intracellular fluorescent proteins at nanometer resolution. Science 313(5793): 1642–1645. https://doi.org/10.1126/science.1127344
2. Manley S, Gillette J, Patterson G et al (2008) High-density mapping of single-molecule trajectories with photoactivated localization microscopy. Nat Methods 5:155–157. https://doi.org/10.1038/nmeth.1176
3. Rust M, Bates M, Zhuang X (2006) Sub-diffraction-limit imaging by stochastic optical reconstruction microscopy (STORM). Nat Methods 3:793–795. https://doi.org/10.1038/nmeth929
4. Heilemann M, van de Linde S, Schüttpelz M et al (2009) Subdiffraction-resolution fluorescence imaging with conventional fluorescent probes. Angew Chem Int Ed Engl 47(33): 6172–6176. https://doi.org/10.1002/anie.200802376
5. Hess ST, Girirajan TP, Mason MD (2006) Ultra-high resolution imaging by fluorescence photoactivation localization microscopy. Biophys J 91(11):4258–4272. https://doi.org/10.1529/biophysj.106.091116
6. Giannone G, Hosy E, Levet F et al (2010) Dynamic superresolution imaging of endogenous proteins on living cells at ultra-high density. Biophys J 99(4):1303–1310. https://doi.org/10.1016/j.bpj.2010.06.005
7. Joensuu M, Padmanabhan P, Durisic N et al (2016) Subdiffractional tracking of internalized molecules reveals heterogeneous motion states of synaptic vesicles. J Cell Biol 215(2): 277–292. https://doi.org/10.1083/jcb.201604001
8. Axelrod D (2001) Total internal reflection fluorescence microscopy in cell biology. Traffic 2(11):764–774. https://doi.org/10.1034/j.1600-0854.2001.21104.x
9. Tokunaga M, Imamoto N, Sakata-Sogawa K (2008) Highly inclined thin illumination enables clear single-molecule imaging in cells. Nat Methods 5:159–161. https://doi.org/10.1038/nmeth1171
10. Jiang A, Kudo K, Gormal RS et al (2024) Dynamin1 long- and short-tail isoforms exploit distinct recruitment and spatial patterns to form endocytic nanoclusters. Nat Commun 15:4060. https://doi.org/10.1038/s41467-024-47677-8
11. Small C, Harper C, Jiang A et al (2024) SV2A controls the surface nanoclustering and endocytic recruitment of Syt1 during synaptic vesicle recycling. J Neurochem 168(9): 3188–3208. https://doi.org/10.1111/jnc.16186
12. Kasula R, Chai YJ, Bademosi AT et al (2016) The Munc18-1 domain 3a hinge-loop controls syntaxin-1A nanodomain assembly and engagement with the SNARE complex during

secretory vesicle priming. J Cell Biol 214(7): 847–858. https://doi.org/10.1083/jcb.201508118
13. Padmanabhan P, Kneynsberg A, Cruz E et al (2024) Single-molecule imaging of tau reveals how phosphorylation affects its movement and confinement in living cells. Mol Brain 17:7. https://doi.org/10.1186/s13041-024-01078-6
14. Cheng S, Nakatani Y, Gagliano G et al (2024) Light sheet illumination in single-molecule localization microscopy for imaging of cellular architectures and molecular dynamics. NPJ Imaging 2:49. https://doi.org/10.1038/s44303-024-00057-9
15. McCann AJ, Lou J, Moustaqil M et al (2021) A dominant-negative SOX18 mutant disrupts multiple regulatory layers essential to transcription factor activity. Nucleic Acids Res 49(19): 10931–10955. https://doi.org/10.1093/nar/gkab820
16. Paakinaho V, Presman D, Ball D et al (2017) Single-molecule analysis of steroid receptor and cofactor action in living cells. Nat Commun 8:15896. https://doi.org/10.1038/ncomms15896
17. Presman DM, Ball DA, Paakinaho V et al (2017) Quantifying transcription factor binding dynamics at the single-molecule level in live cells. Methods 123:76–88. https://doi.org/10.1016/j.ymeth.2017.03.014
18. Mehta GD, Ball DA, Eriksson PR et al (2018) Single-molecule analysis reveals linked cycles of RSC chromatin remodeling and Ace1p transcription factor binding in yeast. Mol Cell 72(5):875–887. https://doi.org/10.1016/j.molcel.2018.09.009
19. Loffreda A, Jacchetti E, Antunes S et al (2017) Live-cell p53 single-molecule binding is modulated by C-terminal acetylation and correlates with transcriptional activity. Nat Commun 8(1):313. https://doi.org/10.1038/s41467-017-00398-7
20. Hipp L, Beer J, Kuchler O et al (2019) Single-molecule imaging of the transcription factor SRF reveals prolonged chromatin-binding kinetics upon cell stimulation. Proc Natl Acad Sci USA 116(3):880–889. https://doi.org/10.1073/pnas.1812734116
21. Callegari A, Sieben C, Benke A et al (2019) Single-molecule dynamics and genome-wide transcriptomics reveal that NF-kB (p65)-DNA binding times can be decoupled from transcriptional activation. PLoS Genet 15(1): e1007891. https://doi.org/10.1371/journal.pgen.1007891
22. Longfield SF, Mollazade M, Wallis TP et al (2023) Tau forms synaptic nano-biomolecular condensates controlling the dynamic clustering of recycling synaptic vesicles. Nat Commun 14(1):7277. https://doi.org/10.1038/s41467-023-43130-4
23. Hao S, Lee YJ, Benhamou Goldfajn N et al (2024) YAP condensates are highly organized hubs. iScience 27(6):109927. https://doi.org/10.1016/j.isci.2024.109927
24. Chen J, Zhang Z, Li L et al (2014) Single-molecule dynamics of enhanceosome assembly in embryonic stem cells. Cell 156(6): 1274–1285. https://doi.org/10.1016/j.cell.2014.01.062
25. Los GV, Darzins A, Karassina N et al (2005) HaloTag™ interchangeable labeling technology for cell imaging and protein capture. Promega Cell Notes 11:2–6
26. Los GV, Wood K (2007) The HaloTag: a novel technology for cell imaging and protein analysis. Methods Mol Biol 356:195–208. https://doi.org/10.1385/1-59745-217-3:195
27. Los GV, Encell LP, McDougall MG et al (2008) HaloTag: a novel protein labeling technology for cell imaging and protein analysis. ACS Chem Biol 3(6):373–382. https://doi.org/10.1021/cb800025k
28. Ohana RF, Encell LP, Zhao K et al (2009) HaloTag7: a genetically engineered tag that enhances bacterial expression of soluble proteins and improves protein purification. Protein Expr Purif 68(1):110–120. https://doi.org/10.1016/j.pep.2009.05.010
29. Encell LP, Friedman Ohana R, Zimmerman K et al (2012) Development of a dehalogenase-based protein fusion tag capable of rapid, selective and covalent attachment to customizable ligands. Curr Chem Genomics 6:55–71. https://doi.org/10.2174/1875397301206010055
30. Keppler A, Gendreizig S, Gronemeyer T et al (2003) A general method for the covalent labeling of fusion proteins with small molecules in vivo. Nat Biotechnol 21(1):86–89. https://doi.org/10.1038/nbt765
31. Juillerat A, Gronemeyer T, Keppler A et al (2003) Directed evolution of O6-alkylguanine-DNA alkyltransferase for efficient labeling of fusion proteins with small molecules in vivo. Chem Biol 10(4):313–317. https://doi.org/10.1016/s1074-5521(03)00068-1
32. Juillerat A, Heinis C, Sielaff I et al (2005) Engineering substrate specificity of O6-alkylguanine-DNA alkyltransferase for specific protein labeling in living cells. Chembiochem

6(7):1263–1269. https://doi.org/10.1002/cbic.200400431

33. Sun X, Zhang A, Baker B et al (2011) Development of SNAP-tag fluorogenic probes for wash-free fluorescence imaging. Chembiochem 12(14):2217–2226. https://doi.org/10.1002/cbic.201100173
34. Gautier A, Juillerat A, Heinis C et al (2008) An engineered protein tag for multiprotein labeling in living cells. Chem Biol 15(2):128–136. https://doi.org/10.1016/j.chembiol.2008.01.007
35. Grimm JB, Xie L, Casler JC et al (2021) A general method to improve fluorophores using deuterated Auxochromes. JACS Au 1(5):690–696. https://doi.org/10.1021/jacsau.1c00006
36. Grimm JB, English B, Choi H et al (2016) Bright photoactivatable fluorophores for single-molecule imaging. Nat Methods 13: 985–988. https://doi.org/10.1038/nmeth.4034
37. Grimm JB, Muthusamy AK, Liang Y et al (2017) A general method to fine-tune fluorophores for live-cell and in vivo imaging. Nat Methods 14(10):987–994. https://doi.org/10.1038/nmeth.4403
38. Grimm JB, Brown TA, Tkachuk AN et al (2017) General synthetic method for Si-Fluoresceins and Si-Rhodamines. ACS Cent Sci 3(9):975–985. https://doi.org/10.1021/acscentsci.7b00247
39. Grimm JB, Tkachuk AN, Xie L et al (2020) A general method to optimize and functionalize red-shifted rhodamine dyes. Nat Methods 17: 815–821. https://doi.org/10.1038/s41592-020-0909-6
40. Erdmann RS, Baguley SW, Richens JH et al (2019) Labeling strategies matter for super-resolution microscopy: a comparison between HaloTags and SNAP-tags. Cell Chem Biol 26(4):584–592. https://doi.org/10.1016/j.chembiol.2019.01.003
41. Banaz N, Mäkelä J, Uphoff S (2019) Choosing the right label for single-molecule tracking in live bacteria: side-by-side comparison of photoactivatable fluorescent protein and halo tag dyes. J Phys D Appl Phys 52(6):064002. https://doi.org/10.1088/1361-6463/aaf255
42. Zheng Q, Ayala AX, Chung I et al (2019) Rational Design of Fluorogenic and Spontaneously Blinking Labels for super-resolution imaging. ACS Cent Sci 5(9):1602–1613. https://doi.org/10.1021/acscentsci.9b00676
43. Bitong L, Wong YY, Flores-Rodriguez N, et al (2024) Anti-phase clustering of regulatory factors shapes gene bursting. bioRxiv 2024.09.10.612362. https://doi.org/10.1101/2024.09.10.612363
44. Corrêa IR Jr, Baker B, Zhang A et al (2013) Substrates for improved live-cell fluorescence labeling of SNAP-tag. Curr Pharm Des 19(30):5414–5420. https://doi.org/10.2174/13816128113193000ll
45. Mollwitz B, Brunk E, Schmitt S et al (2012) Directed evolution of the suicide protein O(6)-alkylguanine-DNA alkyltransferase for increased reactivity results in an alkylated protein with exceptional stability. Biochemistry 51(5):986–994. https://doi.org/10.1021/bi2016537
46. Gronemeyer T, Chidley C, Juillerat A et al (2006) Directed evolution of O6-alkylguanine-DNA alkyltransferase for applications in protein labeling. Protein Eng Des Sel 19(7): 309–316. https://doi.org/10.1093/protein/gzl014
47. Wilhelm J, Kühn S, Tarnawski M et al (2021) Kinetic and structural characterization of the self-labeling protein tags HaloTag7, SNAP-tag, and CLIP-tag. Biochemistry 60(33): 2560–2575. https://doi.org/10.1021/acs.biochem.1c00258

Chapter 8

Single-Molecule Localization Microscopy in vivo at the Drosophila Neuromuscular Junction

Kye Kudo, Patrik Verstreken, Frédéric A. Meunier, and Adekunle T. Bademosi

Abstract

Single-molecule localization microscopy (SMLM) has provided unique insight into the biophysical and nanoscale molecular mechanisms that underpin fundamental cellular functions. Nonetheless, due to technical limitations and challenges, much of the current work on real-time SMLM has relied heavily on in vitro or in cellulo model systems, which may not comprehensively reflect molecular dynamics in vivo. The current chapter acts as a guide for experimenters aiming to extend their research into in vivo or ex vivo SMLM, highlighting the obstacles and challenges that become apparent when attempting to elucidate the nanoscale cellular environment within complex multicellular organisms. Here, we describe our procedure for performing in vivo SMLM at the transgenic *Drosophila melanogaster* larval neuromuscular junction and provide insight into how to overcome these challenges. The protocol described allows direct visualization of molecules involved in synaptic communication within in vivo presynaptic nerve terminals and can be extended to other live multicellular organisms and systems.

Key words *Drosophila*, Single-particle tracking, in vivo, Neuromuscular junction, Presynapse, HILO microscopy

1 Introduction

Super-resolution microscopy techniques have provided extensive insight into both the nanoscale organization and dynamics of biomolecules within biological systems [1–3]. Single-molecule localization microscopy (SMLM) encompasses a group of super-resolution techniques that function by detecting photons emitted from individual sparsely distributed fluorophores to estimate their precise spatial coordinates [4, 5]. When applied to dynamic systems, real-time SMLM, also termed single-particle tracking (SPT), enables the localization and tracking of individual fluorophores over time, providing insight into the nanoscale molecular mobility and biophysical environment. To date, studies that utilize SMLM

Frédéric A. Meunier (ed.), *Super-Resolution Microscopy: Methods and Protocols*, Methods in Molecular Biology, vol. 3034,
https://doi.org/10.1007/978-1-0716-5268-8_8,

remain largely restricted to in vitro settings, including biomolecules on glass coverslips or cell culture systems due to limitations in illumination, sensitivity, and fluorescent probes [5–7]. However, in vitro system models do not completely recapitulate the native environment within live organisms, and it has become apparent that molecular dynamics could be largely distinct between in vitro and in vivo systems [8].

In vivo and ex vivo systems provide unique platforms to study molecular function within a native context, allowing for causal association between nanoscale dynamics and macroscale processes occurring within complex organisms. Real-time SMLM has been applied to a selection of in vivo and ex vivo model systems including *D. rerio* [9–13], *C. elegans* [14–17], *D. melanogaster* [10, 18–22], and *M. musculus* [23, 24] (Table 1). In vivo systems offer the unique opportunity to study complex and specialized biological processes that are often not readily available in vitro. For example, the *C. elegans* tube-shaped gonad contains germ cells and gametes that are sequentially ordered in a temporal gradient, allowing for distinct meiotic stages to be visualized. SMLM within *C. elegans* gonads has been used to characterize the molecular diffusivity of chromatin-associated proteins within nuclei that are actively undergoing meiosis [16]. Additionally, the three-dimensional tissue architecture inherent to in vivo systems provides a platform to visualize the dynamics of extracellular biomolecules in a native multi-dimensional environment [12].

SMLM relies on the highly efficient detection of photons emitted from individual fluorophores and simultaneous mitigation of fluorescence from out-of-focus background emitters. Optical microscope systems with Complementary Metal Oxide Semiconductor (CMOS) or Electron Multiplying Charge-Coupled Device (EMCCD) image sensors are generally used for SMLM to achieve optimal photon detection efficiency and temporal acuity. For in vivo SMLM, obtaining a reliable signal-to-noise ratio is most challenging as fluorescence from fluorescent probes outside of the focal plane and light scattering from the tissue environment is inevitable [25].

Advances in illumination strategies from the conventional epi-fluorescence illumination have been critical for the mitigation of background fluorescence and autofluorescence. These high-contrast imaging techniques include Total Internal Reflection Fluorescence microscopy (TIRF) [26], Highly Inclined and Laminated Optical Sheet microscopy (HILO) [27, 28] and Light Sheet Fluorescence Microscopy (LSFM) [29–31].

TIRF microscopy relies on the production of an exponentially decaying evanescent field that is spatially concentrated approximately 150–350 nm into the sample [26], immediately adjacent to its interface with the glass dish. An evanescent field is generated when the incident illumination beam is totally internally reflected at

Table 1
Applications of real-time SMLM on in vivo *and* ex vivo model systems

	Application/ model	Biomolecule	Fluorophore	System and illumination	Sample mounting	Quantification	References
1	*Danio rerio* embryonic body	Glucocorticoid receptor	eYFP	Custom-built light-sheet fluorescence single-molecule microscope with custom-made mounting system	Contained within a fluorinated ethylene propylene support tube	Demonstrated the presence of two distinct subpopulations of receptors based on diffusivity	[9]
2	*Caenorhabditis elegans* gonads	HORMA domain-containing proteins: High incidence of males (HIM-3) and HIM-3 paralog (HTP-3)	mMaple3 and HaloTag ligands JF549 and PAJF549	Bruker Vutara 352 biplane localization microscope with Hamamatsu ORCA-fusion BT sCMOS camera using epi-fluorescence illumination	Gonads were lightly flattened between glass microscopy slides with tape used as spacers	Demonstrated the presence of distinct subpopulations of meiotic axis proteins defined by molecular motion (diffusive or static states) within the nucleoplasm	[16]
3	*Drosophila melanogaster* larval neuromuscular junction	SNARE complex protein Syntaxin-1a	mEos2	Zeiss Elyra microscope using HILO illumination	Drosophila larvae were dissected on SYLGARD gel and inverted onto a glass-bottom imaging dish	Demonstrated altered mobility of Syntaxin-1a resulting from neuronal stimulation and NSF mutant	[18]

(continued)

Table 1 (continued)

	Application/ model	Biomolecule	Fluorophore	System and illumination	Sample mounting	Quantification	References
4	*Drosophila melanogaster* adult brain	SNARE complex protein Syntaxin-1a	mEos2	Zeiss Elyra microscope using HILO illumination (TIRF angle of 47.3°)	Brains were mounted on a glass slide and sealed with a coverslip; brains were lightly flattened	Demonstrated altered mobility of Syntaxin-1a resulting from neuronal stimulation and general anesthetics	[20, 49]
5	*Drosophila melanogaster* secretory cells of larval salivary glands and peripodial epithelium cells of larval wing imaginal discs	DNA binding protein scalloped (Sd), transcription co-activator Yorkie (Yki), histone H2B	HaloTag ligand JF549	Zeiss Elyra microscope using HILO illumination (TIRF angle of 57° and 60°)	Salivary glands and wing imaginal discs were placed on an Attofluor imaging chamber with hole-punched tape, covered with a permeable membrane	Chromatin-binding dynamics of DNA-binding proteins	[21]
6	*Caenorhabditis elegans* adult neuromuscular synapses	Transmembrane CD4 and voltage-dependent Ca2+ channels (VDCC)	Complementation-activated light microscopy (CALM) split-GFP system	Inverted IX70 Olympus microscope with a QuantEM:512SC EMCCD and an inverted Nikon eclipse Ti-E microscope with an iXon EMCCD using HILO illumination	Worms were placed on an agarose pad, immobilized with polystyrene nanoparticles and covered with a glass coverslip	Diffusion analysis of VDCC	[17]

7	*Caenorhabditis elegans* adult *chemosensory neurons*	Ciliary calcium channel protein OCR-2, kinesin-associated protein (KAP-1)	eGFP	Inverted Nikon eclipse Ti-E microscope with an Andor iXon 897 EMCCD using small-window illumination microscopy (SWIM, epifluorescence illumination with cropped excitation beam)	Worms were sedated, placed on an agarose pad with polystyrene nanoparticles and covered with a glass coverslip	Movement direction of OCR-2-associated vesicles and identification of diffusive and directed states of motion	[14]
8	*Danio rerio* embryo extracellular cavities	TGF-β superfamily ligands cyclops, squint, Lefty1 and Lefty2	HaloTag ligand JF549	Custom-built reflected light-sheet microscope with Nikon TI microscope body and Andor iXon ultra DU 897 U EMCCD	Embryos were mounted onto a glass-bottom dish	Molecular diffusivity in extracellular cavities and cell-cell interfaces, bound-unbound fractions and residence times	[12]
9	*Danio rerio* embryo epidermal cells	C10H-Ras (biologically inert probe within the inner-leaflet of the plasma membrane)	YFP	Custom-built TIRF microscope with Cascade 512B EMCCD using TIRF illumination	Embryos were mounted onto a coverslip, and a sheet of agarose was placed over the tail-region for imaging	Demonstrated confined diffusion within microdomains on plasma membrane	[13]

(continued)

Table 1
(continued)

	Application/ model	Biomolecule	Fluorophore	System and illumination	Sample mounting	Quantification	References
10	*Caenorhabditis elegans* cortex of embryo	Polarity protein PAR-6	GFP	Inverted Nikon Ti-E microscope with Andor iXon3 897 EMCCD using thin laminated optical sheet illumination	Embryos were mounted between glass coverslip and slide containing polystyrene beads to lightly and uniformly compress the embryo surface	Analysis of diffusion coefficients and mobility	[15]
11	*Chironomus tentans* salivary gland cell nuclei	Dextran and RNA-binding protein hrp36	ATTO 647 N	Axiovert 200 inverted microscope with custom built sample chamber and iXon BI DV-860 EMCCD using light-sheet illumination	Placed within custom-built sample chamber	Analysis of effective intranuclear viscosity and mobility fractions	[31]
12	*Danio rerio* embryonic body	TATA-binding protein (TBP) and Sox19b	mEos2	Nikon Ti-E microscope iXon ultra DU 897 U using reflected light sheet microscopy	Embryos mounted on a glass bottom dish	Analysis of bound and unbound fractions of particles and residence times	[50]
13	*Mus Musculus* organotypic hippocampal slices	Glycosylphosphatyidylinositol (GPI)-anchored GFP, Neruoligin1, β-Neurexin	Polyethylene glycol-coated anti-GFP quantum dots	CSU X1 spinning disk confocal system built around upright Olympus BX50 with Andor iXon 897 EMCCD	Organotypic hippocampal slices placed within a perfusion chamber	Analysis of diffusion coefficients and mobility	[23]

the interface between the glass coverslip and the sample media, which occurs when the beam is directed at this interface beyond the critical angle. The spatially concentrated evanescent field produces detections with a high signal-to-noise ratio for fluorescent probes located at (or extremely close to) the coverslip-specimen interface as unwanted excitation of fluorescent probes outside of the focal plane is prevented. However, in vivo SMLM often requires capturing structures situated further into the tissue away from the glass-specimen interface and in such cases HILO illumination can be utilized. HILO involves the generation of a thin angled optical sheet within the specimen by reducing the angle of incidence away from the critical angle [27, 28]. The reduced illumination range and increased light density of the refracted beam result in reduced background fluorescence and greater signal-to-noise ratio in comparison to epi-fluorescence illumination and have been heavily utilized for in vivo SMLM [17, 18, 20, 21].

The *Drosophila melanogaster* is an ideal system for in vivo SMLM due to the optical accessibility to much of its tissue [18–21, 32, 33]. In particular, the highly stereotypic *Drosophila melanogaster* larval neuromuscular junction (NMJ) has been extensively studied and provides a platform for elucidating the molecular mechanics of neurotransmission. The *Drosophila* larval body-wall neuro-muscular system is comprised of distinct efferent neurons whose soma lie within the central nervous system and whose axons sort and extend to specific muscles in a highly stereotypic manner [34]. Type-Ib motor neurons from large varicosities (diameters of 3–5 μm) onto body-wall muscles [35]; each varicosity consists of multiple active-zone structures which are directly aligned with post-synaptic receptors on the muscle [36]. Type-Ib motor neurons form glutamatergic synapses and stimulate muscle contraction [37]. Though these neurons initially establish contacts superficially on the muscle surface, the Type-Ib varicosities later submerge deeper into the muscle during larval development (surrounded by a sub-synaptic reticulum of muscle membrane) [34]. At this stage the Type-Ib NMJ is amenable to SMLM using HILO illumination.

Syntaxin1A, a component of the N-ethylmaleimide-sensitive-factor attachment receptor (SNARE) that is critical for mediating synaptic vesicle fusion, has previously been tracked at single-molecule resolution at the *Drosophila* larval Type-Ib NMJ. Opto- and thermogenetic neuronal stimulation revealed the dynamic activity-dependent release of syntaxin1A from molecular nanoclusters during SNARE complex disassembly at the presynapse [18]. SMLM has also been applied to the voltage-gated Ca^{2+} channel α1 subunit Cacophony at the *Drosophila* larval NMJ. Here, presynaptic homeostatic remodeling was discovered to be associated with the compaction of Cacophony nanodomains, which modulate presynaptic synaptic-vesicle release [38].

Here we describe our protocol for in vivo SMLM at the *Drosophila melanogaster* larval NMJ. As an experimental example, we examine the nanoscale dynamics of the endocytic protein, Endophilin A1, at the presynapse. Previous studies have suggested that Endophilin A1 is sensitive to Ca^{2+} ion influx during depolarization of the presynaptic nerve terminal [33, 39, 40]. We carried out SMLM in transgenic *Drosophila* expressing Endophilin A1-mEos3.1 (EndoA-mEos3.1) in motor neurons using the GAL4/UAS system to investigate the influence of Ca^{2+} ions on Endophilin A1 nanoscale dynamics. We imaged single molecules of EndoA-mEos3.1 in larvae incubated with and without ethylene glycol-bis(β-aminoethyl ether)-N,N,N′,N′-tetraacetic acid (EGTA) to buffer for Ca^{2+} ions. We find that buffering for Ca^{2+} with EGTA serves to restrict Endophilin A1 at the synaptic plasma membrane in vivo.

2 Materials

2.1 Biological Materials

D. melanogaster fruit flies are reared on standard yeast-sugar-agar food vials at room temperature (22–25 °C). The parental generation is flipped out of the original vials following the observation of first-instar larvae to prevent larval crowding. Actively crawling third-instar larvae of either sex are selected and dissected for experiments.

1. Transgenic third-instar *Drosophila melanogaster* larvae expressing the protein of interest tagged with a photoconvertible fluorescent protein (e.g., mEos3.2, mEos4b) or chemical tag (e.g., CLIP, SNAP, Halo) (*see* **Note 1**). Here, we utilize transgenic larvae with UAS-EndoA-mEos3.1 and OK6-GAL4.

2.2 Reagents

1. Modified HL3 Medium: 110 mM NaCl, 5 mM KCl, 10 mM $MgCl_2$, 0.5 mM $CaCl_2$, 10 mM $NaHCO_3$, 5 mM trehalose, 30 mM sucrose, 5 mM HEPES, pH 7.2 [41]. The modified HL3 solution is stored in 50 ml aliquots at −20 °C for up to 3 months.
2. Polydimethylsiloxane (PDMS) base: pour the two-part (10:1) SYLGARD™ 184 Silicone Elastomer Kit (Dow Corning) into a 10 cm petri-dish (see manufacturer's manual). Cure overnight at 60 °C. Cut the resulting gel to form a cube approximately 1 cm × 1 cm × 1 cm. This re-usable PDMS base will serve as the foundation for the larval dissection and imaging. Ensure that the PDMS base can fit within the glass-bottom dish.
3. EGTA (25 μM in modified HL3).

2.3 Dissection Equipment

1. Stereomicroscope (Olympus, SZ51).
2. Vannas scissors (3 mm cutting edge, straight).
3. Dumont tweezer (0.10 mm × 0.06 mm, bent 45°).
4. Dumont tweezer (0.05 mm × 0.01 mm, straight).
5. Dissection (minutien) pins (120 mm length, 0.1 mm diameter).
6. 29 mm glass bottom dishes (20 mm well, #1.5 cover glass)
7. Sterile transfer pipettes.

2.4 Imaging Equipment

In vivo SMLM can be performed with various microscopy systems and illumination methods (*see* Table 1 and **Note 2**). In the current protocol we describe the use of the Zeiss Elyra PS.1 built around a Zeiss Axio Observer.Z1 motorized inverted microscope platform that is equipped with:

1. iXon Ultra 897 EMCCD (Andor Technology),
2. 100× 1.46 NA/Alpha Plan-Apochromat objective (Carl Zeiss)
3. 10× 0.3 NA/Plan-Neofluar objective (Carl Zeiss)
4. BP 570-620 LP750 Filter Cube (Carl Zeiss).
5. BP 495-550 LP650 Filter Cube (Carl Zeiss).
6. 405 nm diode (50 mW)
7. 488 nm OPSL (100 mW)
8. 561 nm OPSL (100 mW)
9. X-Cite® 120PC Q (Excelitas Technologies).
10. WSB PiezoDrive 08 (Carl Zeiss).
11. Definite Focus Controller FC12 (Carl Zeiss).
12. Heating Unit XL S (Carl Zeiss).
13. TempModule S (Carl Zeiss).
14. CO_2 Module S (Carl Zeiss).
15. ZEN 2012 SP5 FP3 (Black) v14.0 (Carl Zeiss).
16. Immersol® 518F/23 °C (Carl Zeiss).

2.5 Software

1. Single-molecule tracking software (PALMTracer, TrackMate, or equivalent, *see* Chapter 6).
2. Fiji (ImageJ2, v2.9.0/1.53 t).
3. SharpViSu (GitHub, v1) (for drift correction, *see* Chapter 6).
4. MATLAB (MathWorks, v2017b) (for drift correction, *see* Chapter 6).

3 Methods

3.1 Dissection of Third-Instar Larvae

1. Remove a crawling third-instar larva from its food vial and place it on the PDMS base with the dorsal side facing upwards. Immobilize the larva by inserting a dissection pin between the two posterior spiracles at the tail of the animal, and another between the two anterior spiracles in front of the mouth hooks (*see* **Note 3**) (Fig. 1a).
2. Add a drop of modified HL3 solution onto the larva and, using dissection scissors, cut longitudinally along the dorsal side of the larvae from above the mouth hooks to the posterior spiracles (Fig. 1b).
3. Remove the internal organs of the larvae using fine forceps and scissors whilst keeping the brain and ventral nerve cord intact (Fig. 1c).
4. Using fine forceps, add two pins on each lateral side of the larva, creating a hexagonal silhouette of the larva preparation, ensuring that the preparation is minimally stretched (Fig. 1d).
5. Push all six inserted dissection pins down such that they are no longer protruding and are flush with the PDMS base (*see* **Note 3**) (Fig. 1e, f).
6. Detach the two brain lobes from the ventral nerve cord to inhibit brain-derived neural activity that can result in contractions whilst imaging.
7. Gently wash the larva preparation with modified HL3 solution to remove any debris remaining on the PDMS base.

3.2 Mounting and Imaging

1. Invert the PDMS base containing the larva and place the base on a glass-bottom dish containing 1 mL of modified HL3 solution, such that the larva is gently resting between the glass bottom dish below and the PDMS base above (Fig. 1g).
2. Mount the glass-bottom dish onto the microscope stage (25 °C) and apply light pressure down onto the PDMS base (either with a weight or the clamp of the mounting frame), such that the gel is stable on top of the glass-bottom dish.
3. Locate the motor neuron of interest using 10× objective, typically Type-Ib motor neurons from abdominal segment A3 and A4, muscle 6/7 or muscle 13. Turn to the 100× objective when the sample position is found.
4. Turn on the 488 nm laser line to visualize the diffraction-limited mEos3.1 within the presynaptic nerve terminal (laser power < 5%, 0.2 W/cm^2). Acquire an image using the 488 nm laser line (Fig. 2a).

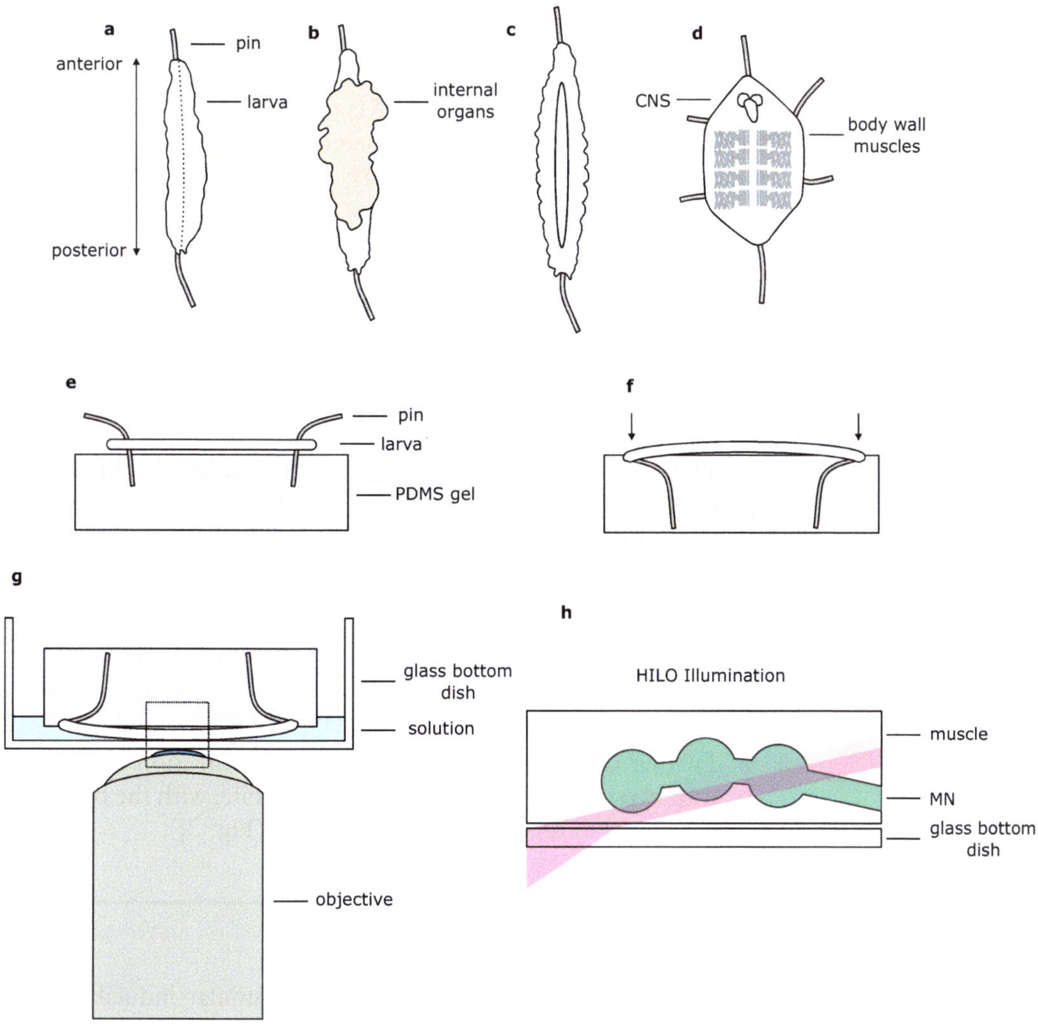

Fig. 1 Drosophila larva dissection and HILO illumination. (**a**) Larva is pinned onto PDMS gel at the posterior and anterior spiracles and an incision is made along the length of the animal (dotted line). (**b**) The internal organs of the animal extrude after the longitudinal incision and are then removed. (**c**) Larva after removal of internal organs. (**d**) Two pins are added on each side of the larva, creating a hexagonal silhouette. (**e** and **f**) Dissection pins are pushed down such that they are no longer protruding and are flush with the PDMS base. (**g**) The PDMS base containing the larva is inverted and placed onto a glass-bottom dish, which is then mounted onto a TIRF microscope. Throughout the dissection and mounting the larva is maintained within modified HL3 solution. (**h**) HILO illumination is used to visualize the neuromuscular junction (MN, motor neuron)

5. Simultaneously turn on the 561 nm (laser power < 25%, 158 W/cm^2) and 405 nm (laser power 0.001% ~ 0.01%, < 0.002 W/cm^2) laser lines (*see* **Notes 4** and **5**).
6. Image the NMJ using oblique illumination in high-power TIRF mode (Fig. 1h). Here we recommend setting the oblique illumination angle at 55.5° with the collimator set at position 2720 (*see* **Note 6**) and capture detections with an acquisition rate between 20 and 50 Hz (*see* **Note 7**).

3.3 Analysis

This protocol briefly summarizes the approach taken to generate the panels for Fig. 2. For detailed information on how to analyze single-molecule imaging data, see Chapters 6 and 13. Here, we find that the addition of EGTA (25 μM) results in reduced mobility of EndophilinA1 as observed by reduced area-under-the-curve of the mean squared displacement (MSD), and a decrease in the ratio of mobile: immobile molecules.

1. Use a single-molecule particle tracking software (e.g., PALM-Tracer [42, 43] or TrackMate [44, 45], *see* Chapter 6) for both the detection and connection of localizations into trajectories. In this instance we used PALMTracer, which utilizes Gaussian fitting for localization of detections and a simulated annealing-based tracking algorithm for tracking.
2. Draw regions of interest around the NMJ using the diffraction-limited 488 acquisition (Fig. 2a) as a reference.
3. Select reconstructed trajectories that last at least eight frames for mobility analysis.
4. Use the output of the tracking software to plot a spatial map of detections, trajectories, and diffusion coefficient (Fig. 2b–f) as well as the MSD, MSD area-under-the-curve and diffusion coefficients for mobility analysis (Fig. 2g–i).
5. Further classify the trajectories based on diffusion coefficients into two populations, immobile and mobile, with the threshold calculated as previously described [46] (Fig. 2j).

4 Notes

1. If using the GAL4-UAS system or a similar inducible over-expression system, fly lines should be maintained such that overexpression is induced only in the final progeny that are used for experiments.
2. In vivo SMLM can be performed with various microscopy systems and illumination methods. The most critical factors are optical accessibility to tissue of interest (whilst mitigating contribution from cellular autofluorescence and light scattering), camera exposure time and camera photon collection efficiency. Since these factors are highly system-dependent, we urge the experimenter to test various microscopy systems if available.
3. During initial insertion of the dissection pins, inserting them not perpendicular to the PDMS gel, but rather at a slight angle towards the center of the larva can help later when pushing the pins in such that they become flush with the PDMS gel. Very subtle protrusion of these pins from the PDMS base can influence mounting and impair optical accessibility of the NMJ and significantly worsen SMLM acquisition quality.

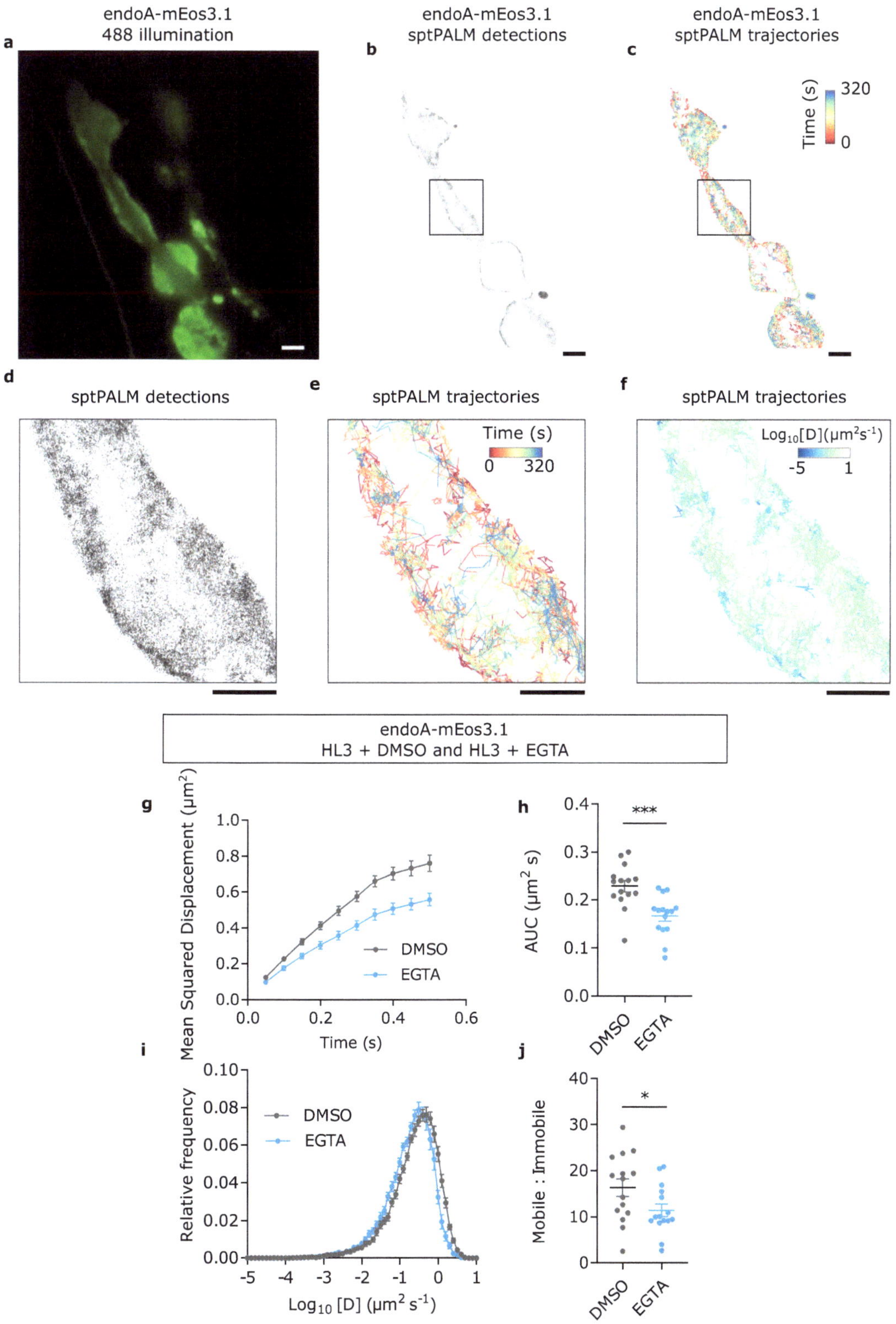

Fig. 2 SMLM of endoA-mEos3.1 at the drosophila NMJ. (**a**) Drosophila Type1b NMJ expressing endoA-mEos3.1, diffraction-limited acquisition made in the 488 channel where single-particles are indistinguishable

4. Auto-photoconversion is apparent within our preparations, and thus 561 nm is often used to first bleach out excess signal. It is critical that the number of molecules detected within the presynapse is sparse enough to ensure reliable tracking of individual molecules between frames—optimal detection density can be achieved through a balance of bleaching with 561 nm and photoconversion with 405 nm.
5. The power of the excitation laser at the specimen plane is a critical factor for reliable and accurate in vivo SMLM, with the significant influence, not only over acquisition quality (probe signal-to-noise ratio, photo-blinking, photo-bleaching) but also sample phototoxicity and heat-influence [47]. This should be empirically measured and optimized [48].
6. The optimal illumination angle and collimator setting are both system and sample-dependent and are found by trial-and-error through optimizing the signal-to-noise ratio (SNR) of detections (using both visual inspection and computational readout of the SNR).
7. Exposure time for the EMCCD camera often becomes a trade-off between localization precision and dynamic information. Increased exposure time increases the photon-count per localization, therefore increasing localization precision, whilst decreased exposure time allows for detection and characterization of faster dynamics. Therefore, exposure time is a critical factor to optimize for the user's specific experiment [48].

Acknowledgments

The single-molecule super resolution imaging was carried out at the Advanced Microscopy Facility within the Queensland Brain Institute (QBI) at the University of Queensland, Australia and at the Advanced Optical Microscopy Centre at the University of Hasselt, Belgium. National Health and Medical Research Council (NHMRC) Fellowship (grant no. 1155794) was awarded to F.A.M. This work was supported by funding from the University

Fig. 2 (continued) from one another. (**b**) Single-molecule localizations made from the NMJ in (**a**), utilizing the 561 channel. (**c**) Trajectories derived from detections in (**b**), colored by time during acquisition when the particle was detected and tracked. (**d**) Scaled image of single-molecule localizations made in box of (**b**). (**e**) Scaled images of single-molecule trajectories made in box of (**c**). (**f**) Single-molecule trajectories from (**c**) colored by the estimated diffusion coefficient. (**g–j**) Comparison of endoA-mEos3.1 mobility parameters after incubation with HL3 solution + DMSO or HL3 solution + EGTA: (**g**) average mean square displacement (MSD) as a function of time, (**h**) corresponding area under the curve (AUC) of the MSD, (**i**) relative frequency plot of the distribution of diffusion coefficients and (**j**) corresponding mobile: immobile ratio. Scalebar represents 2 μm (**a–c**) and 1 μm (**d–f**), p-values $^{*}p < 0.01$, $^{***}p < 0.0001$, data represented as mean ± SEM

of Queensland Strategic Initiatives Fund grant no. DVCR22052A awarded to F.A.M. Research support was provided by ERC and the Chan Zuckerberg Initiative and a Methusalem grant of the Flemish government to PV and a FWO Vlaanderen to PV. ATB was supported by the Race Against Dementia (RAD)—Dementia Australia Research Foundation (DARF) postdoctoral fellowship, Brain Foundation project grant, Motor Neuron Disease Australia Research Australia Innovator Grant and DARF Project Grant. KK is supported by the Research Training Program (RTP) Scholarship. This work is supported by an ARC Linkage Infrastructure, Equipment, and Facilities Grant (LE130100078) to QBI Advanced Microscopy facility. ATB performed drosophila larval dissections and image acquisition. ATB and KK performed the analysis. All authors wrote and edited the manuscript.

Conflict of Interest The authors of this chapter declare no conflict of interest.

References

1. Baddeley D, Bewersdorf J (2018) Biological insight from super-resolution microscopy: what we can learn from localization-based images. Annu Rev Biochem 87:965–989. https://doi.org/10.1146/annurev-biochem-060815-014801
2. Huang B, Bates M, Zhuang X (2009) Super-Resolution Fluorescence Microscopy. Annu Rev Biochem 78:993–1016. https://doi.org/10.1146/annurev.biochem.77.061906.092014
3. Schermelleh L, Ferrand A, Huser T et al (2019) Super-resolution microscopy demystified. Nat Cell Biol 21:72–84. https://doi.org/10.1038/s41556-018-0251-8
4. Betzig E, Patterson GH, Sougrat R et al (2006) Imaging intracellular fluorescent proteins at nanometer resolution. Science 313:1642–1645. https://doi.org/10.1126/science.1127344
5. Lelek M, Gyparaki MT, Beliu G et al (2021) Single-molecule localization microscopy. Nat Rev Methods Primers 1:1–27. https://doi.org/10.1038/s43586-021-00038-x
6. Coelho M, Maghelli N, Tolić-Nørrelykke IM (2013) Single-molecule imaging in vivo: the dancing building blocks of the cell. Integr Biol (Camb) 5:748–758. https://doi.org/10.1039/c3ib40018b
7. Podh NK, Paliwal S, Dey P et al (2021) In-vivo single-molecule imaging in yeast: applications and challenges. J Mol Biol 433:167250. https://doi.org/10.1016/j.jmb.2021.167250
8. Padmanabhan P, Bademosi AT, Kasula R et al (2020) Need for speed: super-resolving the dynamic nanoclustering of syntaxin-1 at exocytic fusion sites. Neuropharmacology 169:107554. https://doi.org/10.1016/j.neuropharm.2019.02.036
9. Bernardello M, Gora RJ, Hage PV et al (2021) Analysis of intracellular protein dynamics in living zebrafish embryos using light-sheet fluorescence single-molecule microscopy. Biomed Opt Express 12:6205–6227. https://doi.org/10.1364/BOE.435103
10. Friedrich M, Nozadze R, Gan Q et al (2009) Detection of single quantum dots in model organisms with sheet illumination microscopy. Biochem Biophys Res Commun 390:722–727. https://doi.org/10.1016/j.bbrc.2009.10.036
11. Gora RJ, de Jong B, van Hage P et al (2022) Analysis of the H-Ras mobility pattern in vivo shows cellular heterogeneity inside epidermal tissue. Dis Model Mech 15:dmm049099. https://doi.org/10.1242/dmm.049099
12. Kuhn T, Landge AN, Mörsdorf D et al (2022) Single-molecule tracking of nodal and lefty in live zebrafish embryos supports hindered diffusion model. Nat Commun 13:6101. https://doi.org/10.1038/s41467-022-33704-z
13. Schaaf MJM, Koopmans WJA, Meckel T et al (2009) Single-molecule microscopy reveals

membrane microdomain organization of cells in a living vertebrate. Biophys J 97:1206–1214. https://doi.org/10.1016/j.bpj.2009.05.044

14. Mitra A, Loseva E, Haasnoot GH et al (2023) A small excitation window allows long-duration single-molecule imaging, with reduced background autofluorescence, in *C. elegans* neurons. Opt Commun 545: 129700. https://doi.org/10.1016/j.optcom.2023.129700
15. Robin FB, McFadden WM, Yao B et al (2014) Single-molecule analysis of cell surface dynamics in Caenorhabditis elegans embryos. Nat Methods 11:677–682. https://doi.org/10.1038/nmeth.2928
16. von Diezmann L, Rog O (2021) Single-molecule tracking of chromatin-associated proteins in the C. elegans gonad. J Phys Chem B 125:6162–6170. https://doi.org/10.1021/acs.jpcb.1c03040
17. Zhan H, Stanciauskas R, Stigloher C et al (2014) In vivo single-molecule imaging identifies altered dynamics of calcium channels in dystrophin-mutant C. elegans. Nat Commun 5:4974. https://doi.org/10.1038/ncomms5974
18. Bademosi AT, Lauwers E, Padmanabhan P et al (2016) In vivo single-molecule imaging of syntaxin1A reveals polyphosphoinositide- and activity-dependent trapping in presynaptic nanoclusters. Nat Commun 7:13660. https://doi.org/10.1038/ncomms13660
19. Greiss F, Deligiannaki M, Jung C et al (2016) Single-molecule imaging in living drosophila embryos with reflected light-sheet microscopy. Biophys J 110:939–946. https://doi.org/10.1016/j.bpj.2015.12.035
20. Hines AD, van Swinderen B (2021) Tracking single molecule dynamics in the adult drosophila brain. eNeuro 8. https://doi.org/10.1523/ENEURO.0057-21.2021
21. Manning SA, Kroeger B, Deng Q et al (2024) The drosophila hippo pathway transcription factor scalloped and its co-factors alter each other's chromatin binding dynamics and transcription in vivo. Dev Cell 59:1640–1654.e5. https://doi.org/10.1016/j.devcel.2024.04.006
22. Mir M, Stadler MR, Ortiz SA et al (2018) Dynamic multifactor hubs interact transiently with sites of active transcription in drosophila embryos. Elife 7:e40497. https://doi.org/10.7554/eLife.40497
23. Biermann B, Sokoll S, Klueva J et al (2014) Imaging of molecular surface dynamics in brain slices using single-particle tracking. Nat Commun 5:3024. https://doi.org/10.1038/ncomms4024
24. Varela JA, Dupuis JP, Etchepare L et al (2016) Targeting neurotransmitter receptors with nanoparticles in vivo allows single-molecule tracking in acute brain slices. Nat Commun 7: 10947. https://doi.org/10.1038/ncomms10947
25. Sokoll S, Prokazov Y, Hanses M et al (2015) Fast three-dimensional single-particle tracking in natural brain tissue. Biophys J 109:1463–1471. https://doi.org/10.1016/j.bpj.2015.07.047
26. Axelrod D (1981) Cell-substrate contacts illuminated by total internal reflection fluorescence. J Cell Biol 89:141–145. https://doi.org/10.1083/jcb.89.1.141
27. Konopka CA, Bednarek SY (2008) Variable-angle epifluorescence microscopy: a new way to look at protein dynamics in the plant cell cortex. Plant J 53:186–196. https://doi.org/10.1111/j.1365-313X.2007.03306.x
28. Tokunaga M, Imamoto N, Sakata-Sogawa K (2008) Highly inclined thin illumination enables clear single-molecule imaging in cells. Nat Methods 5:159–161. https://doi.org/10.1038/nmeth1171
29. Huisken J, Swoger J, Del Bene F et al (2004) Optical sectioning deep inside live embryos by selective plane illumination microscopy. Science 305:1007–1009. https://doi.org/10.1126/science.1100035
30. Ritter JG, Veith R, Siebrasse J-P et al (2008) High-contrast single-particle tracking by selective focal plane illumination microscopy. Opt Express 16:7142–7152. https://doi.org/10.1364/OE.16.007142
31. Ritter JG, Veith R, Veenendaal A et al (2010) Light sheet microscopy for single molecule tracking in living tissue. PLoS One 5:e11639. https://doi.org/10.1371/journal.pone.0011639
32. Bademosi AT, Lauwers E, Amor R et al (2018) In vivo single-molecule tracking at the drosophila presynaptic motor nerve terminal. J Vis Exp:e56952. https://doi.org/10.3791/56952
33. Bademosi AT, Decet M, Kuenen S et al (2023) EndophilinA-dependent coupling between activity-induced calcium influx and synaptic autophagy is disrupted by a Parkinson-risk mutation. Neuron 111:1402–1422. https://doi.org/10.1016/j.neuron.2023.02.001
34. Prokop A (2006) Organization of the efferent system and structure of neuromuscular

junctions in drosophila. Int Rev Neurobiol 75: 71–90. https://doi.org/10.1016/S0074-7742(06)75004-8

35. Atwood HL, Govind CK, Wu C-F (1993) Differential ultrastructure of synaptic terminals on ventral longitudinal abdominal muscles in drosophila larvae. J Neurobiol 24:1008–1024. https://doi.org/10.1002/neu.480240803
36. Muttathukunnel P, Frei P, Perry S et al (2022) Rapid homeostatic modulation of transsynaptic nanocolumn rings. Proc Natl Acad Sci USA 119:e2119044119. https://doi.org/10.1073/pnas.2119044119
37. Johansen J, Halpern ME, Johansen KM et al (1989) Stereotypic morphology of glutamatergic synapses on identified muscle cells of drosophila larvae. J Neurosci 9:710–725. https://doi.org/10.1523/JNEUROSCI.09-02-00710.1989
38. Ghelani T, Escher M, Thomas U et al (2023) Interactive nanocluster compaction of the ELKS scaffold and cacophony Ca2+ channels drives sustained active zone potentiation. Sci Adv 9:eade7804. https://doi.org/10.1126/sciadv.ade7804
39. Chen Y, Deng L, Maeno-Hikichi Y et al (2003) Formation of an endophilin-Ca2+ channel complex is critical for clathrin-mediated synaptic vesicle endocytosis. Cell 115:37–48. https://doi.org/10.1016/s0092-8674(03)00726-8
40. Kroll J, Jaime Tobón LM, Vogl C et al (2019) Endophilin-a regulates presynaptic Ca2+ influx and synaptic vesicle recycling in auditory hair cells. EMBO J 38:e100116. https://doi.org/10.15252/embj.2018100116
41. Khuong TM, Habets RLP, Slabbaert JR et al (2010) WASP is activated by phosphatidylinositol-4,5-bisphosphate to restrict synapse growth in a pathway parallel to bone morphogenetic protein signaling. Proc Natl Acad Sci USA 107:17379–17384. https://doi.org/10.1073/pnas.1001794107
42. Kechkar A, Nair D, Heilemann M et al (2013) Real-time analysis and visualization for single-molecule based super-resolution microscopy. PLoS One 8:e62918. https://doi.org/10.1371/journal.pone.0062918
43. Nair D, Hosy E, Petersen JD et al (2013) Super-resolution imaging reveals that AMPA receptors inside synapses are dynamically organized in nanodomains regulated by PSD95. J Neurosci 33:13204–13224. https://doi.org/10.1523/JNEUROSCI.2381-12.2013
44. Ershov D, Phan M-S, Pylvänäinen JW et al (2022) TrackMate 7: integrating state-of-the-art segmentation algorithms into tracking pipelines. Nat Methods 19:829–832. https://doi.org/10.1038/s41592-022-01507-1
45. Tinevez J-Y, Perry N, Schindelin J et al (2017) TrackMate: an open and extensible platform for single-particle tracking. Methods 115:80–90. https://doi.org/10.1016/j.ymeth.2016.09.016
46. Constals A, Penn AC, Compans B et al (2015) Glutamate-induced AMPA receptor desensitization increases their mobility and modulates short-term plasticity through unbinding from Stargazin. Neuron 85:787–803. https://doi.org/10.1016/j.neuron.2015.01.012
47. Icha J, Weber M, Waters JC et al (2017) Phototoxicity in live fluorescence microscopy, and how to avoid it. Bioessays 39:1700003. https://doi.org/10.1002/bies.201700003
48. Diekmann R, Kahnwald M, Schoenit A et al (2020) Optimizing imaging speed and excitation intensity for single-molecule localization microscopy. Nat Methods 17:909–912. https://doi.org/10.1038/s41592-020-0918-5
49. Hines AD, Kewin AB, Poll MNVD et al (2024) Synapse-specific trapping of SNARE machinery proteins in the anesthetized drosophila brain. J Neurosci 44. https://doi.org/10.1523/JNEUROSCI.0588-23.2024
50. Reisser M, Palmer A, Popp AP et al (2018) Single-molecule imaging correlates decreasing nuclear volume with increasing TF-chromatin associations during zebrafish development. Nat Commun 9:5218. https://doi.org/10.1038/s41467-018-07731-8

Chapter 9

Tracking Synaptic Vesicles in Live Neurons Using Single-Molecule Super-resolution Microscopy

Shanley F. Longfield and Frédéric A. Meunier

Abstract

Neurotransmitter release relies on the regulated fusion of synaptic vesicles (SVs) that are densely packed within the presynaptic bouton of neurons. The mechanisms by which SVs are clustered at the presynapse while dynamically organizing themselves into different SV pools with distinct fusion probabilities remain unknown. The study of SVs has historically been limited to ultrastructural studies of the presynapse. Examining the nanoscale dynamic organization of SVs in live neurons requires the use of innovative optical labelling approaches, super-resolution microscopy techniques, and appropriate stimulation paradigms that can mimic neuronal physiology. In this chapter, we discuss these aspects by highlighting the use of single-particle tracking photoactivated localization microscopy (sptPALM) to resolve the mobility and clustering of the total pool of SVs, Universal Point Accumulation Imaging in Nanoscale Topography (uPAINT) to study the mobility of SV proteins transiting on the plasma membrane, Dual-pulse subdiffractional Tracking of Internalized Molecules (DsdTIM) to simultaneously track the reserve and recycling pool of SVs and electrical field stimulation for depolarizing primary neurons.

Key words Presynapse, Synaptic vesicles, Single particle tracking, Nanobodies, Endocytosis, Exocytosis, Electric field stimulation, TIRF microscopy, Fluorescence microscopy

1 Introduction

Communication between neurons in the central nervous system relies on the sustained production, maintenance, release, and reformation of synaptic vesicles (SVs) at the presynaptic compartment. SVs are small (45 nm) vesicles and can be characterized by their membrane protein composition as well as their neurotransmitter content. Upon the arrival of an action potential (AP) at the presynapse, depolarization of the plasma membrane (PM) triggers the opening of voltage-gated calcium channels, leading to a transient influx of calcium into the presynaptic terminal [1]. The resulting increase in the local calcium concentration is sensed by fusion-ready SVs via Synaptotagmin 1 (Syt1) [2] and triggers SNARE (Soluble N-ethylmaleimide-sensitive factor attachment protein receptor)-

Frédéric A. Meunier (ed.), *Super-Resolution Microscopy: Methods and Protocols*, Methods in Molecular Biology, vol. 3034, https://doi.org/10.1007/978-1-0716-5268-8_9,

mediated fusion of SVs with the PM at the active zone (exocytosis) [3, 4]. Exocytosis mediates the release of neurotransmitters from SVs into the synaptic cleft, a 15–20 nm gap that separates the pre- and postsynapse [5]. Following exocytosis, SVs are reformed and recycled back into the presynaptic compartment, refilled with neurotransmitters and re-incorporated into the SV cluster ready for another round of fusion [6].

SVs can be categorized into different pools based on their release probabilities [7]. The readily releasable pool consists of SVs that are typically docked, primed, and released immediately upon arrival of an AP [8]. The recycling pool includes SVs that are endocytosed upon moderate stimulation and represent typically 10–20% of all SVs [9, 10]. The reserve pool contains the majority of SVs (typically 80–90% of all SVs), the contents of which are only released in response to high-frequency or sustained stimulation leading to partial or full depletion of the recycling pool [7]. Recycling SVs are highly dynamic and constantly move in and out of the presynapse [11–13]. When SVs exit the bouton into the adjoining axon, they are considered to be a part of the superpool of SVs, moving between boutons or towards the cell soma [14]. These mobile recycling vesicles can access a wide range of synaptic terminals over time, acting as a common reservoir to directly support shortages of SVs during sustained neuronal stimulation [15]. As vesicles are functionally heterogeneous, trafficking vesicles capable of fusion across multiple terminals enables rapid responses to synapse-specific homeostatic changes in the SV release probability, positively affecting synaptic performance. Using the techniques highlighted in this chapter, we have recently demonstrated that reserve SVs can also exit the presynapse and become part of this superpool; the functional significance of this is still to be investigated [16].

The study of SVs is both crucial to our understanding of neurotransmission and challenging due to their nanoscopic size. Much of our knowledge about the presynapse comes from electron microscopy [17, 18]. Although this has provided important information on the presynaptic nanoarchitecture, these studies are limited to fixed samples where the dynamic nature of presynaptic proteins and vesicles is lost. The size of the presynaptic hippocampal boutons ranges from 0.5 to ~2 μm, and they are densely packed with proteins both on the PM and in the cytoplasm. SV-associated proteins are found concentrated on the SV membrane, around the SV cluster, at the active and peri-active zones of the PM [19]. This crowded nano-environment, along with the subdiffractional size of the vesicles, makes it challenging to investigate SVs in live neurons using conventional light microscopy techniques. These techniques are limited to the resolution of structures 200 nm or larger, and in some cases, this equals the size of an entire presynaptic compartment. Therefore, to investigate the dynamic nanoscale organization and behavior of SVs and associated presynaptic proteins, one must turn to super-resolution microscopy approaches to overcome the resolution limit of light and isolate single vesicles/molecules within the crowded protein environment of living neurons.

1.1 Visualizing the Total Population of SVs Using sptPALM

There is a subset of key proteins that are resident on all SVs (across all SV pools). As such, a simple approach to visualize and track SVs (total pool of SVs) is by expressing a genetically encoded photo-convertible fluorophore tagged to a compatible SV membrane protein (*see* **Note 1** and Fig. 1a). Controlled stochastic photoconversion of a fluorophore using single particle tracking Photoactivated Localization Microscopy (sptPALM; *see* Chapter 6) enables nanometer resolution tracking of individual SVs, by using the fluorescently tagged SV membrane protein as a marker. EosFP [20] is one of the best-performing photoconvertible fluorophores used in sptPALM, with mEos3.2 [21] currently being the most suitable variation of this protein for protein tracking. mEos3.2 can be visualized as a green fluorescent protein, and excitation of mEos3.2 with UV irradiation (405 nm) leads to its stochastic photoconversion from green to red emission fluorescence [20]. Careful control of mEos3.2 photoconversion with a ~405 nm laser enables the imaging of individual SV proteins over time, from which trajectories can be inferred. This is a powerful technique to investigate SV dynamics in live neurons, with SV membrane proteins vesicular glutamate transporter 1 (vGLUT1) [12] and synaptophysin [22] having been successfully used to track individual SVs. However, exogenous or over-expression of proteins have disadvantages, including the potential for non-specific SV labelling. Hence, the expression of mEos3.2-tagged SV proteins should be carefully controlled so as not to cause interference with the function of the tagged protein or the fluorescent signal being so dense within the small presynaptic compartment that it interferes with the super-resolution tracking of the proteins (*see* Chapter 6).

1.2 Visualizing Surface Pool with uPAINT

Universal Point Accumulation Imaging in Nanoscale Topography (uPAINT) is a super-resolution imaging technique that can be used to track surface proteins using fluorescently tagged ligands. In the study of SVs, uPAINT can be used to track the surface stranded population or pool of SV proteins [11]. This technique does not require stochastic photoactivation of single molecules but instead relies on sparse labelling of surface biomolecular targets [23]. For instance, these ligands can be nanobodies conjugated to highly fluorescent probes. A key benefit of uPAINT is its ability to image real time fluorescent ligand binding to proteins on the PM in comparison to techniques such as sptPALM, where all ligands are already bound but stochastically photoconvert in response to a UV laser. To track the surface pool of an SV protein using uPAINT, the neurons must be genetically encoded to express a pHluorin

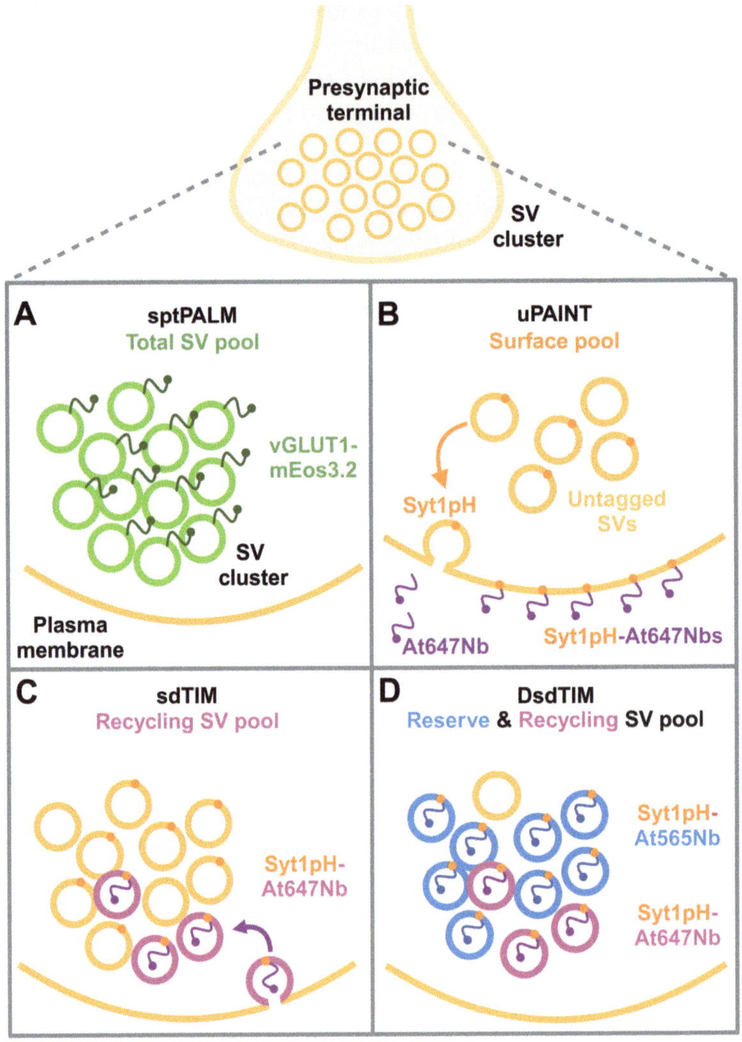

Fig. 1 Graphical illustration of the super-resolution techniques used to image different synaptic vesicle (SV) populations in the presynaptic boutons and axons of live cultured neurons. (**a**) Single particle tracking Photoactivated Localization Microscopy (sptPALM) tracks the total pool of SVs in live cells by imaging an integral SV membrane protein of interest labelled with a photoconvertible fluorescent marker. Example shown: vesicular glutamate transporter 1 (vGLUT1) tagged with mEos3.2 (vGLUT1-mEos3.2). (**b**) universal Point Accumulation Imaging in Nanoscale Topography (uPAINT) tracks the surface pool of SV proteins in live cells by imaging immediately after the addition of nanobodies that recognize the fluorescent tag of the SV protein of interest. Example shown: Synaptotagmin 1 (Syt1) tagged with pHluorin (pH) bound to Atto647N-anti-GFP-nanobodies (Syt1pH-At647Nbs) on the PM. (**c**) Subdiffractional Tracking of Internalized Molecules (sdTIM) tracks the recycling pool of SVs in live cells by imaging a nanobody-bound fluorescently labelled SV protein of interest following its internalization into SVs. Example shown: Syt1pH-At647Nbs after internalization. (**d**) Dual-pulse subdiffractional Tracking of Internalized Molecules (DsdTIM) simultaneously tracks the reserve and recycling pool of SVs in live cells by imaging two spectrally distinct nanobodies bound to a fluorescently labelled SV protein of interest, which are added at different time points to capture SV internalization into different pools. Example shown: Syt1pH-At647Nbs imaged 10 min after labelling to capture the recycling SV pool and Syt1pH-At565Nbs imaged 48 h after labelling to capture the reserve SV pool

(a GFP-variant) attached to the transmembrane SV protein of interest (*see* **Note 2**). External application of anti-GFP nanobodies tagged with Atto565 or Atto647N, in a depolarizing high potassium buffer (termed a pulse), will promote SV fusion with the PM. Upon fusion, SVs expose the luminal pHluorin tag to the synaptic cleft thereby enabling nanobody binding (*see* Fig. 1b). Thus, uPAINT facilitates the nanoscale detection and tracking of individual surface stranded SV proteins as they transit the PM in live neurons.

1.3 Visualizing the Recycling Pool with Subdiffractional Tracking of Internalized Molecules (sdTIM)

Subdiffractional Tracking of Internalized Molecules (sdTIM) is a super-resolution imaging technique that takes advantage of the process of endocytosis to label the endocytic population of fluorescently tagged ligands following their internalization [13]. In this chapter, we have specifically used sdTIM to image recycling SVs [11, 13]. Similar to the uPAINT labelling of an SV protein surface pool, subdiffractional tracking of the recycling SV pool requires the genetic encoding of pHluorin to a transmembrane SV protein of interest, and the external application of anti-GFP nanobodies in a depolarizing high potassium buffer. Upon stimulation, the luminal pHluorin becomes accessible to the anti-GFP nanobodies. Critically, after a wash step to remove unbound anti-GFP nanobodies and a given chase time, the internalized bound anti-GFP nanobodies can be tracked into recycling SVs [11, 13]. Nanobody binding and their selective internalization in recycling SVs facilitates the nanoscale detection and tracking of individual recycling SVs. Further, when the SVs initially fuse with the plasma membrane during the pulse, the pHluorin unquenching can be used to generate regions of interest (ROIs) and identify synapses from nearby axons. As the localization precision of the anti-GFP nanobodies used in uPAINT/sdTIM is typically ~35–40 nm and the diameter of an SV is ~45 nm [24], the nanoscale mobility data from endocytosed ligand reflects the mobility of recycling SVs. Thus, sdTIM is a useful technique to preferentially label and image the actively recycling population of SVs in live primary neurons.

1.4 Visualizing the Reserve Pool with Dual-Pulse Subdiffractional Tracking of Internalized Molecules (DsdTIM)

Dual-pulse subdiffractional Tracking of Internalized Molecules (DsdTIM) is an adaptation of sdTIM [11] (*see* Fig. 1c) to co-track recycling and reserve SVs in the same live neurons. Recycling vesicles have been shown to slowly decrease their mobility at the presynapse as they enter the composition of the reserve pool [16]. Furthermore, SV proteins such as Synaptotagmin 1 (Syt1) age over time, and their co-localization with newly synthesized SV proteins decreases along with their ability to undergo fusion (a key feature of the reserve pool of SVs). By implementing a longer incubation period (48 h chase) and two temporally separated labelling pulses, we were able to co-track the older, reserve SVs in tandem with the newly labelled, actively recycling population of

SVs [16]. Therefore, DsdTIM enables the accurate differentiation of the large and enigmatic "reserve" pool of SVs from the smaller recycling SV pool (*see* Fig. 1d). This chapter describes the DsdTIM protocol and highlights variations of the protocol that are useful to the study of SVs in primary neurons (*see* Subheadings 2 and 3).

1.5 Extracting Information About the Superpool of SVs

The superpool is defined as peri-synaptic SVs found in the axons of neurons. The dynamics of this population can be drawn out in the data analysis portion of the protocol pipeline (*see* Chapter 6). To isolate the mobility patterns of the reserve and recycling SVs within the axons, it is important to ensure that clear axonal regions devoid of excessive overlap between adjacent neurons are imaged and that precise ROIs are drawn around the synaptic and axonal segments of the neurons. Analyzing the data from these ROIs separately (compared to not drawing ROIs or drawing ROIs around whole neurons) will extract key differences between how the SVs or SV-associated proteins behave in axons versus presynapses.

1.6 The Use of Dissociated Primary Neurons in Super-Resolution Microscopy

Each of the above SV tracking techniques utilizes primary hippocampal neurons for obtaining single-molecule trajectory data. The use of cultured cells on glass-bottomed dishes or coverslips remains the gold standard for super-resolution experiments due to limitations in terms of tissue laser penetration depth and microscopy stage equipment. For such cultures, the use of Total Internal Reflection Fluorescence (TIRF) is a key strategy for limiting the detection of fluorescent markers to those on the PM, thus minimizing overlapping detections that can occur with fluorescent markers that are located deeper in the sample. In contrast to traditional widefield imaging (where the laser beam directly illuminates the sample), TIRF imaging uses an angled excitation beam to greatly restrict the evanescent field, thus illuminating only a thin section of the membrane. When the TIRF angle is immediately adjacent to the sample, it allows for selective excitation of the fluorophores found directly on the neuronal PM, as well as a small window immediately adjacent to the PM. In contrast to dissociated cells, brain slices, organoids, and other thicker tissue samples are more difficult to image using many super-resolution approaches. The laser penetration is more limited in thicker samples, and the viability (brain slices) and orientation of neuronal projections (altered in brain organoids) can be more variable than what is usual for cultured neurons. Thus, performing single particle tracking on samples, such as small organisms (e.g., *Drosophila melanogaster*), requires specific sample mounting techniques and acquisition setup, to accommodate the larger distance between the sample and TIRF field and to preserve the tissue integrity for the duration of the acquisition (for more information on small animal preparations, see Chapter 8).

1.7 Different Neuronal Stimulation Paradigms

Cultured neurons are an excellent cellular model that recapitulates many aspects of neuronal communication. Rodent cultured neurons can develop in vitro, mature, form synaptic connections and can be modulated by a range of exogenous stimuli for the purposes of scientific investigation. Understanding how proteins behave during neuronal activity, or in disease conditions, is of great interest to the field of Neuroscience. Inducing neuronal activity can be achieved via several methods in vitro, each with its advantages and disadvantages. Depolarizing (elevated or high K^+) buffers, electric field stimulation and optogenetic stimulation are three key methods to stimulate neurotransmitter release in primary neurons cultured in glass-bottomed dishes or coverslips.

A simple way of promoting neuronal activity is to increase the extracellular potassium concentration of cultured neurons using an isotonic buffer (with an equimolar reduction of Na^+) to induce membrane depolarization. When the extracellular K^+ concentration increases, the K^+ gradient across the membrane decreases, and the potassium equilibrium potential becomes more positive. As external Na^+ continues to leak into the soma, global neuronal depolarization occurs. As calculated by the Goldman equation, a tenfold change in extracellular K^+ from 5 to 50 mM will cause a long-lasting 48 mV membrane depolarization [25]. Electric field stimulation, on the other hand, induces physiologically relevant fast depolarization of cultured neurons [26]. This is achieved through superimposing an electrically induced transmembrane voltage onto the resting membrane voltage proportional to the strength and duration of the field [27, 28]. Exposure of the neurons to this electric field hijacks the resting membrane potential and initiates APs across the neurons. The pros and cons of elevated K^+ buffers vs electric field stimulation is summarized in Table 1 as they are the most versatile for the super-resolution study of SVs. Optogenetic stimulation is another elegant way to induce activity in a subset of neurons, by genetic encoding of light-sensitive ion channels, pumps or enzymes into neurons to induce activity. However, since optogenetic stimulation requires a separate laser activation, care needs to be taken into designing an experiment that does not interfere with the fluorescent activation of super-resolution compatible tags [31].

Elevated extracellular K^+ buffers have been widely used due to their affordable nature and effectiveness in depolarizing live neurons. The body of work generated from their use in vitro has greatly contributed to our understanding of neuronal function by elucidating the signaling, transcriptional, and plasticity-related events that occur in neurons. One should carefully consider the timing and concentration parameters of high K^+ stimulation, bearing in mind that prolonged high K^+ exposure is not directly comparable to neuronal, optogenetic or field stimulation [25]. Below we provide detailed descriptions of both high K^+ and electrical field stimulation (as an alternative method) for stimulating primary neurons for DsdTIM.

Table 1
Comparison between high potassium and electric field stimulation of hippocampal neurons

Elevated extracellular potassium		Electric field stimulation	
Pros	**Cons**	**Pros**	**Cons**
Accessible (affordable)	Not physiological	More physiological methods of neuronal stimulation	Expensive to set up and time-consuming to establish
Reliably depolarizes the whole culture	Strong stimulus that increases the frequency of quantal spontaneous neurotransmitter release—Difficult to modulate the stimulation intensity in the same way that electrical stimulation can	Can be modulated to see the effect of different types of electrical pulses or stimulation trains	Limited to the stimulation of neurons in very small wells (10 mm for the stimulation insert described in this chapter)
Can be used for all cell culture dishes regardless of size	May yield unphysiological results that occur only under episodes of strong tetanic stimulation [25]		Requires a custom microscope stage
Can accentuate the effect size of SV protein transient interactions that would otherwise be difficult to detect	Stimulates alternative signaling pathways inside the neuron relative to synaptic activity–dependent Ca^{2+}influx (e.g., via NMDA receptors) [29]		Requires a way to confirm stimulation has occurred with either a pHluorin SV marker or a calcium sensor (preferably ratiometric)
	Higher K^{+} concentrations induce long-term depolarization. Because voltage-gated Ca^{2+} channels inactivate rapidly, it is difficult to determine the accurate level of stimulation or depletion of SV at the terminals [30]		Insert is fragile and it is liable to breaking

2 Materials

2.1 Primary Hippocampal Neuronal Culture

The protocols described in this chapter use primary hippocampal neurons obtained from C57BL/6J embryonic day (E) 16 mice. The isolation of hippocampi has been previously described [11].

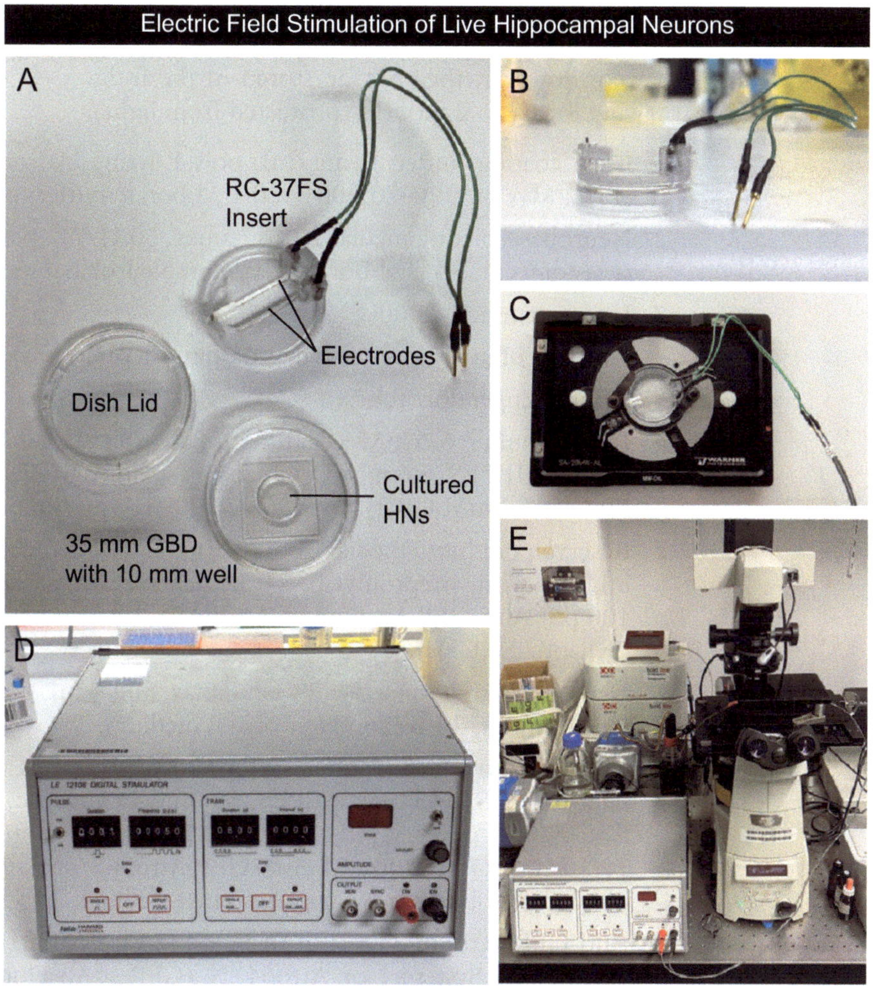

Fig. 2 Equipment needed for electric field stimulation of live hippocampal neurons (HNs). (**a**) RC-37FS Chamber Insert with field stimulation electrodes (Warner Instruments) and a 35 mm glass-bottomed dish (GBD) with a 10 mm micro-well for cultured primary hippocampal neurons. (**b**) Image illustrating how the RC-37FS Chamber Insert should fit into the GBD. (**c**) Image illustrating how the GBD with RC-37FS Chamber Insert is inserted into the Quick Exchange Platform (QE-1; Warner Instruments) and how the RC-37FS field stimulation electrodes are attached to the connecting wires of the digital stimulator. (**d**) PanLab LE12106 1 Single Channel Digital Stimulator (Warner Apparatus) set to a stimulation of 1 ms pulse width, 50 Hz, for 6 s. (**e**) Complete super-resolution microscope set up with an electrical field stimulation apparatus

Note that for this protocol, the primary neurons should be plated in 29 mm or 35 mm glass-bottomed cell culture dishes with 10 mm micro-wells (*see* Fig. 2a), depending on the specific requirements of the experiment.

2.2 Buffers, Media, and Ligands

Prepare all buffers using Milli-Q or Ultrapure water unless otherwise stated. All buffers should be made fresh for each experiment, warmed to 37 °C, and filtered using a 0.22 μm syringe-driven filter unit before use. The Anti-GFP labelled FluoTag-Q single-domain camelid nanobody aliquots should not be repeatedly freeze-thawed; however, they can be stored in the fridge (~4 °C) after thawing for up to 2 months (protected from light).

1. Glass coating buffer: 1 mg/mL poly-L-lysine hydrobromide (PLL; MW 30,000–70,000) in 0.1 M borate buffer (pH 8.4).
2. Neurobasal plating media: 2% Glutamax, 50 U/mL penicillin/streptomycin, 5% fetal bovine serum in Neurobasal medium.
3. Lipofectamine® 2000.
4. Plasmids of interest (e.g., synaptotagmin 1 pHluorin).
5. 1.5 mL Eppendorf tubes.
6. Low K^+ buffer: 5.6 mM KCl, 0.5 mM ascorbic acid, 0.1% BSA, 15 mM HEPES, 5.6 mM D-Glucose, 145 mM NaCl, 0.5 mM $MgCl_2$, 2.2 mM $CaCl_2$, at pH 7.4, 290–310 mOsm.
7. Reserve pool pulse media: 30 mM KCl, 2% Glutamax, 50 U/mL penicillin/streptomycin, 2% B27 in Neurobasal medium, 400 pM Anti-GFP Atto565 or Atto647N-labelled nanobodies (Synaptic Systems).
8. Neurobasal growth media: 2% Glutamax, 50 U/mL penicillin/streptomycin, 2% B27 in Neurobasal medium.
9. Recycling pool pulse buffer: 56 mM KCl, 0.5 mM ascorbic acid, 0.1% BSA, 15 mM HEPES, 5.6 mM D-Glucose, 95 mM NaCl, 0.5 mM $MgCl_2$, 2.2 mM $CaCl_2$, at pH 7.4, 290–310 mOsm, 200 pM Anti-GFP Atto565 or Atto647N-labelled nanobodies (Synaptic Systems).
10. Anti-GFP Atto565 (GFP sdAb-FluoTag-Q, #N0301-At565-L, Synaptic Systems) or Atto647N (GFP sdAb-FluoTag-Q, #N0301-At647N-L, Synaptic Systems)-labelled FluoTag-Q single-domain camelid nanobodies.
11. TetraSpeck™ microspheres (Invitrogen-Thermo Fisher Scientific, 0.1 μm).

2.3 Light and Super-Resolution Microscopy

1. Inverted light microscope.
2. iLas2 ring-TIRF illumination system (Roper Scientific, France) mounted on a Nikon Ti-E inverted microscope with a Nikon CFI Apo TIRF 100/1.49 NA oil-immersion objective and a Perfect Focus System (Nikon Corporation, Japan).
3. Two Evolve 512 Delta EMCCD cameras (Teledyne Photometrics, USA) mounted on a TwinCam LS dual-emission image splitter (Cairn Research, UK).
4. Quad-band dichroic beam-splitter (ZT405/488/561/647rpc, Chroma Technology, USA).

5. Triple-band emission filter on the reflected arm (ZET405/488/561 m, Chroma Technology, USA).
6. Far-red emission filter on the transmission arm (ET690/50 m, Chroma Technology, USA).
7. TetraSpeck™ microspheres (Invitrogen-Thermo Fisher Scientific, 0.1 μm).

2.4 Equipment for Electric Field Stimulation of 35 mm Dishes

1. Silicone Grease Kit: Dow Corning 111® Valve Lubricant & Sealant, and acrylic painter's brushes (Warner Instruments).
2. QE-1—Quick Exchange Platform for 35–40 mm dishes (Warner Instruments) (*see* Fig. 2c).
3. RC-37FS—Chamber Insert for 35 mm cell culture dishes with field stimulation (Warner Instruments) (*see* Fig. 2a).
4. PanLab LE12106 1—Channel Digital Stimulator (Warner Apparatus) (*see* Fig. 2d).

3 Methods

3.1 Primary Hippocampal Neuronal Culture and Transfection

The protocols described in this chapter use primary hippocampal neurons obtained from C57BL/6J embryonic day (E) 16 mice. The isolation of hippocampi and neuronal transfection has been previously described [11]. Here, we provide a brief outline of these methods:

1. Prepare the neurobasal plating and growth media in advance, warm to 37 °C, and filter using a 0.22 μm syringe-driven filter unit directly before use.
2. Dissect the mouse hippocampi and seed these neurons in neurobasal plating medium onto either 29 mm or 35 mm glass-bottomed dishes with the 10 mm micro-wells (pre-coated with 1 mg/mL PLL in borate buffer). We recommend ~40,000 neurons per 10 mm micro-well.
3. Exchange the neurobasal plating media with growth media 5–6 h post-seeding.
4. Supplement the plated neurons with 500 μL fresh growth media every week to maintain the health of the culture.
5. Transfect primary neurons (*with Synaptotagmin 1 pHluorin for DsdTIM*) at DIV14–15 using Lipofectamine® 2000, as per manufacturer's instructions.
6. Image the transfected neurons via super-resolution microscopy 3–7 days post-transfection (up to DIV22).

3.2 Labelling the Reserve Pool of Vesicles Using a High K^+ Buffer

Labelling of the reserve pool of SVs requires sterile technique and must be performed with the intention of returning the pulsed (*see* **Note 3**) synaptotagmin 1 pHluorin (Syt1pH)-transfected hippocampal neurons to the incubator for 48 h before imaging. The reserve pool pulse is carried out from DIV19–20.

1. Calculate the appropriate amount of reserve pool pulse media required for each dish (typically 150 μL per 10 mm well). The reserve pool pulse buffer will be added dropwise to only the 10 mm well, not the entire glass-bottom dish.
2. Filter the pulse media (30 mM KCl, 2% Glutamax, 50 U/mL penicillin/streptomycin, 2% B27 in Neurobasal medium) using a 0.22 μm syringe-driven filter unit before adding the anti-GFP Atto565 nanobodies (At565Nbs) (*see* **Notes 4** and **5**).
3. Taking no more than four glass-bottomed dishes at a time, move the Syt1pH-transfected hippocampal neurons from the incubator to the tissue culture hood.
4. Collect the conditioning media (the supernatant of in vitro cultivated cells that contains all the secreted factors by those cells) from each dish into individual sterile 1.5 mL Eppendorf tubes, leaving enough media in the 10 mm well to cover the transfected primary neurons (*see* **Note 6**).
5. Immediately before adding to the dishes, invert the reserve pool pulse media to ensure homogeneous distribution of the At565Nbs. Add 150 μL of the reserve pool pulse media (dropwise) to each well and incubate the neurons in this media for 1 min.
6. After the 1 min pulse, wash off the excess At565Nbs with warm, filtered neurobasal growth media. We recommend pipetting 500–1000 μL growth media onto the plastic immediately adjacent to the well. This will allow for the gentle flow of media across the neurons without harming the culture. Repeat this wash at least 3 times, then replace the dish's collected conditioning media.
7. Incubate the neurons for 48 h at 37 °C with 5% CO_2.
8. After the 48 h chase (*see* **Note 7**), if you are only interested in the reserve pool of SVs proceed to imaging the neurons with your super-resolution setup. After checking the health of the neurons (using an inverted light microscope; *see* **Note 8**), wash the dish twice with warm low K^+ buffer to remove the culture medium.
9. Place the glass-bottomed dish on the imaging platform of the microscope operating at 37 °C and allow the dish to equilibrate to the platform temperature for at least 5–10 min (*see* **Note 9**).

10. Look for transfected neurons and healthy presynaptic regions of interest (ROIs) using ~491 nm excitation for pHluorin. Ideally, once a transfected neuron is found, we recommend following the neuron down its axon to the cell body to ensure that it is healthy throughout.
11. After identifying an ROI, switch to the red channel (~561 nm excitation for At565Nbs) and acquire. We recommend 8000 frames at 20 ms exposure time (50 Hz).

3.3 Labelling the Reserve Pool of Vesicles Using Electric Field Stimulation

For this protocol, the neurons must be cultured in 35 mm glass-bottomed dishes with a 10 mm micro-well.

1. Sterilize the RC-37FS insert by wiping it down thoroughly with 70% ethanol (v/v) and placing both the top and bottom of the insert underneath UV light for at least 20 min.
2. Set up the PanLab single-channel digital stimulator next to the tissue culture hood, close enough for the connecting wires to reach the sample. For the stimulation parameters, we recommend 100 mA, 1 ms pulse width, 50 Hz for 6 s (300 high-frequency APs).
3. Check the health of the Syt1pH-transfected neurons using an inverted light microscope.
4. Calculate the appropriate amount of neurobasal growth media needed for both the washes (typically 3 mL per dish) and the electrical pulse (typically 1 mL per dish; add 400 pM At565Nbs to this media in order to label the reserve pool of SVs).
5. Connect the digital stimulator to the RC-37FS insert and leave it resting in the tissue culture hood until needed.
6. Take a glass-bottomed dish from the incubator to the tissue culture hood and collect the conditioning media into a sterile 1.5 mL Eppendorf tube, leaving enough media in the 10 mm well to cover the transfected primary neurons.
7. Invert the neurobasal growth media containing 400 pM At565Nbs and add 1 mL of this media to the dish.
8. Place the connected RC-37FS insert inside the 35 mm glass-bottomed dish. Ensure that the electrodes are fully submerged in the media.
9. Stimulate the neurons in the presence of the At565Nbs at 50 Hz for 6 s (*see* **Note 10**).
10. After stimulation, wash off the excess nanobodies with warm, filtered growth media. We recommend pipetting 500–1000 μL wash media onto the plastic immediately adjacent to the well. This will allow for the gentle flow of media across the neurons without harming the culture. Repeat this wash at least 3 times, then return the previously collected conditioning media to the dish.

11. Incubate the neurons for 48 h at 37 °C with 5% CO_2. To image proceed with **steps 8–11** in the above section (*see* Subheading 3.2).

3.4 Labelling the Recycling Pool of Vesicles Using a High K^+ Buffer

If you are interested in simultaneously imaging the reserve and recycling pool of SVs (DsdTIM) or only the recycling pool of SVs (sdTIM), proceed with labelling of the recycling pool. The recycling pool pulse is carried out in Syt1pH-transfected hippocampal neurons from DIV 21–22.

1. Calculate the appropriate amount of recycling pool pulse buffer required per dish (typically 1 mL per dish).
2. After checking the health of the neurons (using an inverted light microscope), wash the dish twice with warm low K^+ buffer to remove all the conditioning media. Immediately before adding to the dish, invert the recycling pool pulse buffer to ensure homogeneous distribution of the At647Nbs. Add 1 mL of the recycling pool pulse buffer to the dish and incubate the neurons for 5 min.
3. After the 5 min pulse, wash off the excess At647Nbs with a warm, low K^+ buffer. We recommend pipetting 1000 μL imaging buffer onto the plastic immediately adjacent to the well. This will allow for the gentle flow of media across the neurons without harming the culture. Repeat this wash at least 3 times.
4. Incubate the neurons for at least 10 min at 37 °C with 5% CO_2 before imaging to ensure complete reinternalization of the SVs. We recommend moving the dish directly to the microscope imaging platform and using this time to search for transfected neurons.
5. After the chase, if you would only like to image the recycling pool of SVs, switch to the far-red channel (~647 nm excitation for At647Nbs) and acquire. We recommend 8000 frames at 20 ms exposure time (50 Hz).

3.5 Labelling the Recycling Pool of Vesicles Using Electric Field Stimulation

This protocol describes a variation on the sdTIM protocol where high K^+ stimulation is replaced with electric field stimulation to trigger internalization of the Syt1pH-bound anti-GFP At647Nbs. The neurons must be cultured in 35 mm glass-bottomed dishes with a 10 mm micro-well for this field stimulation protocol.

1. Check the health of the Syt1pH-transfected neurons using an inverted light microscope.
2. Connect the PanLab single-channel digital stimulator as per the manufacturer's instructions and ensure that its position to the microscope stage is close enough for the connecting wires to reach the sample (*see* Fig. 2e). Set the stimulation parameters, we recommend 100 mA, 1 ms pulse width, 50 Hz for 6 s.

3. Using the silicone grease kit brush, scoop a generous amount of silicone grease from the tub and apply it to the bottom of the RC-37FS chamber insert for 35 mm cell culture dishes. Spread the grease around the edges of the insert well (ensuring an even spread on both sides) and place it upside down on the table on a paper towel or tissue (silicone grease facing upward).
4. Retrieve a dish from the incubator and wash twice with warm low K^+ buffer to remove the conditioning media.
5. Add to the dish 1 mL of low K^+ buffer containing 200 pM At647Nbs.
6. Place the RC-37FS insert into the culture dish, press down firmly, and hold it in place for 10 s.
7. Place the glass-bottomed dish with the insert on the imaging platform (*see* Fig. 2b, c) of the microscope operating at 37 °C.
8. Connect the digital stimulator to the RC-37FS insert and stimulate the dish at 50 Hz for 6 s. Immediately following stimulation, wash off the excess At565Nbs or At647Nbs with warm low K^+ buffer (8–10 mL), leaving at least 1 mL in the dish.
9. During the 10-min chase, look for transfected neurons and healthy presynaptic ROIs using ~491 nm excitation for pHluorin. Ideally, once a transfected neuron is identified, we recommend following the neuron down its axon to the cell body to ensure that it is healthy throughout.
10. After the chase, identify an ROI, switch to the far-red channel (~647 nm excitation for At647Nbs) and acquire. We recommend 8000 frames at 20 ms exposure time (50 Hz).

3.6 Simultaneously Imaging the Reserve and Recycling Pool of SVs (Dual-Color Imaging)

1. Before imaging, ensure that the EMCCD cameras are aligned and the TIRF angle is calibrated (*see* **Note 11**).
2. After labelling both the reserve and recycling pool of SVs in the same hippocampal neurons (*see* **Note 12**), identify an ROI using ~491 nm excitation for pHluorin. Switch to the dual color setup using the red (~561 nm excitation) and far-red (~647 nm excitation) channels and acquire. We recommend 8000 frames at 20 ms exposure time (50 Hz).

3.7 Simultaneously Imaging the Reserve Pool of SVs and a Presynaptic Protein Tagged with mEos

1. On DIV 14–15, co-transfect the primary cultures with an mEos-tagged candidate protein and Syt1pH using Lipofectamine® 2000 (Invitrogen), as per the manufacturer's instructions.
2. On DIV 19–20, label the reserve pool of SVs using At647Nbs (*see* Subheading 3.2).

3. After the 48 h chase, check the health of the neurons using an inverted light microscope. If the neurons are healthy enough to image, wash the dish twice with warm low K^+ buffer to remove the conditioning media.
4. Place the glass-bottomed dish on the imaging platform of the microscope operating at 37 °C and allow the dish to equilibrate to the platform temperature for at least 5–10 min.
5. Look for transfected neurons and healthy presynaptic ROIs using ~491 nm excitation for pHluorin and mEos. Ideally, once an ROI has been identified, we recommend checking for photo conversion in the red channel (~561 nm excitation) and single-molecule signal in the far-red channel (~647 nm excitation) at a low laser power to confirm that the candidate neuron is co-transfected with both plasmids.
6. Once a candidate neuron has been identified, switch to the dual color setup using the red (~561 nm excitation) and far-red (~647 nm excitation) channels and acquire. We recommend 8000 frames at 20 ms exposure time (50 Hz).

3.8 Electric Field Stimulation of Live Neurons Transfected with an mEos-Tagged SV Protein

This protocol assumes the experiment aims to track the mobility of an mEos2-tagged SV-associated protein before and after high-frequency stimulation of the transfected neurons. Aspects of the protocol should be altered to suit specific experimental needs (i.e., *stimulation frequency settings and pulse timing*). The neurons must be cultured in 35 mm glass-bottomed dishes with a 10 mm micro-well for this field stimulation protocol.

1. Prepare the microscope for imaging and check the health of the transfected neurons (using an inverted light microscope).
2. Connect the PanLab single-channel digital stimulator as per the manufacturer's instructions and ensure that its position to the microscope stage is close enough for the connecting wires to reach the sample. Set the stimulation parameters, we recommend 100 mA, 1 ms pulse width, 50 Hz for 6 s.
3. Using the silicone grease kit brush, scoop a generous amount of silicone grease from the tub and apply it to the bottom of the RC-37FS chamber insert for 35 mm cell culture dishes. Spread the grease around the edges of the insert well (ensuring an even spread on both sides) and place it upside down on the table on a paper towel or tissue (silicone grease facing upwards).
4. Retrieve a dish from the incubator and wash twice with warm imaging buffer to remove the culture medium.
5. Leaving 1 mL of imaging buffer in the 35 mm dish, firmly press the RC-37FS insert into the culture dish and hold it in place for 10 s.

6. Place the glass-bottomed dish with the insert on the imaging platform of the microscope operating at 37 °C and allow the dish to equilibrate to the platform temperature for at least 5–10 min.
7. Look for transfected neurons and healthy presynaptic ROIs using ~491 nm excitation for mEos. Ideally, once a transfected neuron has been identified, we recommend following the neuron down its axon to the cell body to ensure that it is healthy throughout.
8. Once a candidate neuron has been identified, switch to the red channel (~561 nm excitation) and acquire the mEos signal in resting (unstimulated) conditions. We recommend 8000 frames at 20 ms exposure time.
9. Connect the digital stimulator to the RC-37FS insert and stimulate the dish for 50 Hz for 6 s. Immediately following stimulation, image the same neuron (for paired analysis) or another transfected neuron (unpaired analysis), acquiring the mEos signal post-stimulation (stimulated conditions). We recommend 8000 frames at 20 ms exposure time (50 Hz).

4 Notes

1. To image the total pool of SVs using sptPALM, we recommend using vGLUT1 mEos3.2, which functions as a key glutamate/proton exchanger and controls SV acidification at hippocampal synapses [32].
2. When choosing which SV-protein to tag with pHluorin, it is important to note that the time proteins spend stranded on the membrane is dependent on the type and copy number of the given SV protein. Synaptobrevin (VAMP2) has a relatively long residence time on the plasma membrane (partially due to its abundance), whereas vGLUT1 resides on the membrane for significantly less time, returning to the SVs upon endocytosis [33]. These are also important considerations to make when designing uPAINT experiments and interpreting super-resolution trajectory data.
3. *Pulse:* depolarizing the neurons in the presence of anti-GFP nanobodies. This can be achieved through the use of high K^+ or electrical field stimulation. This process aims to promote the externalization of the luminal pHluorin tagged to the SV transmembrane protein of interest (i.e., Syt1pH) upon exocytosis of the vesicle. This allows for the binding of the anti-GFP/pHluorin fluorescent nanobody to the plasma membrane and the internalization of the complex in recycling SVs.

4. The Anti-GFP Atto565 or Atto647N-labelled FluoTag-Q single-domain camelid nanobodies can be used interchangeably depending on the design of the experiment and whether neurons are co-transfected with other fluorescent markers. For example, in the effect of an effector protein on SV mobility in knockout conditions, one can track the mobility of the reserve pool of SVs when re-expressing an mEos-tagged effector protein of interest. In this case, the reserve pool can be labelled with the At647Nbs, leaving the red channel open to visualize the mEos-tagged effector protein [16].
5. We recommend keeping all media in the incubator with the lid open enough for the CO_2 to equilibrate, and filter sterilize immediately before use. Primary hippocampal neurons are sensitive to temperature changes—the addition of media that is at a significantly lower temperature than that of the neurobasal media the neurons are in (~37 °C) can affect the health of the primary cultures.
6. Primary hippocampal neurons are sensitive to media volume fluctuations. Therefore, never remove media from the central well in which the neurons are cultured. Instead, remove media from the plastic surrounding the well by pressing the pipette tip into the side of the dish, gently removing excess media on the plastic surface of the dish. This will ensure that the media directly above the neurons does not fall below a critical volume and ensures the health of the culture.
7. *Chase:* rest time afforded to the neurons after the removal of excess anti-GFP nanobodies. This will allow for uninterrupted compensatory endocytosis and for the incorporation of SVs labelled with nanobody-bound Syt1pH into the SV cluster.
8. After labelling the reserve pool of SVs, it is not uncommon to see a small portion of cell death (10–20%). If there is a significant amount of cell death (>50% of the neurons), discard the dish and move on to a healthier culture. Imaging neurons with a significant amount of cell death around them (even if there is one or two live neurons to image) can create variability in the data set. If the culture is unhealthy, almost all the neurons will be affected by this cell death and the proteins of interest will display different mobility patterns to physiological conditions, potentially reflecting a neuronal survival state.
9. The temperature of the glass-bottomed dish will fluctuate as it is removed from the tissue culture incubator, washed, and placed in the microscope imaging platform. This fluctuation in temperature (from 37 °C to room temperature back to 37 °C) will cause nanoscale contraction and relaxation of the culture dishes themselves. This fluctuation in dish size can lead to significant drift in the acquisitions if one were to immediately start imaging. Drift can be identified as the smearing of

neuronal trajectories after the acquisitions have been processed and is caused by the location of the ROI shifting several hundred nanometers as the dish expands in the microscope incubator. If this drift is too severe, it will cause inaccuracies in the localizations of individual trajectories, which can render the data difficult to analyze. To avoid this, the dish should be incubated for at least 5–10 min in the heated microscope stage to allow for the plastic/glass to reach the correct temperature thus minimizing drift.

10. One way to confirm that the field stimulation electrodes are working is the presence of bubbles forming at the tip immediately post-stimulation. The higher the stimulation frequency, the more bubbles.
11. In order to make a 2D spatial comparison between SV trajectories from the reserve or recycling pools, it is imperative to precisely align the two cameras before each imaging session. This can be achieved using TetraSpeck™ microspheres (Invitrogen-Thermo Fisher Scientific, 0.1 μm) diluted in PBS (*see* Chapter 6).
12. It is important to note that the reserve pool pulse contains a higher concentration of nanobodies to compensate for the degradation of Syt1-tagged proteins over time (48 h) [34]. In order to make comparisons between the number of labelled reserve pool SVs and recycling pool SVs, one should calculate the ratio of SVs in a presynaptic bouton and its immediately adjacent axonal compartment and compare these ratios for the reserve and recycling SV pools.

Acknowledgments

The Synaptotagmin1-pHluorin plasmid was kindly provided by Prof Volker Haucke (Freie Universität Berlin). Super-resolution imaging was carried out at the Queensland Brain Institute's (QBI's) Advanced Microscopy Facility. We thank the Clem Jones Centre for Ageing Dementia Research (CJCADR) for its support. S.F.L. is supported by the Research Training Program (RTP) Scholarship. S.F.L. performed the culturing of the primary neurons, and the simultaneous dual-color DsdTIM and electric field stimulation experiments. F.A.M. supervised the experiments, was involved in funding acquisition and led the project administration. All authors wrote and edited the manuscript. We thank Dr. Alex McCann and Dr. Rachel Gormal for critically appraising and editing this book chapter.

Conflict of Interest The authors of this chapter declare no conflict of interest.

References

1. Sudhof TC (2004) The synaptic vesicle cycle. Annu Rev Neurosci 27:509–547. https://doi.org/10.1146/annurev.neuro.26.041002.131412
2. Chapman ER (2002) Synaptotagmin: a Ca2+ sensor that triggers exocytosis? Nat Rev Mol Cell Biol 3:498–508. https://doi.org/10.1038/nrm855
3. Yoon TY, Munson M (2018) SNARE complex assembly and disassembly. Curr Biol 28(8): R397–R401. https://doi.org/10.1016/j.cub.2018.01.005
4. Jahn R, Fasshauer D (2012) Molecular machines governing exocytosis of synaptic vesicles. Nature 490:201–207. https://doi.org/10.1038/nature11320
5. Palay SL, Palade GE (1955) The fine structure of neurons. J Biophys Biochem Cytol 1(1): 69–88. https://doi.org/10.1083/jcb.1.1.69
6. Saheki Y, De Camilli P (2012) Synaptic vesicle endocytosis. Cold Spring Harb Perspect Biol 4(9):a005645. https://doi.org/10.1101/cshperspect.a005645
7. Rizzoli SO, Betz WJ (2005) Synaptic vesicle pools. Nat Rev Neurosci 6:57–69. https://doi.org/10.1038/nrn1583
8. Kaeser PS, Regehr WG (2017) The readily releasable pool of synaptic vesicles. Curr Opin Neurobiol 43:63–70. https://doi.org/10.1016/j.conb.2016.12.012
9. Rizzoli SO (2014) Synaptic vesicle recycling: steps and principles. EMBO J 33(8):788–822. https://doi.org/10.1002/embj.201386357
10. Krishnan S, Klingauf J (2023) The readily retrievable pool of synaptic vesicles. Biol Chem 404(5):385–397. https://doi.org/10.1515/hsz-2022-0298
11. Joensuu M, Padmanabhan P, Durisic N et al (2016) Subdiffractional tracking of internalized molecules reveals heterogeneous motion states of synaptic vesicles. J Cell Biol 215(2): 277–292. https://doi.org/10.1083/jcb.201604001
12. Longfield SF, Mollazade M, Wallis TP et al (2023) Tau forms synaptic nano-biomolecular condensates controlling the dynamic clustering of recycling synaptic vesicles. Nat Commun 14(1):7277. https://doi.org/10.1038/s41467-023-43130-4
13. Joensuu M, Martínez-Mármol R, Padmanabhan P et al (2017) Visualizing endocytic recycling and trafficking in live neurons by subdiffractional tracking of internalized molecules. Nat Protoc 12:2590–2622. https://doi.org/10.1038/nprot.2017.116
14. Staras K, Branco T, Burden JJ et al (2010) A vesicle superpool spans multiple presynaptic terminals in hippocampal neurons. Neuron 66(1):37–44. https://doi.org/10.1016/j.neuron.2010.03.020
15. Darcy KJ, Staras K, Collinson LM et al (2006) Constitutive sharing of recycling synaptic vesicles between presynaptic boutons. Nat Neurosci 9(3):315–321. https://doi.org/10.1038/nn1640
16. Longfield SF, Gormal RS, Feller M et al (2024) Synapsin 2a tetramerisation selectively controls the presynaptic nanoscale organisation of reserve synaptic vesicles. Nat Commun 15: 2217. https://doi.org/10.1038/s41467-024-46256-1
17. Siksou L, Triller A, Marty S (2009) An emerging view of presynaptic structure from electron microscopic studies. J Neurochem 108(6):1336–1342. https://doi.org/10.1111/j.1471-4159.2009.05888.x
18. Harris KM, Weinberg RJ (2012) Ultrastructure of synapses in the mammalian brain. Cold Spring Harb Perspect Biol 4(5):a005587. https://doi.org/10.1101/cshperspect.a005587
19. Wilhelm BG, Mandad S, Truckenbrodt S et al (2014) Composition of isolated synaptic boutons reveals the amounts of vesicle trafficking proteins. Science 344(6187):1023–1028. https://doi.org/10.1126/science.1252884
20. Wiedenmann J, Ivanchenko S, Oswald F et al (2004) EosFP, a fluorescent marker protein with UV-inducible green-to-red fluorescence conversion. Proc Natl Acad Sci USA 101(45): 15905–15910. https://doi.org/10.1073/pnas.0403668101
21. Zhang M, Chang H, Zhang Y et al (2012) Rational design of true monomeric and bright photoactivatable fluorescent proteins. Nat Methods 9:727–729. https://doi.org/10.1038/nmeth.2021
22. Chanaday NL, Kavalali ET (2018) Optical detection of three modes of endocytosis at hippocampal synapses. eLife 7:e36097. https://doi.org/10.7554/eLife.36097
23. Giannone G, Hosy E, Sibarita J-B et al (2013) High-content super-resolution imaging of live cell by uPAINT. In: Sousa AA, Kruhlak MJ (eds) Nanoimaging: methods and protocols. Methods in molecular biology. Humana, Totowa, pp 95–110

24. Qu L, Akbergenova Y, Hu Y et al (2009) Synapse-to-synapse variation in mean synaptic vesicle size and its relationship with synaptic morphology and function. J Comp Neurol 514(4):343–352. https://doi.org/10.1002/cne.22007
25. Rienecker KDA, Poston RG, Saha RN (2020) Merits and limitations of studying neuronal depolarization-dependent processes using elevated external potassium. ASN Neuro 12: 1759091420974807. https://doi.org/10.1177/1759091420974807
26. Hodgkin AL, Huxley AF (1952) A quantitative description of membrane current and its application to conduction and excitation in nerve. J Physiol 117(4):500–544. https://doi.org/10.1113/jphysiol.1952.sp004764
27. Ye H, Steiger A (2015) Neuron matters: electric activation of neuronal tissue is dependent on the interaction between the neuron and the electric field. J Neuroeng Rehabil 12:65. https://doi.org/10.1186/s12984-015-0061-1
28. Marszalek P, Liu DS, Tsong TY (1990) Schwan equation and transmembrane potential induced by alternating electric field. Biophys J 58(4):1053–1058. https://doi.org/10.1016/s0006-3495(90)82447-4
29. Bading H, Ginty DD, Greenberg ME (1993) Regulation of gene expression in hippocampal neurons by distinct calcium signaling pathways. Science 260(5105):181–186. https://doi.org/10.1126/science.8097060
30. Grubb MS, Burrone J (2010) Activity-dependent relocation of the axon initial segment fine-tunes neuronal excitability. Nature 465(7301):1070–1074. https://doi.org/10.1038/nature09160
31. Altahini S, Arnoux I, Stroh A (2024) Optogenetics 2.0: challenges and solutions towards a quantitative probing of neural circuits. Biol Chem 405(1):43–54. https://doi.org/10.1515/hsz-2023-0194
32. Martineau M, Guzman RE, Fahlke C et al (2017) VGLUT1 functions as a glutamate/proton exchanger with chloride channel activity in hippocampal glutamatergic synapses. Nat Commun 8:2279. https://doi.org/10.1038/s41467-017-02367-6
33. Chanaday NL, Kavalali ET (2021) Synaptobrevin-2 dependent regulation of single synaptic vesicle endocytosis. Mol Biol Cell 32(19):1818–1823. https://doi.org/10.1091/mbc.E21-04-0213. Epub 2021 Jun 30. PMID: 34191540; PMCID: PMC8684713
34. Truckenbrodt S, Viplav A, Jähne S et al (2018) Newly produced synaptic vesicle proteins are preferentially used in synaptic transmission. EMBO J 37:e98044. https://doi.org/10.15252/embj.201798044

Chapter 10

Single-Molecule Imaging of Endogenous Proteins

Anusha Malapaka, Frédéric A. Meunier, and Rachel S. Gormal

Abstract

Single-molecule imaging is a technique of choice to investigate the dynamic nanoscale organization of proteins of interest, revealing a wealth of new information on how proteins perform their biological function. However, overexpression, which is often used in this context, can alter biological functions. It is therefore useful to implement tools to visualize and track endogenous proteins in living cells and observe their dynamic clustering and motion within their native environment. Herein, we describe two approaches: Fluorescent intrabody Localization Microscopy (FiLM) and gene-editing (CRISPR/Cas9) and provide a workflow including the design of plasmid backbones used to track endogenous proteins by either i) introducing nanobodies that target endogenous proteins to perform single-particle tracking or ii) gene-editing cells in culture to express tagged endogenous proteins at biologically relevant expression levels.

Key words Nanobodies, Endogenous protein tracking, CRISPR, Viral transduction, Plasmid construction, sptPALM

1 Introduction

Intracellular antibodies (intrabodies) are recombinant antibody fragments that are expressed and bind to intracellular target epitopes. Nanobody antibody fragments (aka. Single-domain antibodies), used as intrabodies, have recently been adapted for use in both fixed [1] and living tissues [2–4] for the purpose of performing single-molecule localization microscopy (SMLM). Nanobodies have several advantages over conventional antibodies for super-resolution approaches, particularly their small size, superior stability, and solubility (Fig. 1). Nanobodies can also be directed against endogenous targets in living cells, with a 1:1 protein: target binding ratio and a strong affinity to their target, in a similar affinity range to traditional antibodies (such as immunoglobulin antibodies, e.g. IgGs) [5, 6]. Another advantage of using nanobodies (over IgGs) is that they do not contain an Fc region (when derived from camelids) and therefore have a lower

Frédéric A. Meunier (ed.), *Super-Resolution Microscopy: Methods and Protocols*, Methods in Molecular Biology, vol. 3034, https://doi.org/10.1007/978-1-0716-5268-8_10,

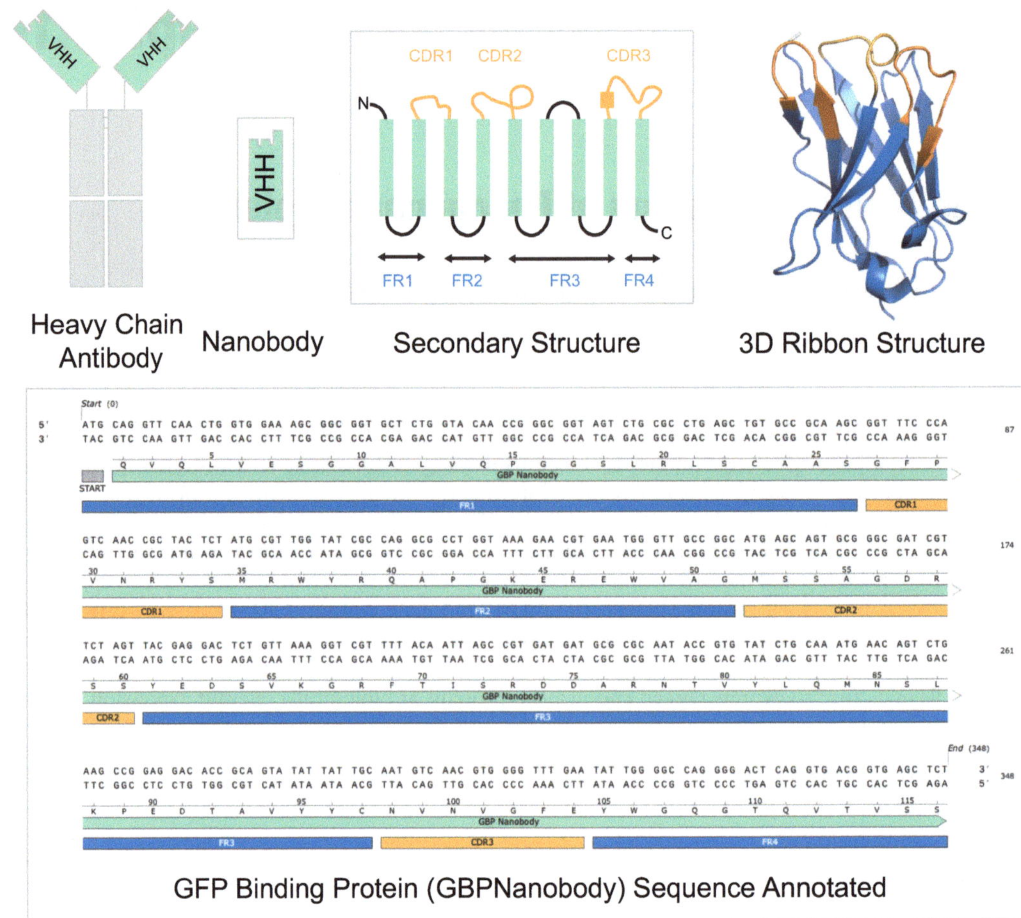

Fig. 1 Nanobodies domain structure. The Camelid Heavy Chain Antibody (HCA) structure is shown. These nanobodies comprise the variable heavy chain domains (VHHs) from HCAs. The secondary structure and 3D ribbon structure of the GBP Nanobody are shown. Binding to GFP is mediated by its CDR1, CDR2 and CDR3 loops (in orange), which are linked by a series of beta-sheet frameworks (FR1–4). The annotated GBP Nanobody sequence is shown (bottom panel)

immunogenic response for both live cell and therapeutic applications [7, 8].

Antibody fragments can be produced using strategies such as phage display, surface display, and ribosome display [9]. Phage display is a molecular biology technique by which bacteriophages are genetically engineered to express viral coat proteins fused to other proteins or peptides of interest, displaying them on their surface. Phage libraries can contain $>10^{11}$ unique species. These surface antigens are then "panned" against an antigen of interest, and several unique clones can be isolated and subsequently validated for their affinity and specificity to their target. By cloning the cDNA corresponding to the nanobody sequence derived from these methods or existing hybridoma clones, one can conveniently create an expression vector to express intrabodies in various model systems. One advantage of using nanobodies over standard expression models is that, where they exist, nanobodies targeting specific conformations of proteins can also be used to study a protein as it performs a discrete biological function. A not-for-profit research repository, Addgene, has recently expanded its nanobody collection and currently has hundreds of nanobody-containing expression plasmids available.

Where no targeting nanobodies exist (nor easily generated), gene-editing approaches are an excellent endogenous labeling alternative. CRISPR, which stands for "clustered regularly interspaced short palindromic repeats," is a gene-editing approach that is simple, quick, and cost-effective and can be applied in almost all animal and cellular models [10]. The CRISPR and CRISPR-associated proteins (Cas) system can selectively target and cut DNA for various applications, such as knock-out, knock-in, base editing, activation, transcriptional repression, or insertion of additional/foreign genetic material. Recent applications have commonly used CRISPR in cell biology to knock down and upregulate protein expression, introducing site-directed mutations (i.e., to mimic disease) and to integrate fluorescent tags (such as GFP) to genes of interest for direct visualization. The CRISPR/Cas9 combination that is often referred to as the "genome engineering powerhouse" is guided by SpCas9 from *S. pyogenes,* a nuclease responsible for generating blunt DNA double-stranded breaks, guided by an RNA molecule (gRNA) to a unique site within the genome. Target recognition of gRNA relies on a protospacer adjacent motif (PAM)—short, conserved sequences (2–5 bp) located adjacent to the target RNA/DNA to be edited. Cas9 is a versatile enzyme that recognizes various PAM sites (3′-NGG). The combination of the CRISPR system with different Cas proteins (Cas3, Cas9, Cas14, etc.) can be applied/adapted to suit various requirements, such as different PAM motif sites, for the alteration of RNA, ssDNA vs dsDNA and/or to alter small vs large stretches of genetic material [11, 12]. Harold D. MacGillavry and his team used CRISPR/Cas9 genome editing technology to develop a system called ORANGE (Open Resource for the Application of Neuronal Genome Editing). ORANGE allows scientists to tag endogenous proteins (i.e., with GFP), making it possible to study these proteins at their native expression levels. Importantly, the ORANGE system enables the creation of knock-in constructs for expression in cultured neurons [13]. Other gene editing approaches have also been used to generate gene-edited model systems, such as zinc-finger nucleases (ZFNs) [14] and transcription activator-like effector nucleases (TALENs) [15]. For this book chapter, the methodologies explained will be limited to the FiLM and CRISPR applications to perform single-particle tracking (spt) of endogenous proteins.

2 Materials

2.1 Cloning Intrabodies (or gRNA) into the Plasmid Backbone

1. Insert Vector/nanobody vector—Phagemid Vector—pOPINE-GFP nanobody (GFP Nanobody is also referred to as GFP Binding Protein (GBP) (or similar nanobody plasmid backbone).
2. Recipient Vector: pmEos3.1-N1 (CMV-MCS-mEos3.1).
3. Restriction enzymes (XhoI, AgeI-HF® and rCutstmart Buffer, New England Biosciences).
4. Plasmid viewing software (Example: SnapGene).
5. PCR primers for Amplification (Custom order—Sigma Aldrich or IDT).
6. Forward 5′-TCAGATCTCGAGCTCATGCAGGTTCAACTG GTG-3′ (Custom).
7. Reverse 5′-GGCGACCGGTGGATCTTTAGAGCTCACCGT CACCTG-3′ (Custom).
8. InFusion®HD Cloning Kit (Clontec).
9. DNAse/RNAse Free Tubes (0.2 thin-walled PCR tubes and 1.5 mL Microtubes).
10. dNTP Mix,
11. Taq Polymerase (and Taq Buffer).
12. Dpnl Enzyme.
13. 6x DNA Loading Dye (Blue)
14. DNA Ladder (Axygen DNA Ladder 100 bp plus).
15. 50x TAE Buffer (2 M Tris, 1 M Acetic Acid and 50 mM EDTA)
16. 1x TAE buffer (20 mL 50x TAE + 980 mL MilliQ H_2O) (1x Conc: 40 mM Tris, 20 mM acetic acid, and 1 mM EDTA)
17. UltraPure Agarose.
18. PCR Cleanup Kit.
19. *E. coli* competent cells (DH5α).
20. 14-ml round-bottomed tube
21. SOC Media.
22. T4 Ligase Enzyme (and T4 Buffer).
23. LB-Agar Plates with Antibiotic (Kanamycin, 50 μg/mL).
24. LB-Broth (Sterile).
25. Antibiotic (Kanamycin, 50 mg/mL).
26. Plasmid DNA Extraction Kit (Mini, Macherey-Nagel).
27. Sanger Sequencing Service (Custom Service, AGRF).

28. Sequencing Primers:
29. mEos-N (Reverse Primer, antisense strand) 5′-TGTACCATC TCCGTCGATCA-3′,
30. CMV For (Forward Primer) 5′-CGCAAATGGGCGGTAGGC GTG-3′.
31. Spectrophotometer (NanoDrop).
32. Water bath.
33. DNA Electrophoresis Tank, Trays and well combs.
34. Blue Box, or UV illuminator (BioRad).
35. Thermal Cycler (PCR).
36. Bacterial Shaking Incubator 37 °C.
37. Bacterial Incubator 37 °C.

2.2 Cloning CRISPR Probe in the Plasmid Backbone

1. Insert Vector: Template for knock-in construct (pmEos3.2-N1), or gDNA block containing target sequences and mEos3.2 knock-in.
2. Backbone Vector: pORANGE template vector (Addgene ID:131471).
3. Restriction enzymes (Bbsl-HF, HindIII-HF, XhoI, and rCutstmart Buffer, New England Biosciences).
4. PCR primers for First target sequence (Custom order—Sigma Aldrich or IDT).
5. Forward 5′CACCGCAAAACGCGCGCTAGGGCGC3′ (Custom).
6. Reverse 5′TTTGCGCGGGATCGCGCGCAAAACG3′ (Custom).
7. PCR primers for Amplification of target sequences and mEos3.2 (Custom order—Sigma Aldrich or IDT).
8. Forward:

5’ATAAAGCTTCAAAACGCGCGCTAGGGCGCGGGAGCCTCGAGGG

TGGAGTGAGCAAGGGCGAGGAG3’

(Custom).

9. Reverse:

5‘ATACTCGAGTCCGCGCCCTAGCGCGCGTTTTG**TTA**CTTGTACAGC

TCGTCCATG3’

(Custom).

10. Annealing Buffer (100 mM Potassium Acetate, 30 mM HEPES-KOH pH 7.4, 2 mM Magnesium Acetate).

11. 6x DNALoading Dye (Blue)
12. DNA Ladder (Axygen 100 bp Plus DNA Ladder).
13. 50x TAE Buffer (2 M Tris, 1 M Acetic Acid, and 50 mM EDTA)
14. 1x TAE buffer (20 mL 50x TAE + 980 mL MilliQ H_2O) (1x Conc: 40 mM Tris, 20 mM acetic acid, and 1 mM EDTA)
15. Ultrapure Agarose.
16. PCR Cleanup Kit.
17. *E. coli* competent cells (DH5α).
18. 14-ml round-bottomed tube
19. T4 Ligase Enzyme and Buffer.
20. SOC Media.
21. LB-Agar Plates with Antibiotic (Ampicillin, 100 μg/mL).
22. LB-Broth (Sterile).
23. Antibiotic (Ampicillin, 100 μg/mL).
24. Plasmid DNA Extraction Kit (Mini, Macherey-Nagel).
25. Sanger Sequencing Service (Custom Service, AGRF).
26. Sequencing Primers:
27. LKO.1 5′ (hU6) (Forward Primer) GACTATCATATGCT TACCGT.
28. CMV Enhancer (Reverse Primer) CGGGCCATTTACCG TAAGTT.
29. Spectrophotometer (NanoDrop).
30. Water bath.
31. DNA Electrophoresis Tank, Trays and well combs.
32. Blue Box, or UV illuminator (BioRad).
33. Thermal Cycler (PCR).
34. Bacterial Shaking Incubator 37 °C.
35. Bacterial Incubator 37 °C.

2.3 Transfection of Intrabody Vectors and CRISPR Knock-in Vectors into Cultured Cells (PC12)

1. Nanobody plasmid backbone or CRISPR knock-in construct (Transfection quality).
2. Pheochromocytoma (PC12) Cells (60% confluent) in 35 mm Nunclon dishes.
3. Lipofectamine LTX and Plus Reagent.
4. Opti-MEM (Reduced Serum Media).
5. Normal Culture Media (DMEM, FBC, HS, GlutaMax).
6. 35 mm Tissue Culture Dishes (Nunclon)
7. Poly-D-Lysine (0.1 mg/mL).

8. Glass-bottom dishes (Cellvis or Mattek).
9. Class II Biosafety Cabinet.
10. Tissue Culture Incubator (37 °C and 5% CO_2).

3 Methods

The procedure described below is an example of endogenous labelling of a protein of interest: by either Subheading 3.1 using nanobodies as intrabodies tagged with a fluorescent protein to perform Fluorescent intrabody Localization Microscopy (FiLM) or Subheading 3.2 by CRISPR/Cas9, to insert a photoconvertible tag (mEos3.1) to perform endogenous single particle tracking Photoactivated Localization Microscopy (sptPALM) (*see* **Note 1**). A pipeline of the protocol is shown in Fig. 2.

3.1 Fluorescently Tagged Nanobody (Intrabody) Expression and Plasmid Construction

1. Identify the nanobody clone to insert into your expression plasmid backbone. Ideal nanobodies should be selective for your target, bind strongly (high affinity, low nM range), and bind to a cytoplasmic epitope of the protein of interest (*see* **Note 2**). Also *see* **Notes 3–5** for more information on the use of nanobodies in mammalian cells.
2. Open the electronic map of the nanobody in SnapGene (or similar sequence visualization software) and identify the open reading frame (ORF). The open reading frame contains the entire conserved nanobody structure. Refer to Fig. 3.
3. Identify any peptides that must be removed for downstream applications. For example, the pelB leader sequence must be removed before many cell biology applications (commonly present in phage display). Cleavage sites, GST, or HISx6 tags used for protein purification may also be unnecessary in downstream applications. Some of the common features of the bacterial expression plasmids have been listed in Table 1.
4. Design a set of PCR amplification primers to add compatible restriction sites at the beginning and end of the Nanobody ORF, for insertion into the multiple cloning site (MCS) of the recipient backbone. The example uses XhoI and AgeI restriction sites (*see* Fig. 3). Use the InFusion®HD Cloning tool and create primers with at least 20 nt overhang to allow for efficient recombination. Standard-desalted primer quality will be suitable for this application. Longer primers may require higher-quality synthesis. When tagging Nanobodies on their C-terminus, you must design appropriate PCR primers to remove the STOP codon (if present) so that the translation will read through the linker and create a fusion protein with the fluorescent protein mEos3.1.

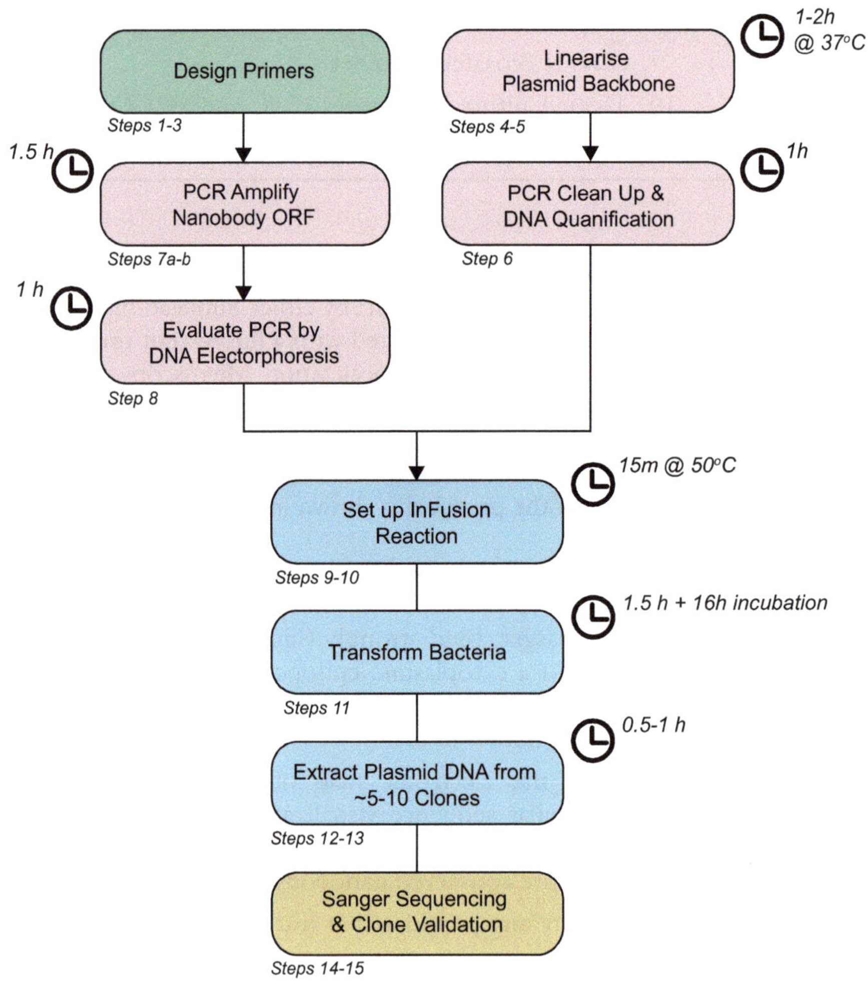

Fig. 2 Procedural Pipeline. The Schematic workflow shows the steps and estimated work time required for each wet lab step. The steps include the (green) Design process, (pink) PCR amplification of Nanobody Insert, vector restriction digestion, (blue) InFusion Reaction, Transformation, DNA extraction and (yellow) clone validation process

5. Set up vector backbone digestion to linearize the backbone of the recipient vector in DNAse/RNAse-free 1.5 mL Tubes as per Table 2.
6. Digest for 1 h at 37 °C in a water bath. You may need to optimize the time and temperature for double digestions by referring to the NEB Double Digest finder (link: https://nebcloner.neb.com/#!/redigest). This resource can also be used to determine whether heat inactivation of the selected enzymes is required and will show the activity of both enzymes in various buffers (*see* **Note 6**).
7. Perform PCR Cleanup of the linearized backbone and measure DNA concentration and quality using a spectrophotometer (NanoDrop).

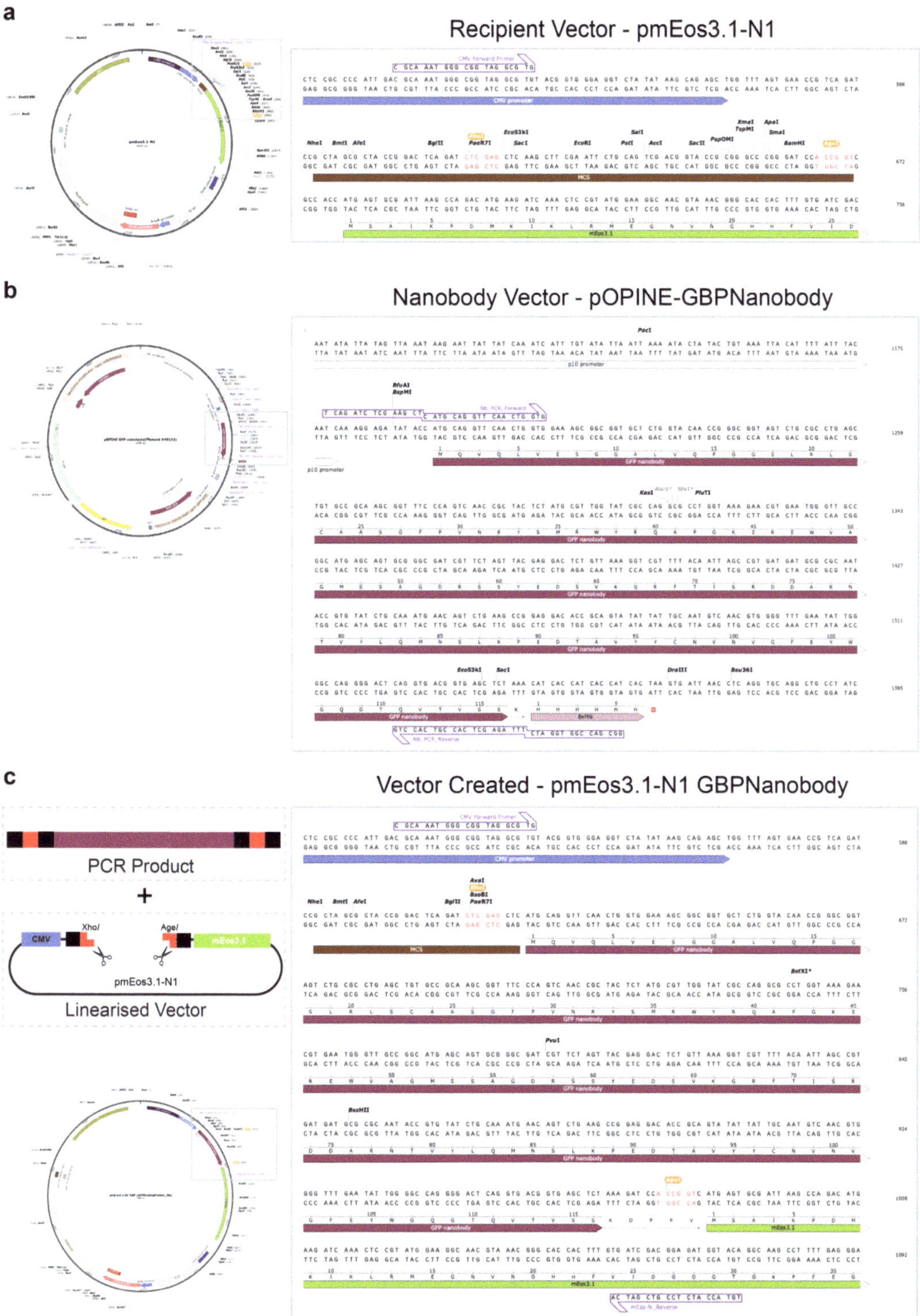

Fig. 3 Plasmid maps and ORF sequences for plasmid construction. (**a**) Full Plasmid map and part of the multiple cloning site (MCS) sequence of the pmEos3.1-N1 (Empty) mammalian expression vector backbone to be used for cloning. Annotated sites include the selected restriction digestion sites (XhoI and AgeI in yellow; red text shows the enzyme recognition site), forward and reverse primers in purple. (**b**) Full Plasmid map and part of the sequence showing the promoter and full open reading frame (ORF) of the GFP binding protein

8. Set up PCR Amplification in 0.2 mL thin-walled PCR tubes (prepare on ice) as per Table 3 (*see* **Note 7**).
9. Set up PCR per cycle per Table 4 on the Thermal Cycler with the heated lid set to 105 °C (*see* **Note 8**).
10. Make a 2% Agarose gel with TAE Buffer (50 mL 1 × TAE buffer +1 g of Agarose) for evaluating the PCR amplification product yield. Run ~10% of the sample with running dye (diluted 1x), alongside a ladder in separate lanes (such as 100 bp Axygen ladder), to evaluate the amplicon size.
11. Examine the agarose gel under UV or blue light to visualize the DNA fragments and confirm the success of the restriction digestion (*see* **Note 9**).
12. Proceed with InFusion Cloning of the remaining PCR product (90%) by adding 2 μL of Cloning enhancer (CE) to 5 μL PCR product.
13. Alternatively, if the PCR yield is low (as confirmed by DNA electrophoresis) perform a PCR cleanup before InFusion® Recombination, which can be used to purify the PCR product and quantify the concentration.
14. Set up the DNA recombination reaction using the InFusion® Kit as per Table 5 and incubate the InFusion®HD reaction for 15 min at 50 °C.
15. Transform 3 μL of the InFusion® Reaction into 50 μL of an appropriate DNA amplifying *E.coli* source (such as NEB5alpha) in a 14 mL Round Bottom tube and incubate on ice for 30 min.
16. Heat-shock the bacteria by dipping the tip of the tube in a water bath (set at 42 °C) for exactly 30 s and return the tube immediately onto ice for 90 s.
17. Add 1 mL of prewarmed SOC Media and incubate at 37 °C with shaking (250 rpm) for 1 h.
18. Spread 100 μL of transformant onto LB-Agar plates with appropriate selection antibiotic (Kanamycin 50 μg/mL) and incubate at 37 °C overnight in an incubator (non-shaking). Do not let the culture exceed 18 h of incubation to prevent the

Fig. 3 (continued) Nanobody (GBPNanobody). The forward and reverse PCR primers designed to amplify the GBP (without a stop codon) and facilitate cloning into the empty vector are annotated in purple. The complete sequence of the GBP is translated and shown in mauve. (**c**) The PCR product and linearized vector generate the new vector, pmEos3.1-N1 GBP Nanobody. The full Plasmid map and part of the plasmid sequence are annotated to show the insertion of the GBP Nanobody into the recipient vector. The GBP is now in frame with mEos3.1, with no stop codon and a small linker of 5 amino acids. Note: Only part of the mEos3.1 sequence is shown with the sequencing primers in purple

Table 1
Common features in bacterial expression plasmids

Feature	Function	Example/protein sequence	Length
Promoter	Region of DNA where transcription of a gene is initiated	Common prokaryotic promoters: T7, Sp6, Lac	Various
pelB leader sequence	Directs the expressed protein into the bacterial periplasm	MKYLLPTAAAGLLLLAAQPAMA	22 amino acids (aa)
Purification "tags"	N- or C-terminal tag used for subsequent protein isolation approaches	6xHis: HHHHHH FLAG: DYKDDDDK Myc: EQKLISEEDL ALFA: SRLEEELRRRLTE GST:MSPILGYWKIKGLVQPTRLLLEY LEEKYEEHLYERDEGDKWRNK KFELGLEFPNLPYYIDGDVKLTQSMAII RYIADKHNMLGGCPKERAEIS MLEGAVLDIRYGVSRIAYSKDFETLKVD FLSKLPEMLKMFEDRLCHK TYLNGDHVTHPDFMLYDALDVVLYMDP MCLDAFPKLVCFKKRIEAIP QIDKYLKSSKYIAWPLQGWQA TFGGGDHPPKSD	Various
Polypeptide linker	A short stretch of amino acids to separate two ORFs. It serves to prevent folding issues of two adjacent proteins or to incorporate cleavage sites between them.	Flexible: GGGS(n) Rigid: EAAAK(n) Thrombin cleavage: LTPRGVRL TEV cleavage: ENLYFQS	Various

Table 2
Restriction digestion

Components	Per reaction
Plasmid backbone (pmEos3.1-N1)	1–4 μg
XhoI	1 unit/μg DNA
AgeI-HF	1 unit/μg DNA
rCutSmart buffer (10x)	1x
DNAse/RNAse-free H_2O	Top up to total
	Total volume = 50 μL

Table 3
PCR amplification of nanobody

Components	Per reaction
Insert backbone (pOPINE-GBP)	50–100 ng
Forward primer	0.2 μM
Reverse primer	125 ng
dNTP mix	200 μM
10x *Taq* buffer	1x
Taq DNA polymerase	1–2 units
DNAse/RNAse free H_2O	Top up to total
	Total volume = 50 μL

Table 4
PCR cycles

	Cycles	Temperature °C	Time
Initial denaturation	1	95 °C	5 min
Denature Anneal Extension	30x	95 °C 60 °C[a] 68 °C	30 s 60 s 60 s per kb
Final extension	1x	68 °C	10 min
HOLD 4–10 °C			

[a]Temperature should be ~5 °C less than Tm. Gradient Cyclers can also be used to determine the best annealing temperature

Table 5
InFusion®HD reaction

Components	Per reaction
CE-PCR product or Purified PCR product	1–4 μL 10–200 ng
Linearized backbone (X ng)	50–200 ng
InFusion enzyme (5x)	1x
DNAse/RNAse-free H_2O	Up to total
	Total volume = 10 μL

generation of satellite colonies and potential mutations (*see* **Note 10**).

19. Select 5–10 individual colonies for amplification using a sterile pipette tip (or small inoculation loop to isolate them) and culture in 2 mL of LB containing the appropriate antibiotic (Kanamycin 50 μg/mL) in 14 mL round-bottom tubes. Incubate overnight (~18 h) at 37 °C and 250 rpm in a shaking incubator.
20. Extract plasmid DNA from each clone, using the Mini Spin Plasmid Extraction Kit (Macherey-Nagel) and quantify the quality and concentration of DNA using a spectrometer.
21. Send each clone with the appropriate sequencing primer for Sanger sequencing to confirm the correct insertion of the nanobody.
22. Assess each clone to confirm the insertion of the nanobody into the correct position. Ensure that the entire reading frame of the nanobody and fluorescent protein is covered to ensure no mutations exist. Ensure that the nanobody is in frame with the fluorescent tag to translate as one fusion protein rather than two independent proteins.

3.2 Designing of Probes—CRISPR/Cas9-Based mEos2 Tagging

This method uses CRISPR/Cas9 to add a photoconvertible tag (Example: mEos3.2) to a protein within its natural environment. This allows for single particle tracking Photoactivated Localization Microscopy (sptPALM) to study the protein's movement and behavior. The technique was developed by Harold D. MacGillavry and colleagues. It uses a single CRISPR/Cas9 knock-in template vector and takes advantage of the cell's natural Non-homologous End Joining (NHEJ) repair pathway [16]. This ensures efficient and accurate insertion of the tag into the protein's DNA sequence. The steps involved in making the ORANGE (Open Resource for the Application of Neuronal Genome Editing) constructs are described in Fig. 4 and are divided into 3 sections: Target sequence design (Part 1), Primer design (Part 2), and Cloning design (Parts 3 and 4).

3.2.1 Target Sequence Design

This protocol outlines the steps to design the pORANGE Munc13a-mEos3.2 knock-in construct.

1. The first step is crucial: Identify the location within the protein sequence where you wish to insert the fluorescent tag (*see* **Note 11**). This ensures the protein's function remains intact, an essential requirement for successful tagging. For Munc13a, tagging the protein on its C-terminal does not affect the protein function.
2. Identify the target sequence within the Munc13a gene, *Unc13a* (*see* Fig. 5).

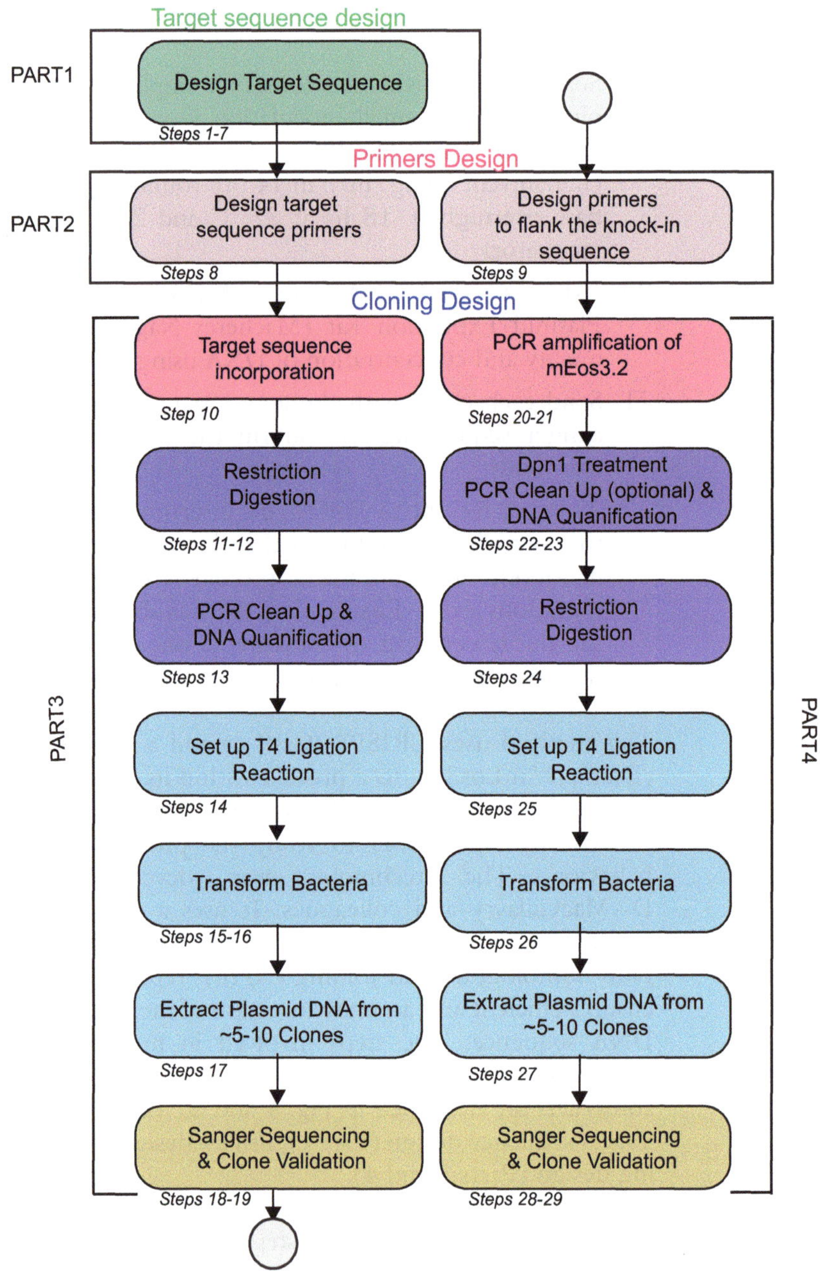

Fig. 4 Pipeline of pORANGE-knock-in design. A schematic workflow showing the steps to incorporating a pORANGE knock-in (KI) design. The main sections are Target Sequence Design (PART 1), Primers Design (PART 2), Cloning Design—Target sequence incorporation (PART 3) and Cloning Design—mEos3.2 tag knock-in (PART 4). The steps involved in the incorporation of the target sequence (PART 3) are: target sequence incorporation, restriction digestion, PCR cleanup and quantification, T4 ligation, Bacterial transformation, DNA extraction and clone validation process. The steps involved in the incorporation of the mEos3.2 tag (PART 4) are: PCR amplification of mEos3.2, PCR cleanup and Dpn1 treatment followed by DNA quantification, restriction digestion, T4 ligation, bacterial transformation, DNA extraction, and clone validation

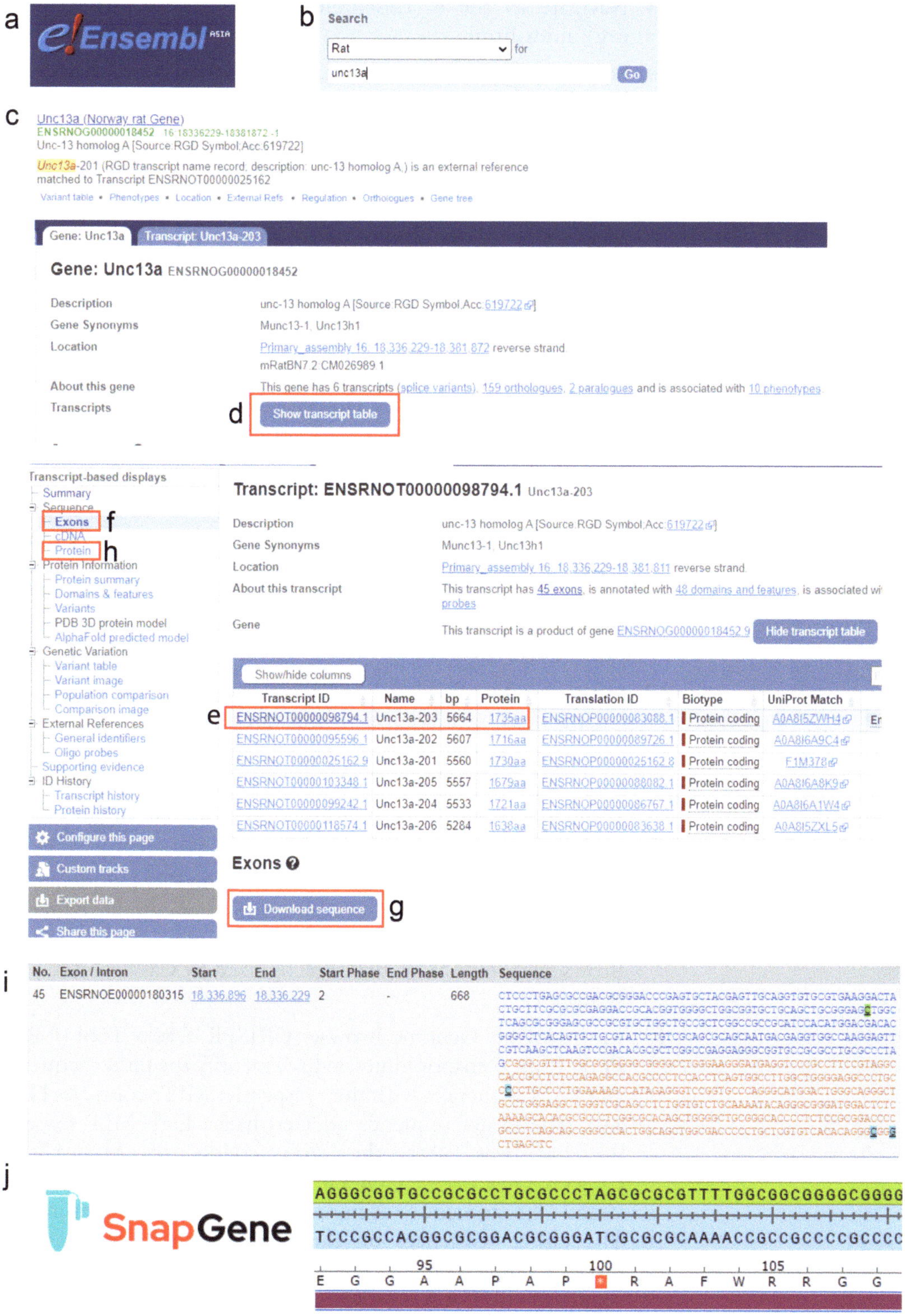

Fig. 5 Guide to identifying the target sequence exon from the Ensemble website. A graphical representation of the steps (**a**–**j**) involved in identifying the exon for the genomic locus of integration (the target sequence) from the Ensemble website

3. Navigate to the e! Ensemble website https://asia.ensembl.org/index.html
4. Select the species you want to knock in from the dropdown box. RAT (Norway Rat).
5. Search for your gene of interest — Unc13a.
6. Select the "Show transcript table" in the Unc13a gene tab.
7. Select the longest protein sequence transcript (ENSRNOT00000098794.1) from the transcript table (*see* **Note 12**).
8. Select "Exons" on the top left selection panel (under Transcript-based displays/Sequence).
9. Download the complete sequence, including exons and introns, by selecting the "Download sequence" option.
10. Download the protein sequence for the transcript by selecting the "Protein" option on the top left selection panel (under Transcript-based displays/Sequence).
11. Identify the start and end of the protein open reading frame, which can be identified in the exon table. The genomic coordinates can help us identify the last exon of the protein. As Munc13a needs to be tagged at the C-terminus, locate the last exon (Exon 45).
12. Open the downloaded FASTA file using SnapGene software (snapgene.com or equivalent software) and locate Exon 45.
13. Copy the coding region sequence in the approximate location you wish to insert the mEos3.2 tag into (for Munc13a, Exon 45 marks the coding region of interest for tagging with mEos3.2).
14. To identify a target sequence with a flanking PAM sequence use a tool that helps design, evaluate, and clone guide sequences for the CRISPR/Cas9 System (for example, UCSC Genome Browser) (the PAM sequence recognized by Cas9 is NGG) (*see* **Note 13**).
15. In the UCSC Genome Browser CRISPR Tracks Tool (CRISPOR, http://crispor.gi.ucsc.edu/) identify the target sequence and PAM sequences with their respective MIT scores (*see* **Note 14**). The target sequence selected had a high MIT score of 98 and was near the stop codon of Munc13a - CCTGCGCCCTAGCGCGCGTTTTG in the template strand (the stop codon is in red; the PAM site is underlined).
16. The target sequences on the template strand need to be reverse-complemented. The reverse complement of the sequence is CAAAACGCGCGCTAGGGCGCAGG in the coding strand (*see* **Note 15**).

17. The target sequence must be inserted into three locations within the pORANGE plasmid backbone. The first is before the Cas9 gRNA scaffold guide, which cleaves the DNA at the desired location. The second and third target sequences flank the knock-in sequence of interest (in this case, the mEos3.2 cDNA) (*see* **Note 16**).

3.2.2 Primer Design

1. Primers must be designed to insert the first target sequence (20 bp) at the Bbsl restriction site before the Cas9 guide in the plasmid backbone pORANGE-cloning template vector, Addgene #131471, with sticky ends of BbsI on each end (*see* **Note 17**).

   5′-CACCGNNNNNNNNNNNNNNNNNN-3′

   3′-CNNNNNNNNNNNNNNNNNNNCAAA-5′

2. The next set of primers must be designed to incorporate the fluorescent tag, flanked by two Cas9 target sites identical to the genomic target site and the PAM sequence. Primers must contain all the following features

(restriction sites (Hind*III* (for), Xho*I* (rev)), Target sequence (reverse complemented), PAM, linker (to correct the frame) and the first few base pairs of the donor tag and overhang)

(*see* Fig. 6).

3.2.3 Cloning Design—Incorporate Initial Target Sequence

1. Anneal the first set of primers by setting up the following reaction (Table 6) and incubating at 95 °C for 4 min, 70 °C for 10 min and then decreasing the temperature to 4 °C (−0.1 °C/min). Hold at 4 °C for a further 10 min. *See* Fig. 7 for an overview of plasmid construction.
2. Set up a vector digestion (Table 7) to linearize the backbone for the first Target insertion.
3. Digest for 1 h at 37 °C in a water bath. Perform heat inactivation (if applicable for the enzyme used). (*see* **Note 6**).
4. Perform PCR Cleanup to remove the enzymes and buffers or perform Gel extraction of the linearized backbone (*see* **Note 9**). Measure DNA concentration and quality using a spectrophotometer (NanoDrop). (*see* **Note 18**).
5. Set up the T4 reaction to ligate the annealed primers (at equimolar concentration) and linearized vector. Mix T4 Reaction per Table 8 and incubate overnight at 4 °C.

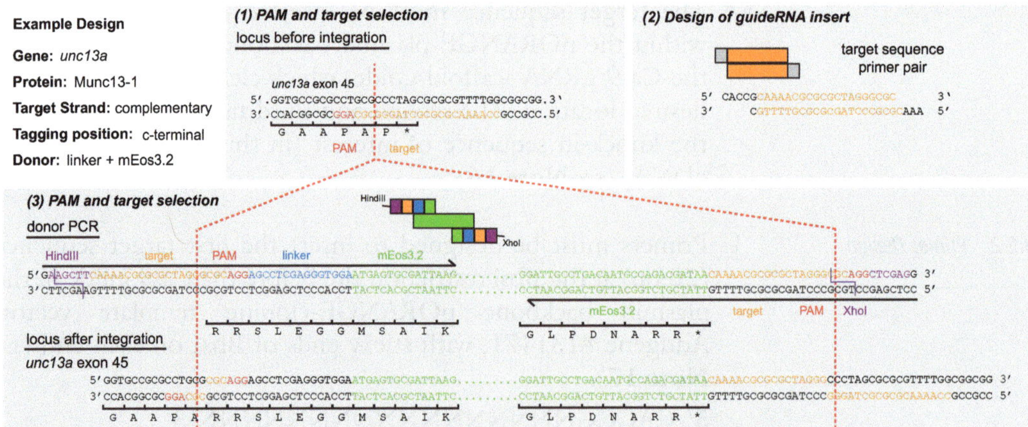

Fig. 6 Guide RNA (gRNA) design for CRISPR/Cas9 Fluorescent Protein insertion. Example of a knock-in construct design for Munc13-1, which has the target sequence on the complementary genomic strand. The target sequence is highlighted in orange, the PAM sequence in red, the linker sequence in blue and the primers used in the amplification of the donor DNA in green. The amino acid sequences are represented under the nucleotide sequence. The asterisk denotes the stop codon. The red dotted line specifies the Cas9 cleavage position and the mEos3.2 tag integration site. The purple solid lines identify the sites of restriction enzymes used for cloning into the pORANGE backbone template vector. Munc13–1, Mammalian uncoordinated-13-1; ORANGE, Open Resource for the Application of Neuronal Genome Editing; PAM, protospacer adjacent motif

Table 6
Primer annealing (first target incorporation)

Components	Per reaction
Forward primer (60 pmol/μL)	1 μL
Reverse primer (60 pmol/μL)	1 μL
Annealing Buffer	48. μL

6. Transform 3 μL of T4 ligation® Reaction into 50 μL of an appropriate DNA amplifying *E.coli* source (such as NEB5alpha) in a 14 mL Round Bottom tube and incubate on ice for 30 min.
7. Heat shock bacteria by dipping the tip of the tube in a water bath (set at 42 °C) for exactly 30 s and return the tube immediately onto ice for 90 s.
8. Add 1 mL of prewarmed SOC Media and incubate at 37 °C with shaking (250 rpm) for 1 h.
9. Spread 100 μL of transformant onto LB-Agar plates with appropriate selection antibiotic (Ampicillin 100 μg/mL) and incubate at 37 °C overnight in an incubator (non-shaking). Do not let the culture exceed 18 h of incubation to prevent the generation of satellite colonies and potential mutations (*see* **Note 10**).

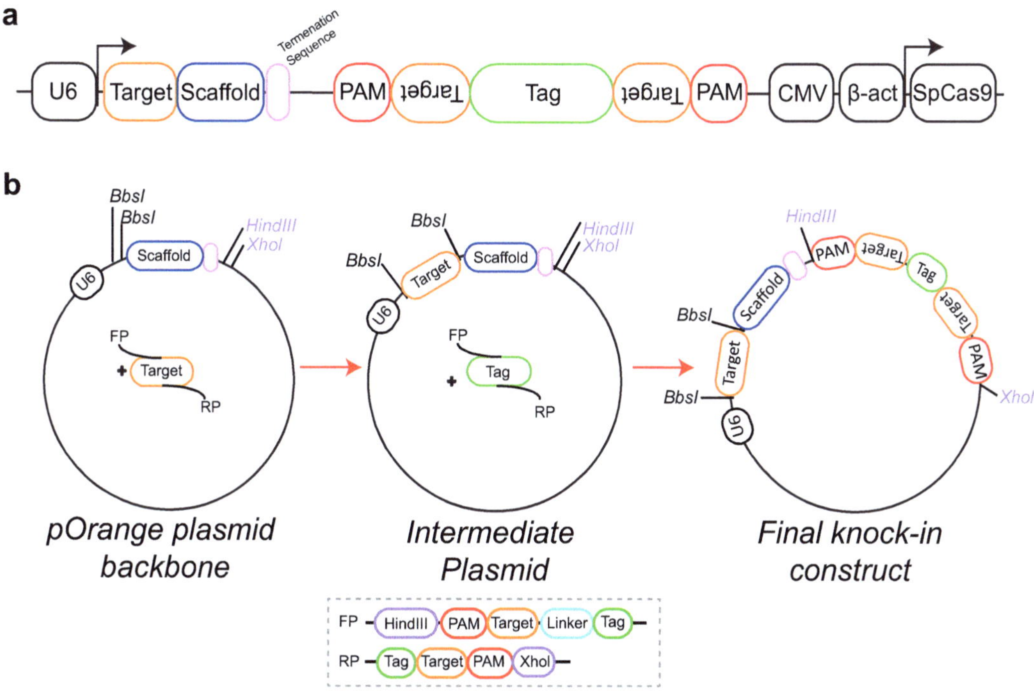

Fig. 7 Plasmid construction diagram. (**a**) An overview of the pORANGE knock-in (KI) construct. (**b**) The first step to clone a knock-in construct is the insertion of a 20 bp target sequence for the genomic locus of the gene of interest into the guide RNA cassette, generating an intermediate construct with the target sequence ligated before the guide RNA. In the second step, the donor sequence containing the tag of interest is generated by PCR and inserted into the donor DNA cassette, generating the final knock-in construct. PAM, protospacer adjacent motif; β-act, β-actin; SpCas9, *streptococcus pyrogenes* Cas9 ORANGE, Open Resource for the Application of Neuronal Genome Editing

Table 7
Restriction digestion (first target incorporation)

Components	Per reaction
Plasmid backbone (pORANGE-cloning template vector)	1–4 μg
BbsI-HF	1 unit/μg DNA
rCutSmart buffer (10x)	1x
DNAse/RNAse-free H_2O	Top up to total
	Total volume = 50 μL

10. Select 5–10 individual colonies for amplification using a sterile pipette tip (or small inoculation loop to isolate them) and culture in 2 mL of LB containing the appropriate antibiotic (Ampicillin 100 μg/mL) in 14 mL round-bottom tubes. Incubate overnight (~18 h) at 37 °C and 250 rpm in a shaking incubator.

Table 8
T4 Ligation

Components	Per reaction
Annealed primers (first target)	~60 fmol
Linearized backbone (BsbI)	~60 fmol
T4 ligation enzyme	0.5–1 μL
T4 enzyme buffer	1x
DNAse/RNAse-free H_2O	Up to total
	Total volume = 10 μL

Table 9
PCR Amplification of mEos3.2

Components	Per reaction
Insert backbone (pmEos3.2-N1)	50 ng
Forward primer	125 ng
Reverse primer	125 ng
dNTP mix	200 μM
10x *Taq* buffer	1x
Taq DNA polymerase	1–2 units
DNAse/RNAse-free H_2O	Top up to total
	Total volume = 50 μL

11. Extract plasmid DNA from each clone, using the Mini Spin Plasmid Extraction Kit (Macherey-Nagel) and quantify the quality and concentration of DNA using a spectrometer.
12. Send each clone with the appropriate sequencing primer (pLKO1.5 for and CMV Enhancer rev) for Sanger sequencing to confirm the correct insertion of the target sequence.
13. Assess each clone to confirm the insertion of the target sequence into the correct position before proceeding with the tag knock-in step.

3.2.4 Cloning Design—Second-Step Target and mEos3.2 Knock-in (KI)

1. After selecting a clone where the target sequence is inserted in the BsbI site in the correct orientation, set up a PCR reaction to amplify mEos3.2 with target and restriction site overhangs in 0.2 mL thin-walled PCR tubes (prepare on ice) as per Table 9. (*see* **Note** 7).

Table 10
PCR cycles (mEos3.2)

	Cycles	Temperature °C	Time
Initial denaturation	1	95 °C	5 min
Denature Anneal Extension	30x	95 °C ~72 °C[a] 68 °C	30 s 60 s 60 s per kb
Final extension	1x	68 °C	10 min
HOLD 4–10 °C			

[a]Temperature should be ~5–10 °C less than Tm. Tm calculators can be used to estimate the ideal Annealing Temperature. Gradient Cyclers can also be used to perform various temperatures simultaneously

Table 11
Restriction digestion vector (for mEos3.2 tag knock-in)

Components	Per reaction
Plasmid backbone (pORANGE first target clone)	1–4 μg
HindIII-HF	1 unit/μg DNA
XhoI	1 unit/μg DNA
rCutSmart buffer (10x)	1x
DNAse/RNAse-free H_2O	Top up to total
	Total volume = 50 μL

2. Set up PCR per cycle as per Table 10 on a Thermal Cycler with the heated lid set to 105 °C.
3. Make a 2% Agarose gel with TAE Buffer (50 mL 1 × TAE buffer +1 g of Agarose) for evaluating the PCR amplification product yield. Run ~10% of the sample with running dye (diluted 1x), alongside a ladder (such as 100 bp Axygen ladder), to evaluate the amplicon size. Ensure the PCR fragment is correct (~750 bp).
4. To increase the positive screening in the subsequent steps, treat cells with 2 μl of DpnI enzyme at 37 °C to digest the template DNA (i.e., methylated DNA from most *E. coli* sources) in the sample. (*see* **Note 19**).
5. Set up two [2] restriction digestion reactions to linearize the backbone (Table 11) and create sticky ends on the PCR fragment (Table 12) (*see* **Note 20**).

Table 12
Restriction digestion PCR fragment (for mEos3.2 tag knock-in)

Components	Per reaction
PCR fragment (target-mEos3.2-target)	1–4 μg
HindIII-HF	1 unit/μg DNA
XhoI	1 unit/μg DNA
rCutSmart buffer (10x)	1x
DNAse/RNAse-free H_2O	Top up to total
	Total volume = 50 μL

Table 13
T4 Ligation

Components	Per reaction
Digested PCR product (mEos3.2-knock-in)	50 ng
Linearized backbone (HindIII/XhoI)	250 ng
T4 ligation enzyme	0.5–1 μL
T4 enzyme buffer	1x
DNAse/RNAse-free H_2O	Up to total
	Total volume = 10 μL

6. Set up a T4 reaction to ligate the digested PCR product and linearized vector.as per Table 13 and incubate overnight at 4 °C.
7. Transform 3 μL of T4 ligation reaction into 50 μL of an appropriate DNA amplifying *E.coli* source (such as NEB5alpha) in a 14 mL Round Bottom tube and incubate on ice for 30 min.
8. Heat shock the bacteria by dipping the tip of the tube in a water bath (set at 42 °C) for exactly 30 s and return the tube immediately onto ice for 90 s.
9. Add 1 mL of prewarmed SOC Media and incubate at 37 °C with shaking (250 rpm) for 1 h.
10. Spread 100 μL of transformant onto LB-Agar plates with appropriate selection antibiotic (Ampicillin 100 μg/mL) and incubate at 37 °C overnight in an incubator (non-shaking). Do not let the culture exceed 18 h of incubation to prevent the generation of satellite colonies introduced with potential mutations (*see* **Note 10**).

11. Select 5–10 individual colonies for amplification using a sterile pipette tip (or small inoculation loop to isolate them) and culture in 2 mL of LB containing the appropriate antibiotic (Ampicillin 100 μg/mL) in 14 mL round-bottom tubes. Incubate overnight (~18 h) at 37 °C and 250 rpm in a shaking incubator.
12. Extract plasmid DNA from each clone, using the Mini Spin Plasmid Extraction Kit (Macherey-Nagel) and quantify the quality and concentration of DNA using a spectrometer.
13. Send each clone with the appropriate sequencing primer (pLKO1.5 for and CMV Enhancer rev) for Sanger sequencing to confirm the correct insertion of the target sequence.
14. Assess each clone to confirm the insertion of the target sequences and mEos3.2 into the correct position.

3.3 Transfection of Nb-mEos3.2 or pORANGE-unc13-mEos3.2-KI Vector into Cultured Cells

1. Passage a flask of PC12 cells onto 35 mm tissue culture Petri dishes so that cells will be ~60% confluent in 24 h (*see* **Note 21**). In a sterile Class II Biological cabinet, culture cells in the required media and incubate at 37 °C with 5% CO_2.
2. Prepare the transfection reaction per Table 14, then incubate at room temperature (RT) for 5 min (*see* **Note 22**).
3. Add 6.75 μL of Lipofectamine LTX to the transfection reaction and incubate for 25 min at RT.
4. Add the transfection mixture dropwise to the cells. Perform a media change no sooner than 6 h post-transfection and leave in culture for 24 h.
5. Prepare 0.1 mg/mL Poly-D-Lysine and incubate on glass-bottom dishes for 30 min at 37 °C. Wash 3 times with sterile MilliQ H_2O (do not allow the glass surface to dry out) and add media to plates ready for cells to be seeded into (keep at 37 °C until needed). (*see* **Note 23**).
6. Flush transfected cells off 35 mm dishes by replacing the media with fresh warm media and triturating with a 1 mL Pipette 10 times. Replating 1x 35 mm into 3–5x glass-bottom dishes is ideal for imaging individual cells 24–48 hrs later. (*see* **Note 24**).

Table 14
Transfection reaction

Components	Per reaction
Opti-MEM (reduced serum media)	500 μL
Plasmid DNA (1–3 μg)[a]	1–3 μg
Plus reagent (1 μL per μg)	1–3 μL

[a]Plasmid DNA concentration should be tested carefully

7. Visualize cells to check for cell health and morphology before proceeding with single-particle tracking.
8. Consider additional validation of new intrabody applications, such as complementary assessment of transfection efficiency, nanobody expression level (western-blotting) and subcellular localization assessment (immunocytochemistry) (*see* **Note 25**).
9. Consider additional validation of CRISPR incorporation, such as qPCR, whole genome sequencing, and/or FACS sorting of CRISPR-edited cells to isolate positive clones for expansion. Additional notes on validation testing are mentioned in Notes.

4 Notes

1. Gene-editing applications in model organisms must only be made after the relevant biosafety and ethical clearances have been obtained.
2. Nanobodies that have not previously been used as intrabodies should be validated to ensure they retain their solubility and targeting ability *in cellulo.*
3. Approaches to remove disulfide bonds from nanobodies have been shown to improve solubility when expressed in the cytosol of mammalian cells [17].
4. Nanobodies can also be modified to be "destabilized" so that when they are not bound to their target, they are degraded via the proteasome (see [18] for further information). This can reduce non-specific background from some intrabody applications.
5. Nanobodies in single-particle tracking applications (e.g., FiLM and uPAINT) that are unbound to their target and moving freely in the cytosol (or extracellular space for uPAINT) are moving extremely fast (estimated ~10–20 $\mu m^2 s^{-1}$ within the cytosol). As a result, they will not be tracked effectively using standard imaging speeds and tracking approaches.
6. High Fidelity (HF) enzymes are preferred as they are quick, 100% active in rCutSmart buffer and do not exhibit any STAR activity (digestion of DNA at positions other than the recognized site).
7. Using a good-quality *Taq* polymerase with proofreading properties will reduce the likelihood of mutations (errors in the PCR product).
8. When amplifying the ORF of intrabodies by PCR, optimizing the annealing temperature may be necessary for efficient amplification. Try using a gradient Thermocycler and a ± 10 °C range around the estimated annealing temperature of your primer pairs.

9. Limit the exposure of DNA fragments to UV light during DNA electrophoresis applications if they will be subsequently purified (e.g., if gel Extraction will be used in place of PCR Cleanup). UV will damage DNA. Use a Blue light box for gel excise, if available.
10. Seal plates and place at 4 °C if not preceded by culture immediately. Plates should be kept at 4 °C for no longer than 1 month.
11. It is essential to consider where a tag can be placed in the protein. For example, for some proteins, tagging on their C-terminus may interfere with their function.
12. If your protein of interest has multiple splice variants, insertion at a particular exon may not tag all splice variants. Factor this into your choice of where to place the tag of choice.
13. The PAM sequence identified before the target sequence can be interchangeable so long as it is in the NGG format (recognized by Cas9), where "N" is interchangeable between A, T, C, and G.
14. MIT score, also known as the Multiplexed insertion or deletion targeting efficiency score, is a score used to evaluate the efficiency of CRISPR-Cas9 genome editing (a score that summarizes the off-targets into a single number between 0 and 100; the higher the number, the fewer off-target effects expected predicting if the guide leads to relatively strong or weak cleavage) [19].
15. For CRISPR target sequences that target the complementary strand, the guide sequence must be reverse complemented in the pORANGE backbone in all three locations (For target sequences on the coding strand, only the target sequences flanking the KI DNA should be reverse complemented).
16. As an alternative, instead of a two-step process to make a knock-in construct, a single-step process could be to obtain a gDNA block and ligate the block into the restriction sites Bbsl and Xhol (Noting that long sequences or those containing secondary structures may not be possible to generate using gene block technology methods).
17. The BbsI recognition site is found twice within the pORANGE template plasmid backbone. BbsI is a Type IIS restriction enzyme that cleaves outside a non-palindromic recognition sequence. Digestion and incorporation of the target sequence as designed (i.e., with sticky ends designed to ligate within the linearized backbone at these sites) will ensure that the correct ligation orientation is within the plasmid backbone.
18. CIP treatment may also be performed on linearized vector. CIP refers to the use of Calf Intestinal Alkaline Phosphates, an enzyme that removes the phosphate groups from the 5′ end DNA and RNA, may increase the insertion efficiency and prevent vector re-linearization.

19. A DpnI treatment may not be necessary if the template DNA used is less than 50 ng.
20. Most PCR buffers are suitable for direct restriction digestion, and enzymes will remain active. If the PCR buffer is incompatible with your enzyme of choice, perform a PCR cleanup before restriction digestion of the PCR fragment.
21. Splitting cells during the log growth phase will ensure healthier cultures with more successful transfection outcomes.
22. The concentration of the plasmid transfected should be tested carefully. Choose a concentration that does not interfere with protein function nor impact cell viability. Modulate expression levels in your chosen model system by carefully selecting the promoter (for strength and cell specificity) or using an inducible system.
23. PC12 cells are loosely adherent on glass surfaces, and subsequent imaging applications benefit from adhesion growth substrate coating.
24. CRISPR incorporation may take longer to incorporate and produce sufficient protein levels for imaging. Strongly adherent cell lines will require trypsinization for removal and replating.
25. Successful transfection of the construct and intrabody expression should be confirmed. The Intrabody expression can be validated by various methods such as (i) assessing the correct localization of the nanobody by immunofluorescence (including negative and positive controls), (ii) by western blotting to assess expression level, and (iii) as well as by co-immunoprecipitation experiments to validate binding to its target.

Acknowledgments

The authors thank the University of Queensland Advanced Imaging Facility for their guidance and support. R.S.G. is supported by an Australian Research Council Discovery Project grant (DP220110334) awarded to F.A.M., F.A.M is supported by an NHMRC Fellowship SRFB (GNT1155794). A.M. is supported by a UQ postgraduate scholarship. This work was also supported by an ARC Large Infrastructure Equipment Facilities Grant (LE130100078) grant as well as funding from The University of Queensland Strategic Intitatives Fund grant no. DVCR22052A awarded to F.A.M.

Conflict of Interest The authors declare no conflict of interest.

References

1. Ries J, Kaplan C, Platonova E et al (2012) A simple, versatile method for GFP-based super-resolution microscopy via nanobodies. Nat Methods 9:582–584. https://doi.org/10.1038/nmeth.1991
2. Gormal RS, Padmanabhan P, Kasula R et al (2020) Modular transient nanoclustering of activated β2-adrenergic receptors revealed by single-molecule tracking of conformation-specific nanobodies. Proc Natl Acad Sci 117: 30476–30487. https://doi.org/10.1073/pnas.2007443117
3. Wagner TR, Rothbauer U (2020) Nanobodies right in the MIDDLE: INTRABODIES as toolbox to visualize and modulate antigens in the living cell. Biomolecules 10:1–19
4. Traenkle B, Rothbauer U (2017) Under the microscope: single-domain antibodies for live-cell imaging and super-resolution microscopy. Front Immunol 8:1030
5. Kubala MH, Kovtun O, Alexandrov K et al (2010) Structural and thermodynamic analysis of the GFP:GFP-nanobody complex. Protein Sci 19:2389–2401. https://doi.org/10.1002/pro.519
6. Asaadi Y, Jouneghani FF, Janani S et al (2021) A comprehensive comparison between camelid nanobodies and single chain variable fragments. Biomark Res 9:87. https://doi.org/10.1186/s40364-021-00332-6
7. Ackaert C, Smiejkowska N, Xavier C et al (2021) Immunogenicity risk profile of Nanobodies. Front Immunol 12:632687
8. Bao G, Tang M, Zhao J et al (2021) Nanobody: a promising toolkit for molecular imaging and disease therapy. EJNMMI Res 11:6. https://doi.org/10.1186/s13550-021-00750-5
9. Pirkalkhoran S, Grabowska WR, Kashkoli HH et al (2023) Bioengineering of antibody fragments : challenges and opportunities. Bioengineering 10(2):1–17
10. Doudna JA, Charpentier E (2014) Genome editing. The new frontier of genome engineering with CRISPR-Cas9. Science 346(6213): 1258096. https://doi.org/10.1126/science.1258096
11. Xu Y, Li Z (2020) CRISPR-Cas systems: overview, innovations and applications in human disease research and gene therapy. Comput Struct Biotechnol J 18:2401–2415. https://doi.org/10.1016/j.csbj.2020.08.031
12. Nidhi S, Anand U, Oleksak P et al (2021) Novel crispr–cas systems: an updated review of the current achievements, applications, and future research perspectives. Int J Mol Sci 22: 1–42. https://doi.org/10.3390/ijms22073327
13. Willems J, de Jong APH, Scheefhals N et al (2020) Orange: a CRISPR/Cas9-based genome editing toolbox for epitope tagging of endogenous proteins in neurons. PLoS Biol 18(4):e3000665. https://doi.org/10.1371/journal.pbio.3000665
14. Urnov FD, Rebar EJ, Holmes MC et al (2010) Genome editing with engineered zinc finger nucleases. Nat Rev Genet 11:636–646. https://doi.org/10.1038/nrg2842
15. Joung JK, Sander JD (2013) TALENs: a widely applicable technology for targeted genome editing. Nat Rev Mol Cell Biol 14:49–55. https://doi.org/10.1038/nrm3486
16. Song B, Yang S, Hwang GH et al (2021) Analysis of nhej-based dna repair after crispr-mediated dna cleavage. Int J Mol Sci 22(12): 6397. https://doi.org/10.3390/ijms22126397
17. Dingus JG, Tang JCY, Amamoto R et al (2022) A general approach for stabilizing nanobodies for intracellular expression. elife 11:e68253. https://doi.org/10.7554/eLife.68253
18. Ariotti N, Rae J, Giles N et al (2018) Ultrastructural localisation of protein interactions using conditionally stable nanobodies. PLoS Biol 16:1–11. https://doi.org/10.1371/journal.pbio.2005473
19. Concordet J-P, Haeussler M (2018) CRISPOR: intuitive guide selection for CRISPR/Cas9 genome editing experiments and screens. Nucleic Acids Res 46(W1):W242–W245. https://doi.org/10.1093/nar/gky354

Chapter 11

Pan-Optical Shadow Imaging of Brain Microanatomy

Yulia Dembitskaya, Kosuke Okuda, Thibault Brugiere, Jordan Girard, Pranay Gaikwad, and U. Valentin Nägerl

Abstract

Understanding the fine-scale organization of brain tissue is critical for elucidating the principles of neural circuit function and dysfunction. The SUper-resolution SHadow Imaging (SUSHI) technique was developed to provide panoramic yet high-resolution visualization of brain microarchitecture by labeling the extracellular space with membrane-impermeant fluorescent dyes, rendering surrounding cellular structures visible in a pan-optical way with negative contrast. Initially established in combination with STED microscopy, SUSHI enables unbiased nanoscale imaging of live brain tissue without relying on cell-type-specific labeling. Recent adaptations have extended the method to in vivo applications using confocal and light-sheet microscopy, broadening its accessibility and utility.

This chapter presents comprehensive protocols for implementing shadow imaging in acute brain slices, organotypic slice cultures, and the intact mouse brain in vivo, including both acute and chronic preparations. We provide detailed instructions on sample preparation, dye delivery, imaging setup, and data acquisition using multiple imaging modalities (2P, STED, confocal). Additionally, we describe shadow imaging in fixed tissue using extracellular matrix labeling. Together, these protocols offer a flexible and scalable approach to resolve the structural interplay between neurons, glia, and extracellular space in diverse experimental models, including those relevant to neurological diseases.

Key words Shadow imaging, Pan-optical tissue imaging, Super-resolution microscopy, Two-photon microscopy, Brain extracellular space, Brain microstructure, In vivo brain imaging, Organotypic and acute brain slices, Microglia imaging

1 Introduction

Understanding the organization of brain tissue at the micro- and nanoscale is essential for unraveling how neural circuits function in health and disease. In particular, the fine-grained spatial relationships between astrocytes, neurons, and the extracellular space form a complex microarchitecture that remains challenging to resolve with conventional imaging techniques [1]. The motivation behind developing the SUSHI (SUper-resolution SHadow Imaging) technique [2] was to create an unbiased and comprehensive method to

Frédéric A. Meunier (ed.), *Super-Resolution Microscopy: Methods and Protocols*, Methods in Molecular Biology, vol. 3034, https://doi.org/10.1007/978-1-0716-5268-8_11,

visualize the extracellular landscape and tissue organization in a panoramic/pan-optical yet high-resolution manner, across multiple experimental contexts.

SUSHI achieves this by fluorescently labeling the extracellular space using a membrane-impermeant dye, thereby rendering cellular structures visible in negative contrast. This straightforward approach allows for seamless visualization of the brain's densely packed microenvironment without relying on cell-type-specific labeling, which can introduce bias or obscure certain components. Combined with live Stimulated Emission Depletion (STED) microscopy, SUSHI offers super-resolution imaging capabilities under physiological conditions.

Initially introduced in Tønnesen et al. [2], SUSHI provided the first live-cell nanoscale images of the extracellular space in the living native state. In Dembitskaya et al. [3], the technique was extended to the intact mouse brain in vivo using cranial windows and was applied with both confocal and light-sheet microscopy. This expanded the accessibility of shadow imaging beyond the STED platform, allowing broader application across imaging modalities. The technique also has proven itself for imaging the fine morphology of cultured neurons, whose fine processes are difficult to image without inflicting photodamage if the fluorescent marker is inside of the cells [4].

The integration of SUSHI with STED microscopy, as detailed in Arizono et al. [5, 6], highlights its particular utility for investigating tripartite synapses at the nanoscale. This protocol outlines how combining super-resolution imaging with extracellular labeling provides a powerful tool for dissecting the fine structure and dynamics of astrocyte–synapse interactions.

This book chapter presents a comprehensive protocol for implementing SUSHI in three major experimental preparations: acute brain slices, organotypic slice cultures, and the intact brain in vivo (*see* Table 1 for comparisons). It includes practical guidance on sample preparation, dye selection and delivery, imaging parameters for optimal acquisition across modalities, and post-processing for quantification. By equipping researchers with a robust methodology for unbiased, super-resolved visualization of brain microarchitecture, SUSHI opens new avenues for studying the dynamic interplay of cellular and extracellular elements in the living brain.

2 Materials

2.1 Buffers and Solutions for Organotypic Brain Slices

1. Millicell Cell Culture Insert (0.40 μm pore size).
2. PTFE Membrane Filter (0.45 μm pore size).
3. Heat-inactivated horse serum: Thaw horse serum and inactivate it at 56 °C for 30 min. Store aliquots at −20 °C.

Table 1
Comparison of different preparations/imaging modalities (pros and cons)

	Live organotypic brain slices	Live acute brain slices	Fixed brain slices	Acute shadow imaging in vivo	Chronic shadow imaging in vivo
Age of the animal	Restricted to P3-P6 (with possible exceptions)	Unrestricted	Unrestricted	Unrestricted (typically >2 months old)	Unrestricted (typically >2 months old)
Label type	Cell-nonpermeable: Calcein, Alexa Fluor (488/514/532/594)	Cell-nonpermeable: Calcein, Alexa Fluor (488/514/532/594)	ECM-binding label—WGA with fluorescent reporter (488/514/532/594)	Cell-nonpermeable: Calcein, Alexa Fluor (488/514/532/594)	Cell-nonpermeable: Calcein, Alexa Fluor (488/514/532/594)
Possibility to study deep brain structures	Yes	Yes	Yes	Possible (requires cortex removal)	Possible (requires cortex removal)
Dynamic processes	Yes	Yes	No	Yes	Yes
Longitudinal studies (> 1 day)	No	No	No	No	Yes
Facility of applying pharmacological agents	High	High	Limited	Limited	Limited
Optical/electrical stimulation	Yes	Yes	No	Limited	Limited
"Natural" stimulation	No	No	No	Yes	Yes
Disease models	Limited	Yes	Yes	Yes	Yes
Preparation complexity	Low	Average	Low	High	Very high

4. Glucose-solution: Dissolve 55 g D(+)-Glucose* H_2O in 50 mL H_2O. Sterile filter and store in 500 μL aliquots at −20 °C.
5. Kynurenic acid: Dissolve 946 mg Kynurenic acid in 5 mL of 1 M NaOH solution, and add 45 mL H_2O. Sterile filter and store aliquots at −20 ° C.
6. Culture medium: 100 mL Basal Medium Eagle (BME) without Glutamine, 50 mL Hank's Balanced Salt Solution (HBSS), 50 mL heat-inactivated Horse serum, 1 mL of 200 mM Glutamine, 2 mL Glucose-solution. Sterile filter and store at 4 °C.
7. Microglial-supportive growth medium: 10 mL BME, 7.3 mL HBSS, 2 mL heat-inactivated Horse Serum, 400 μL B27+, 200 μL Glucose-solution, 100 μL of 200 mM Glutamine. Sterile filter and store at 4 °C.
8. Gey's Balanced Salt Solution (GBSS): 0.220 g $CaCl_2$ * 2 H_2O, 0.740 g KCl, 0.030 g KH_2PO_4 0.210 g $MgCl_2$ * 6 H_2O, 0.070 g $MgSO_4$ * 7 H_2O, 8.000 g NaCl, 0.227 g $NaHCO_3$ 0.120 g Na_2HPO_4 1.100 g D(+)-Glucose * H_2O. Dissolve in 1000 mL H_2O. Sterile filter and store at 4 °C.
9. Dissection solution: 98 mL GBSS, 1 mL Kynurenic acid, 1 mL Glucose-solution. Adjust pH with 1 M HCl to 7.2. Sterile filter and store at 4 °C. The dissection solution must be freshly prepared on the day of the dissection.
10. HEPES-based Artificial cerebrospinal fluid (ACSF). 125 mM NaCl, 2.5 mM KCl, 1.25 mM NaH_2PO_4, 1 mM $MgCl_2$, 2 mM $CaCl_2$, 30 mM D-glucose, 10 mM HEPES.

2.2 Buffers and Solutions for Acute Brain Slices

1. $NaHCO_3$-based ACSF (in mM): 125 NaCl, 2.5 KCl, 11 glucose, 25 $NaHCO_3$, 1.25 NaH_2PO_4, 2 $CaCl_2$, 1 $MgCl_2$, at 295–300 mOsm, pH 7.4.
2. NMDG solution (in mM): 1.25 NaH_2PO_4-H_2O, 2.5 KCl, 7 $MgCl_2$-H_2O, 20 HEPES, 0.5 $CaCl_2$- 2 H_2O,28 $NaHCO_3$,8 D-Glucose, 5 L (+)-Ascorbate, 3 Na-Pyruvate, 2 Thiourea, 93 N-methyl-D-glucamine (NMDG), at pH 7.4, 290–310 mOsm. Solutions must be freshly prepared on the day of the dissection.

2.3 Consumables for Acute Brain Slices

1. Whatman filter paper disk.
2. Recovery chambers.
3. Dissection tools: Scalpel, razor blades, fine scissors, forceps (#5 or #7 forceps (fine-tip)), broad-tip forceps, spatulas/spoons.
4. Perfusion tubes.
5. Imaging chamber.
6. Platinum Harp.

Table 2
Fluorescent dye preparation and working concentrations

Type of preparation	Fluorescent dyes	Stock solution (mM)	Solvent for stock solution	Final solution (μM)
Organotypic brain slices	Calcein	2	ACSF	200
	Alexa 488/532 Fluor carboxylic acid	20	H_2O	10–200
	Atto-488/514 carboxy	20	H_2O	10–200
Acute brain slices	Calcein	20	ACSF or H_2O	200–400
	Alexa 488/532 Fluor carboxylic acid	20	ACSF or H_2O	50–200
	Atto-488 carboxy	20	ACSF or H_2O	50–200
Brain in vivo	Alexa 488/532 Fluor carboxylic acid	50	ACSF (Mg^{2+}, Ca^{2+} – free)	≈400–800
Fixed brain slices	WGA-CF488	1 mg/mL	PBS	5 μg/mL

2.4 Fluorescent Dyes for Shadow Imaging

The solvent used to prepare the dye solution and the final concentration of the dye solution depend on the preparation type (*see* Table 2).

1. Calcein (Dojindo Molecular Technologies, # C001).
2. Alexa 488 Fluor Carboxylic Acid, tris(triethylammonium) salts (Thermo Scientific, #A33077).
3. Alexa 532 Fluor Carboxylic Acid, tris(triethylammonium) salts (Thermo Scientific, #A33078).
4. Atto-488 carboxy (Atto-tech, #AD 488-25).
5. WGA-CF488A (Wheat Germ Agglutinin, CF®488A Conjugate (Biotium, #29022-1).
6. Dilution buffer: Mg^{2+}, Ca^{2+}—free ACSF (in mM): 125 NaCl, 2.5 KCl, 25 $NaHCO_3$, 1.25 H_2PO_4, at 280–290 mOsm, pH 7.4. Alexa Fluor™ 488 dye (50 mM solution) in Mg^{2+}, Ca^{2+}-free ACSF.
7. PBS.
8. H_2O.

2.5 Materials for Surgery, Intracerebroventricular Dye Injection, and Imaging In Vivo

2.5.1 Anesthesia and Pain Management

1. Isoflurane vaporizer with an anesthetic induction chamber and a nose mask.
2. Oxygen or air delivery system with flow regulators.
3. Isoflurane.
4. Buprenorphine (0.1 mg/kg).
5. Ketamine/xylazine (100/10 mg/kg).
6. Sterile 0.8 M NaCl solution (for dilution of buprenorphine, ketamine, and xylazine).

2.5.2 Surgical Supplies

1. 70% ethanol or 4% Betadine solution (for disinfection)
2. Medical-grade glue (e.g., Superbond C&B).
3. Dental cement (e.g., Superbond C&B).
4. Glass coverslip (#1 thickness, 4 mm diameter).

2.5.3 Dye Injection Supplies

1. Hamilton syringe (10 μL capacity) with a 34 G needle.
2. Motorized syringe pump with a compatible holder for the Hamilton syringe.

2.5.4 Surgical Setup

1. Stereotaxic frame with ear bars.
2. Heated surgical blanket or heating pad with adjustable temperature control.
3. High-speed surgical drill with appropriate burrs (for craniotomy).

2.5.5 Surgical Instruments and Consumables

1. Shaving razor or hair removal cream.
2. Sterile gauze pads.
3. Sterile gloves.
4. Fine forceps.
5. Scissors.
6. Scalpel and blades.
7. Microspatula or retractor.
8. Absorbent pads or bench covers (to maintain a clean surgical area).

2.5.6 Additional Materials for Chronic Shadow Imaging

1. PDMS (Polydimethylsiloxane) membrane (Sterne Elastomere, 250 μm thickness (custom-made)). Prepare adapted round pieces of ≈2 mm diameter.

2.6 Materials and Solutions for Preparation of Fixed Samples

1. Fixative: 4% of paraformaldehyde (PFA) in 30% solution of sucrose.
2. Phosphate-buffered saline (PBS) 10X stock solution containing in mM: 137 NaCl, 2.7 KCl, 10 Na_2HPO_4, and 1.8 KH_2PO_4. Dilute 10 times for 1X working concentration.
3. Permeabilizing solution: 5% normal goat serum and 0.5% Triton X-100 in 1X PBS.
4. Glass slides.
5. Coverslip.
6. CitiFluor™ CFM-3 Mounting Medium (E17979–30).
7. Immersion media (adapted for objective type).

2.7 Equipment for Slice Preparation

2.7.1 For Organotypic Slice Preparation

1. Chopper (Mcllwain tissue chopper).
2. Binocular (Zeiss Stemi 508).
3. Incubator (Heracell).

2.7.2 For Acute Slice Preparation

1. Water bath (Grant instruments JB Aqua 5 Plus).
2. Vibratome (Leica VT1200S).
3. Peristaltic pump (Gilson minipuls 3).
4. Perfusion tubes (Tygon).

2.7.3 For Fixed Slice Preparation

1. Vibratome (Leica VT1200S).

2.8 Equipment for Shadow Imaging

2.8.1 For Two-Photon Shadow Imaging in Organotypic and Acute Slices

1. Two-photon microscope (Prairie Technologies).
2. Temperature controller.
3. 40X water-immersion objective with an NA of 1.0 (Plan-Apochromat, Zeiss).

2.8.2 For Shadow Imaging in Intact Brain In Vivo

1. 2 Photon-STED Microscope (custom-built)
2. Upright research microscope (e.g., BX51WI, Olympus) equipped with epifluorescence light source—CoolLED, Femtosecond mode-locked fiber laser (e.g., Alcor 920, Spark Lasers) with 920 nm wavelength, 700 fs pulse width at 80 MHz repetition rate.
3. Pockels cell (e.g., 302 RM, Conoptics) for power modulation.
4. Spatial Light Modulator (SLM, e.g., Aberrior Instruments) for aberration correction.
5. Telecentric scanner (e.g., Yanus IV, TILL Photonics).
6. Lens combinations for beam conjugation and alignment. Appropriate dichroic mirror and filters for laser/signal separation.
7. 60x silicone oil immersion objective lens (e.g., UPlanSAPO, NA 1.3, Olympus)
8. Piezo-actuator for z-axis focusing (e.g., Pifoc 725.2CD, Physik Instrumente).
9. Avalanche photodiode (e.g., SPCM-AQRH-14-FC, Excelitas).
10. Data Acquisition Card—PCIe-6259 (National Instruments) or equivalent.
11. Software for imaging control and data acquisition (e.g., Imspector scanning software, Abberior Instruments).

12. Heating pad or temperature control unit for maintaining the animal's body temperature.
13. Stage for animal stabilization.

2.8.3 For Fixed Slices

1. Confocal STED microscope.
2. 2-Photon or STED microscope as described in the section above.

3 Methods

3.1 Organotypic Slice Culture Preparation

The experimental workflow is presented in Fig. 1a.

3.1.1 Plate Preparation (2 h Before Dissection)

1. Prepare culture medium: add 1 mL of culture medium into each well of a 6-well plate.
2. Place cell culture inserts into each well. Add sterilized membranes: place 3–4 sterilized PTFE membranes (cut into small squares) onto each insert.
3. Place the prepared plate in the incubator at 35 °C to equilibrate for at least 2 h before use.

3.1.2 Brain Dissection

1. Brain extraction: carefully extract the brain and place it into a cold dissection solution at 4 °C for 2–3 min to cool it down. Use WT animals for single-channel recording, or transgenic CX3CR1-EGFP mice for two-channel recordings.
2. Hippocampus dissection: using a binocular microscope, dissect the hippocampus and transfer it to a tissue chopper.
3. Slice preparation: cut the hippocampus into slices of 350 μm thickness using the chopper.
4. Slice Incubation: transfer the slices into a fresh dish containing cold dissection solution and incubate them at 4 °C for 30–60 min.
5. Transfer to membranes: place the hippocampal slices onto the PTFE membranes in the prepared plate. Place the plate back into the incubator at 35 °C to maintain optimal culture conditions.

3.1.3 Maintenance of Slices

1. Culture medium exchange: change the culture medium in the plate every 3–4 days to ensure the health and viability of the slices.

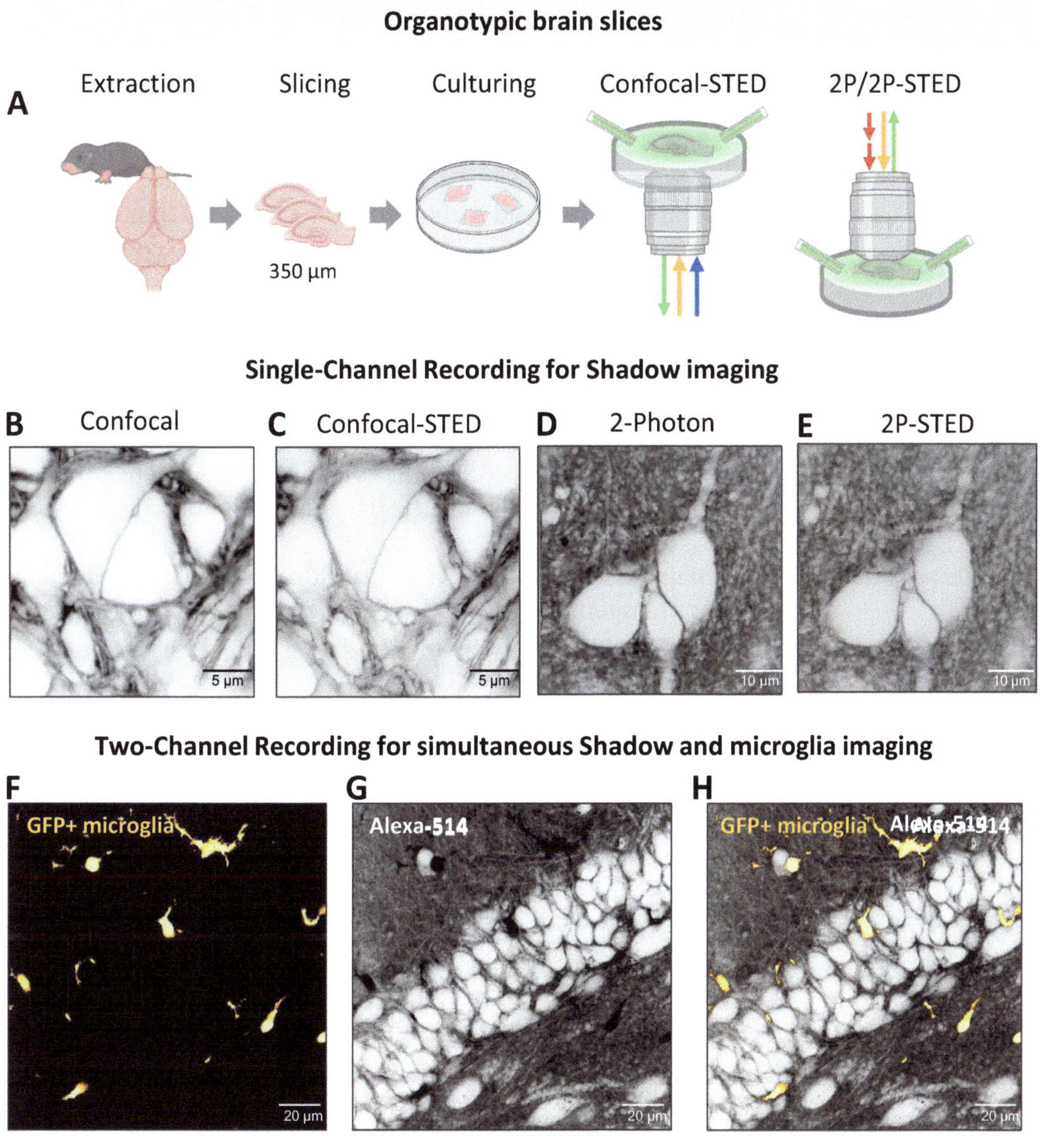

Fig. 1 Shadow imaging in organotypic hippocampal slice cultures
(**a**) Schematic overview of organotypic slice preparation, including brain extraction, slicing into 350 μm sections, culturing, and imaging in different modalities: confocal/confocal-STED, two-photon (2P/2P-STED)
(**b**–**e**) Single-channel shadow imaging of the extracellular space using Alexa Fluor 514 in organotypic hippocampal slices, imaged with: confocal (**b**), confocal-STED (**c**), two-photon (2P) (**d**), and 2P-STED (**e**) microscopy
(**f**–**h**) Dual-channel shadow imaging showing GFP-positive microglia (**f**), extracellular space labeled with Alexa Fluor 514 (**g**), and merged imaging (**h**), enabling visualization of microglia in the context of surrounding cellular structures

3.2 Acute Slices Preparation

The experimental workflow is presented in Fig. 2a.

3.2.1 Preparation for Brain Slicing

1. Cool 200 mL of NMDG solution in the freezer, monitoring closely to ensure it begins to chill but does not freeze completely. Once it reaches a slushy consistency, transfer the solution on ice and carbonate it for at least 10 min.
2. Prepare a recovery chamber with 100 mL of NMDG solution in a 250 mL beaker. Carbonate the solution continuously at 34 °C.
3. Prepare another recovery chamber with 100 mL of $NaHCO_3$--based ACSF in a separate 250 mL beaker. Carbonate the solution continuously at room temperature.
4. Vibratome Setup. Set the vibratome parameters: Speed: 0.12 mm/s, Amplitude: 2, Slice thickness: 300–350 μm.

3.2.2 Dissection and Slicing

1. Brain dissection. Use WT animals for single-channel recording, or transgenic CX3CR1-EGFP mice for two-channel recordings. Perform cervical dislocation and extract the brain. Immediately place the brain into an ice-cold NMDG solution to cool. (*see* **Note 1**).
2. Slicing of the brain. Remove the excess solution with soft tissue and glue the brain to the vibratome sample holder. Place the brain and sample holder into the vibratome chamber. Make 300–350 μm-thick slices and collect them. (*see* **Note 2**).

3.2.3 Slice Recovery

1. Initial Recovery. Transfer the slices into the NMDG solution in the recovery chamber at 34 °C and incubate for 10 min.
2. Final Recovery. Transfer the slices into the $NaHCO_3$-based ACSF and incubate for 40 min at room temperature. Ensure the recovery chambers are continuously carbonated throughout the process.

3.3 Single-Channel TUSHI Imaging and Dual-Channel TUSHI/Microglia Imaging in Organotypic and Acute Brain Slices

3.3.1 Preparation of Microscope Setup

1. Preheat Stage Heater or Incubator. Set the microscope stage heater to 32 °C. Begin preheating at least 30 min before the experiment to ensure stable conditions during imaging.
2. (For acute brain slices) Set up the perfusion system with $NaHCO_3$-based ACSF at a flow rate of 2 mL/min. Ensure that the system provides a constant and stable inflow and outflow of solution throughout the recording. (*see* **Note 3**)

3.3.2 Organotypic Brain Slices

1. Place the organotypic brain slice in a dry chamber.
2. Add 2–3 mL of 10–200 mM fluorescent dye (e.g., calcein, Alexa-488 or Atto514) dissolved in HEPES-based ACSF into the chamber.

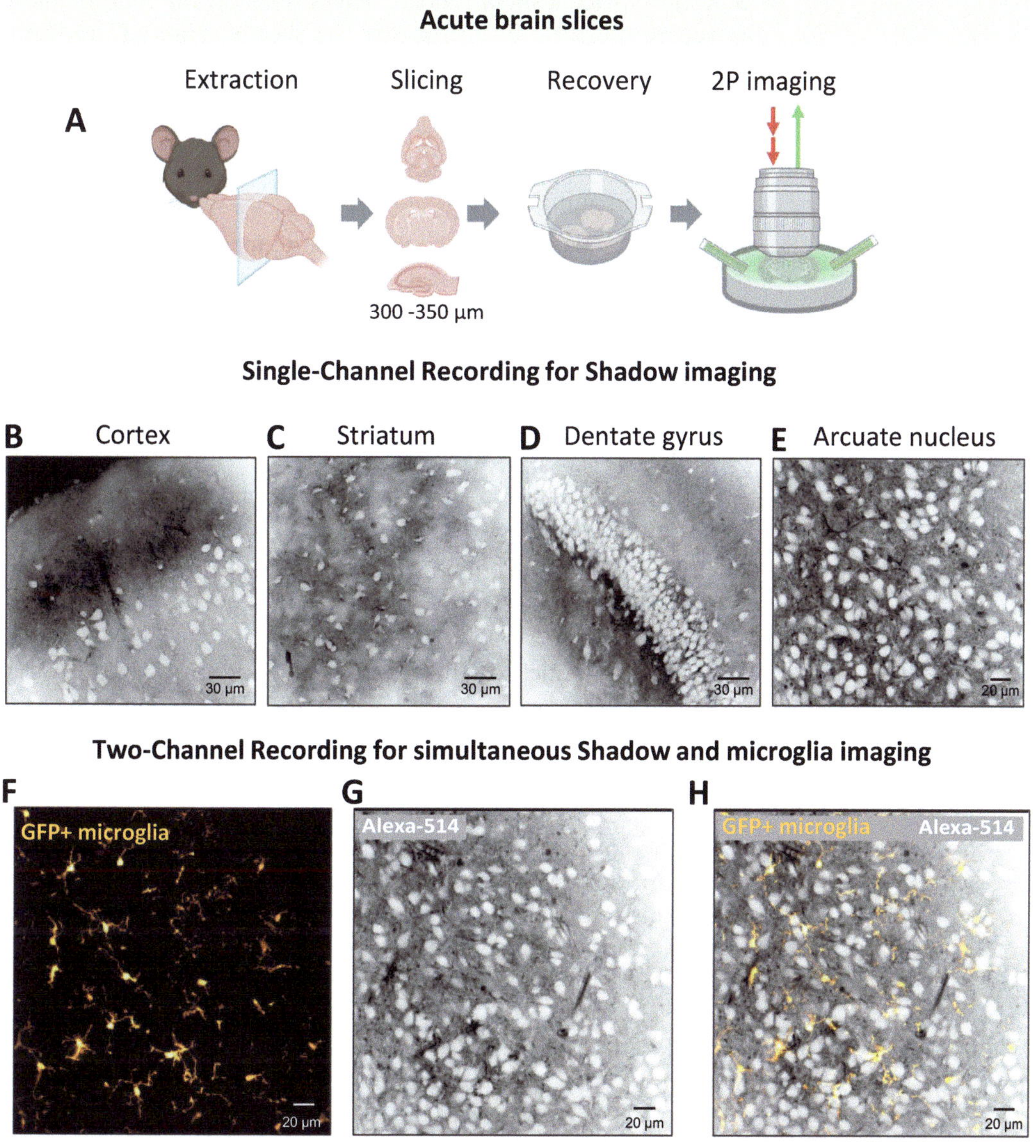

Fig. 2 Two-photon shadow imaging in acute brain slices
(**a**) Schematic overview of acute brain slice preparation, including brain extraction, slicing into 300–350 μm sections, recovery, and subsequent 2P shadow imaging
(**b**–**e**) Single-channel 2P shadow imaging of the extracellular space using Alexa Fluor 514 in acute brain slices from different brain regions: cortex (**b**), striatum (**c**), dentate gyrus (**d**), and arcuate nucleus (**e**)
(**f**–**h**) Dual-channel 2P shadow imaging showing GFP-positive microglia (**f**), Alexa Fluor 514-labeled extracellular space (**g**), and merged image (**h**), enabling visualization of microglia within the surrounding cellular structures and extracellular environment in arcuate nucleus (maximum projection of a 20 μm-thick Z-stack)

3. Gently place a small metallic flat washer on the membrane to secure the slice, ensuring that the slice is centered inside the washer.

3.3.3 Acute Brain Slices

1. Place the acute brain slice in the perfusion chamber and secure the slice by placing a harp on top of it to prevent movement during imaging.
2. Use a binocular microscope or a large magnification objective under visible light to locate the general area of interest. Use a camera to refine the selection and focus on the precise area of interest.
3. Add 50–200 mM fluorescent dye (e.g., calcein, Alexa-488 or Alexa-532) to the chamber to achieve a final concentration of 50 μM.
4. Allow the dye to diffuse and stain the tissue for optimal imaging contrast. (*see* **Note 4**).
5. Proceed to Subheading 3.3.5 for two-channel recording using two-photon imaging.

3.3.4 Confocal/Confocal-STED/Two-Photon (2P)/2P-STED Imaging (For Single-Channel Recording)

1. Use one channel to visualize a single fluorophore (e.g., Alexa-488 for shadow imaging). Adjust the detector and emission filter to maximize the signal of the target fluorophore in confocal/confocal-STED or 2P/2P-STED mode.
2. Adjust scanning speed and resolution to balance image quality and acquisition time. Set the pixel size to approximately half of the maximum achievable resolution, in accordance with the Nyquist theorem for proper sampling: For confocal: ~ 120 nm, confocal-STED ~50 nm, 2P: ~ 150 nm. For 2P-STED: ~75 nm.
3. Use software to record data and ensure proper image acquisition for subsequent analysis. (*see* **Notes 5** and **6**). Refer to representative images of hippocampal tissue in organotypic brain slices obtained in confocal/confocal-STED or 2P/2P-STED mode (Fig. 1b–e). Refer to representative images of hippocampal tissue in acute brain slices obtained in 2P mode in cortex, striatum, dentate gyrus and arcuate nucleus (Fig. 2b–e).

3.3.5 Two-Photon (2P) Imaging (For Two-Channel Recording)

1. Configure Channel 1 for GFP and Channel 2 for Alexa-532 to enable simultaneous shadow imaging and microglia visualization. Ensure sufficient spectral separation by optimizing emission filters and detectors for both channels. (*see* **Note 7**).
2. Adjust scanning speed and resolution to balance image quality and acquisition time. Set the pixel size to approximately half of the maximum achievable resolution, in accordance with the Nyquist theorem for proper sampling: For 2P: ~ 150 nm.

3. For high-contrast, multichannel imaging, verify that emission spectra are well-separated and avoid cross-talk between channels.
4. Use software to record data and ensure proper image acquisition for subsequent analysis. (*see* **Note 5** and **6**). Refer to representative images of hippocampal tissue and GFP-positive microglia in organotypic brain slices obtained in 2P mode (Fig. 1f–h). Refer to representative images of hippocampal tissue and GFP-positive microglia in acute brain slices obtained in 2P mode in arcuate nucleus (Fig. 2f–h).

3.4 In Vivo Labeling of ECS

The protocol described in this chapter was applied to adult (>2 months old) C57BL/6 J mice and can be adapted to different mouse strains, ages or species. The experimental workflow is presented in Fig. 3a.

3.4.1 Surgical Procedures for Acute Shadow Imaging In Vivo

1. Administer buprenorphine (0.1 mg/kg) diluted in sterile NaCl (0.8 M) solution intraperitoneally for pain relief approximately 30 min before surgery.
2. Prepare isoflurane anesthesia equipment, including mask for the nose of the mouse. Maintain the anesthesia levels at 1.5–2.5% during surgery while constantly monitoring proper anesthetic effect—normal respiration, absence of pinch reflexes, lack of gasping or irregular breathing.
3. Place the mouse on a heated blanket to maintain body temperature in a stereotaxic frame and ensure stability and symmetric head fixation. Avoid insufficient or excessive heating. (*see* **Note 8**).
4. Shave and disinfect the surgical area using ethanol (70%) or Betadine solution (4%).
5. Carefully remove a sufficient amount of skin to expose the area above the somatosensory cortex. Avoid removing excessive skin or damaging posterior neck muscles.
6. Prepare a 50 mM solution of Alexa Fluor™ 488 dye. Load 5–8 μL of the dye into a Hamilton syringe fitted with a 34G needle. Secure the syringe in a motorized syringe pump holder, ensuring there are no air bubbles in the solution.
7. Make a ~ 4 mm circular craniotomy over the somatosensory cortex while keeping the dura mater intact. Use a high-speed drill with care to avoid damaging underlying tissues.
8. Locate the lateral ventricle using the following coordinates relative to bregma: Medial/Lateral (M/L): + or − 1.1 mm; Anterior/Posterior (A/P): −0.52 mm; Dorsal/Ventral (D/V): −2.3 mm. Insert the injection needle gently to avoid additional damage. (*see* **Note 9**).

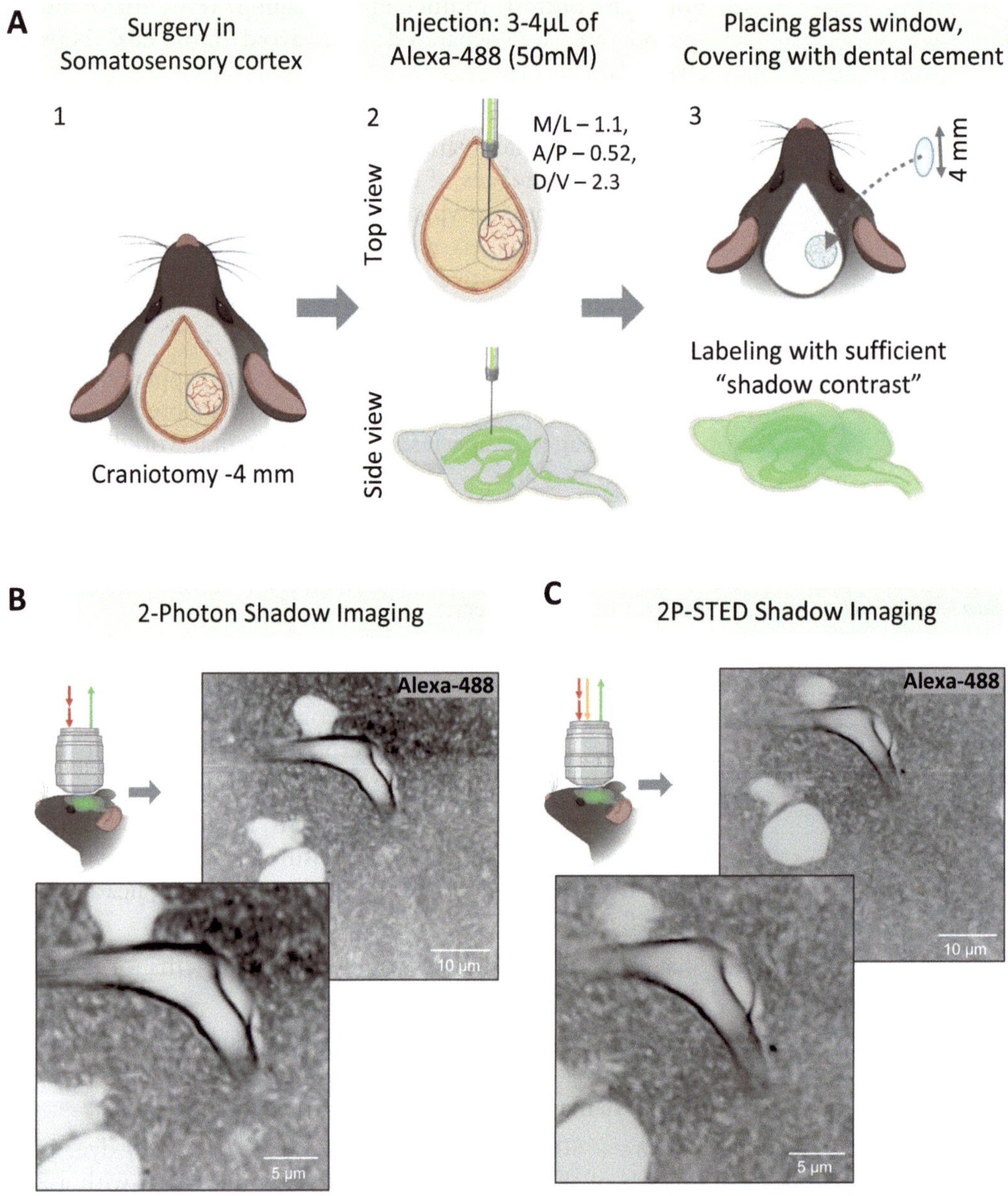

Fig. 3 Acute in vivo shadow imaging in the mouse brain

(**a**) Schematic of surgical preparation for acute in vivo 2P shadow imaging in the somatosensory cortex: [1] preparation of 4 mm craniotomy, [2] injection of 3–4 μL Alexa Fluor 488 (50 mM) at stereotaxic coordinates (M/L: ±1.1 mm, A/P: −0.52 mm, D/V: −2.3 mm) into lateral ventricles (top and side view), and [3] sealing of craniotomy with glass coverslip

(**b**) 2P shadow imaging of the extracellular space following Alexa Fluor 488 injection

(**c**) High-resolution 2P-STED shadow imaging of the same region, providing enhanced nanoscale visualization of the extracellular space and fine cellular structures

9. Gently insert the injection needle into the ventricle to avoid tissue damage. Inject 3–4 μL of the dye at a rate of 500 nL/min using the motorized syringe pump. After injection, leave the needle in place for 3 min to prevent dye leakage. Carefully remove the needle.
10. Cover the exposed craniotomy area with a #1 thickness glass coverslip (4 mm diameter). Secure the coverslip with medical-grade glue.
11. Seal the craniotomy with dental cement (e.g., Superbond C&B).
12. Remove the mouse from the stereotaxic frame. Place the mouse in a heated recovery cage to maintain body temperature. Allow the dye to evenly diffuse for 20–30 min. Transfer the mouse for imaging.

3.4.2 Surgical Procedures for Chronic Shadow Imaging In Vivo

The experimental workflow is presented in Fig. 4a.

1. From Subheading 3.4.1 "Surgical procedures for acute shadow imaging" skip **steps 6**, **8**, **9** and repeat the **steps 1–5**, 7, **10**.
2. On the contralateral side perform a second craniotomy of ≈1 mm diameter with the center at coordinates relative to bregma: Medial/Lateral (M/L): +1.1 mm; Anterior/Posterior (A/P): −0.52 mm; Dorsal/Ventral (D/V): −2.3 mm.
3. Seal it with ~2 mm in diameter PDMS membrane, and secure it with medical-grade glue and dental cement (e.g., Superbond C&B).
4. Seal the craniotomy with dental cement.
5. Remove the mouse from stereotaxic frame. Place the mouse in a heated recovery cage to maintain body temperature.
6. Transfer the mouse to a home cage and monitor its health for at least 2 days after surgery. Allow full recovery from inflammation for 2–3 weeks before imaging.

3.5 Acute In Vivo TUSHI/SUSHI Imaging

3.5.1 Microscope Preparation and Optical Alignment

1. Use a femtosecond mode-locked laser for 2-photon excitation at ~920 nm (80 MHz repetition rate) (e.g., Alcor 920, Spark Lasers).
2. For STED imaging, employ a second laser tuned to ~592 nm with pulse duration ~700 ps (e.g. Onefive KATANA 06 HP from NKT Photonics) synchronized with excitation laser with a pulse delay generator (e.g., Tombak from AlphaNov).
3. Turn on the system components: laser controller, scanner power supply, Pockels cells, piezo Z-stage, SLM controller, motorized sample stage, and beam alignment optics.
4. Verify system calibration using a standard fluorescent calibration slide to ensure uniform illumination.

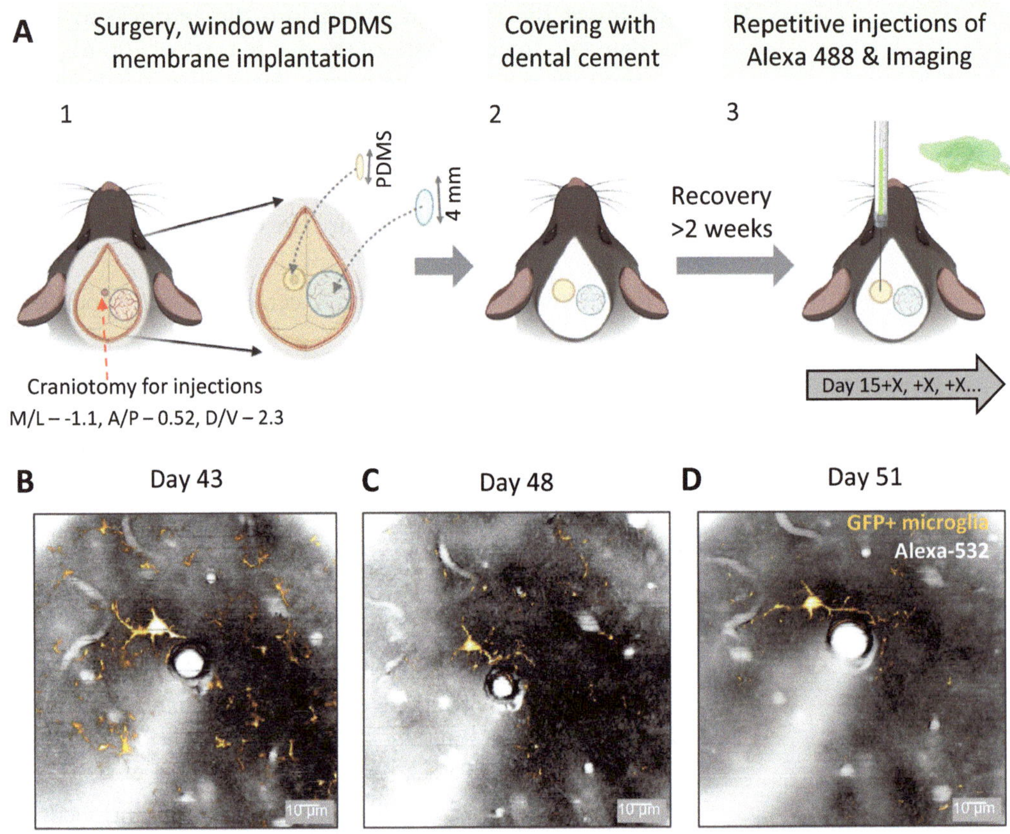

Fig. 4 Chronic in vivo shadow imaging via PDMS window
(**a**) Schematic of surgical preparation for chronic 2P shadow imaging in vivo in the somatosensory cortex: [1] implantation of a 4 mm glass coverslip over a craniotomy for imaging, and a second ~1 mm craniotomy on the contralateral side at stereotaxic coordinates (M/L: +1.1 mm, A/P: −0.52 mm), sealed with a ~2 mm PDMS membrane for repeated dye injections; [2] sealing of both craniotomies and allowing the mouse to recover for >2 weeks; [3] repetitive injections of 3–4 μL Alexa Fluor 488 or Alexa Fluor 532 (50 mM) into the lateral ventricle via the PDMS window 30 min before each imaging session
(**b**–**d**) Dual-channel 2P shadow imaging at different time points post-surgery:
(**b**) Day 43, (**c**) Day 48, (**d**) Day 51, showing Alexa Fluor 532-labeled extracellular space (grayscale) and GFP-positive microglia (orange)

5. For 2P-STED: Use the SLM to correct common optical aberrations (e.g., coma, astigmatism, spherical aberration) and set the beam shape for 2D and 3D STED using beam reflections on 150 nm gold nanospheres. Adjust scanning patterns in XY, XZ, and YZ planes to achieve symmetrical and optimal beam profiles. Verify proper alignment of the 2P and STED beams in X, Y, and Z using fluorescent beads. Optimize defocus settings in the SLM to match 2P and STED imaging planes.
6. Assess the point spread function (PSF) with 170 nm fluorescent nanospheres to ensure proper resolution in both 2P and 2P-STED modes. Measure Full Width at Half Maximum

(FWHM) for validation. Expected values of FWHM: 2P: ~350 nm in XY, ~1200 nm in Z; STED: ~170 nm in XY for 2D, <400 nm in Z for 3D.

7. Ensure proper optical alignment to avalanche photodiodes (APDs) detectors. Maximize signal collection efficiency and reduce background fluorescence. Refer to representative images of brain tissue in vivo in somatosensory cortex obtained in 2P/2P-STED mode (Fig. 3b and c).

3.5.2 Shadow Imaging In Vivo

1. Administer ketamine/xylazine (100/10 mg/kg) diluted in sterile HaCl (0.8 M) solution intraperitoneally approximately 10 min before imaging. Alternatively, the mouse can be kept under isoflurane anesthesia during the imaging session. Ensure anesthetic effect—normal respiration, absence of pinch reflexes, lack of gasping or irregular breathing.
2. Place the anesthetized mouse fixed in an adapted miniature stereotaxic frame under the objective. Ensure the craniotomy area aligns with the field of view. Use an epifluorescence signal (CoolLED, 488 nm) to locate the region of interest (ROI).
3. Switch to 2-photon mode. Use fast scanning to navigate at μm-presicion within your ROI. Use a galvanometric scanner for x-y axis scanning and a piezo actuator for z-axis focusing.
4. Adjust Imaging Parameters. Set appropriate laser power (5–20 mW for 2P; 30–80 mW for STED). Adjust scanning speed and resolution to balance image quality and acquisition time. Set the pixel size to approximately half of the maximum achievable resolution, in accordance with the Nyquist theorem for proper sampling: For 2P: ~ 150 nm. For 2P-STED: ~75 nm. (*see* **Note 10**).
5. Use Imspector scanning software with a data acquisition card to control imaging parameters and record data. Use z-stack acquisition for 3D imaging or time-lapse for dynamic measurements as required. (*see* **Note 11**).

3.6 Chronic TUSHI Imaging In Vivo

Microscope adjustments are described in Subheading 3.5 "Microscope preparation and optical alignment."

3.6.1 Dye Injection for Chronic Shadow Imaging In Vivo

1. Prepare isoflurane anesthesia equipment, including mask for the nose of the mouse. Maintain the anesthesia levels at 1.5–2.5% during surgery while constantly monitoring proper anesthetic effect—normal respiration, absence of pinch reflexes, lack of gasping or irregular breathing.
2. Place the mouse (WT for single-channel recording, or transgenic CX3CR1-EGFP mice for two-channel recording) on a heated blanket to maintain body temperature in a stereotaxic frame and ensure stability and symmetric head-fixation. Avoid insufficient or excessive heating.

3. Prepare a 50 mM solution of Alexa Fluor™ 488 (for WT animals) or 532 (for CX3CR1-EGFP transgenic mice) dye. Load 3–4 μL of the dye into a Hamilton syringe fitted with a 34 G needle. Secure the syringe in a motorized syringe pump holder, ensuring there are no air bubbles in the solution.
4. Locate the center of PDMS membrane, insert the injection needle gently at 2.55 mm depth (to account for the thickness of the PDMS membrane +0.25 mm). Inject 3–4 μL of the dye at a rate of 500 nL/min using the motorized syringe pump. After injection, leave the needle in place for 3 min to prevent dye leakage. Carefully remove the needle.
5. Allow dye to evenly diffuse for 20–30 min. Transfer the mouse for imaging.

3.6.2 Chronic Shadow Imaging In Vivo (Two-Channel Recording)

1. Imaging procedures are identical to those described in Subheading 3.5 "Imaging *in vivo*." Refer to representative images of brain tissue and GFP-positive microglia from CX3CR1-EGFP transgenic mouse in vivo in somatosensory cortex obtained in 2P/2P-STED mode at day 43, 48 and 51 after surgery (Fig. 4b–d).
2. After the imaging session, remove the mouse from the stereotaxic frame. Place the mouse in a heated recovery cage to maintain body temperature. Once the mouse is fully awake, transfer the mouse to the home cage and monitor its health.

3.7 COSHI/TUSHI/SUSHI in Fixed Slices

The experimental workflow is presented in Fig. 5a.

3.7.1 Sample Preparation

1. Fixation of Mouse Brain. Carefully extract the mouse brain and immediately place it into a solution of 4% paraformaldehyde (PFA) in PBS. Incubate the brain in the PFA solution overnight at 4 °C to ensure proper fixation. (*see* **Note 12**).
2. Brain Slicing. Using a vibratome, cut the fixed brain into slices with a thickness of 70–100 μm. Place the prepared brain slices into a 12-well plate in PBS for further processing.
3. Permeabilization. Incubate the brain slices in blocking buffer (5% normal goat serum and 0.5% Triton X-100 in 1X PBS) at room temperature (RT) for 1 h to permeabilize the tissue. Rinse the slices by washing them in 1X PBS for 5 min, repeating this process 3 times.
4. WGA-CF488 Staining. Incubate the slices in PBS containing 5 μg/mL WGA-CF488. Allow the staining to proceed overnight at 4 °C with agitation for 16 s to ensure even labeling.
5. Washing. After staining, wash the slices in 1X PBS for 30 min, repeating this process 3 times to remove excess stain.

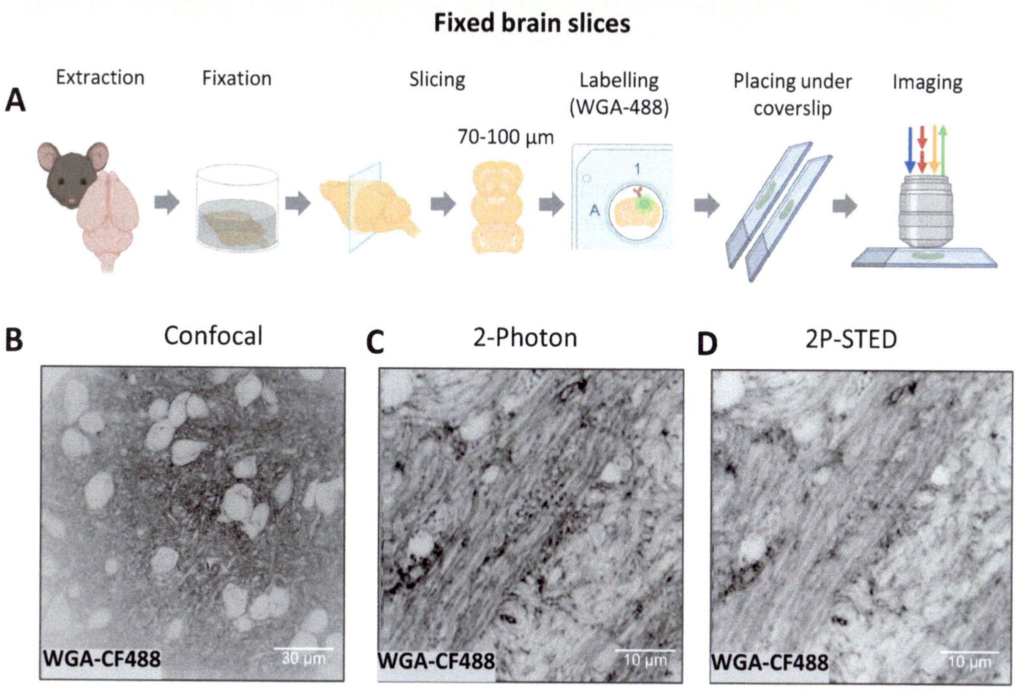

Fig. 5 Shadow imaging of the extracellular space in fixed brain slices
(**a**) Schematic of sample preparation for shadow imaging in fixed brain slices: [1] brain extraction and fixation in 4% paraformaldehyde (PFA), [2] slicing into 70–100 μm sections using a vibratome, [3] staining of the extracellular matrix with WGA-CF488, [4] mounting under a coverslip with antifade medium for imaging
(**b**–**d**) Single-channel shadow imaging of the extracellular space labeled with WGA-CF488, imaged using: (**b**) confocal microscopy, (**c**) 2P microscopy, (**d**) 2P-STED microscopy

6. Mounting. Carefully mount the stained slices onto glass slides. Apply a suitable mounting medium (e.g., CFM3 or similar) and seal the slides with a coverslip to preserve the sample. (*see* **Note 13**).

3.7.2 Shadow Imaging in Fixed Slices

1. Place the glass slide with the fixed brain slice under the objective of the microscope (Fig. 5a).
2. Use the epifluorescence signal (CoolLED, 488 nm) to locate your region of interest (ROI). Adjust the focus and field of view under low magnification, then refine under higher magnification.
3. Microscope Selection for Imaging Modes. Choose the Appropriate Microscope: Confocal Microscope: For COSHI (Confocal Shadow Imaging); 2-Photon Microscope: For TUSHI (2-Photon Shadow Imaging); STED Microscope: For SUSHI (Super-resolution Shadow Imaging).
4. Use fast scanning to navigate at μm-precision within your ROI. Use a galvanometric scanner for x-y axis scanning and a piezo actuator for z-axis focusing.

5. Perform fast scanning to identify and refine your ROI with micron-level precision.
6. Adjust Imaging Parameters. Set the laser power appropriate for your sample to avoid photobleaching while ensuring a strong signal.
7. Adjust scanning speed and resolution to balance image quality and acquisition time. Set the pixel size to approximately half of the maximum achievable resolution, in accordance with the Nyquist theorem for proper sampling: For Confocal: $\approx$ 125 nm; For 2P: $\approx$ 150 nm; For 2P-STED: $\approx$ 75 nm.
8. Image Acquisition. Control imaging parameters with Imspector (or relevant imaging software) and a data acquisition card. Configure and record data efficiently using the software interface. For 3D imaging, enable z-stack acquisition to capture multiple layers of the sample. (Fig. 5b–d).

4 Notes

1. Intracardial perfusion with cold NMDG solution before brain extraction can improve the quality of acute brain slices.
2. Use a wet Whatman paper disk with $NaHCO_3$-based ACSF to cut part of the brain to access specific structures and define slicing orientation as required.
3. Mechanical stability of the brain slice is critical for high-quality imaging. Stable perfusion is necessary to minimize motion artifacts. Alternatively, a HEPES-based solution can be used for recording without perfusion, depending on the specific experimental requirements.
4. Acute slices must be in good condition to produce strong contrast for shadow imaging. Handle slices gently to avoid damage. Imaging should begin at a depth of 50 μm from the surface to prevent interference from dead cells on the surface of the slice.
5. Adjust the fluorescent dye concentration for shadow imaging depending on the brightness of the expression of the genetically encoded indicator. This ensures equal 2P excitation and comparable fluorescence levels from both channels.
6. Optimize the excitation wavelength to favor the genetically encoded indicator and reduce photobleaching while maintaining a strong signal.
7. If improved spectral separation is needed, use Alexa-594 instead of Alexa-532.
8. Apply ophthalmic ointment to prevent corneal drying during anesthesia.

9. The precision of coordinates can vary depending on the size of the animals. Coordinates must be adjusted in case of other species, e.g., rats.
10. For 2P-STED adjust the balance of laser powers between 2P and STED lasers to achieve maximum signal-to-noise ratio and to avoid damaging the tissue.
11. Maintain sample sterility and minimize motion artifacts by ensuring firm fixation of the mouse head.
12. Intracardiac perfusion with cold ACSF followed by 4% PFA solution before brain extraction can improve the quality of fixed brain slices.
13. This protocol is compatible with additional fluorescent labeling. Ensure dye spectra do not overlap and use a multichannel acquisition setup when needed.

Acknowledgments

This work received funding from the European Research Council (ERC) under the European Union's Horizon 2020 research and innovation program (Grant agreement No. #951294) and the Human Frontier Science Program (Grant No. RGP0036/2020) to UVN.

References

1. Arizono M, Idziak A, Quici F et al (2023) Getting sharper: the brain under the spotlight of super-resolution microscopy. Trends Cell Biol 33:148–161. https://doi.org/10.1016/j.tcb.2022.06.011
2. Tonnesen J, Inavalli V, Nagerl UV (2018) Super-resolution imaging of the extracellular space in living brain tissue. Cell 172:1108–1121. https://doi.org/10.1016/j.cell.2018.02.007
3. Dembitskaya Y, Boyce AKJ, Idziak A et al (2023) Shadow imaging for panoptical visualization of brain tissue in vivo. Nat Commun 14:6411. https://doi.org/10.1038/s41467-023-42055-2
4. Inavalli VVGK, Lenz MO, Butler C et al (2019) A super-resolution platform for correlative live single-molecule imaging and STED microscopy. Nat Methods 16:1263–1268. https://doi.org/10.1038/s41592-019-0611-8
5. Arizono M, Inavalli VVGK, Bancelin S et al (2021) Super-resolution shadow imaging reveals local remodeling of astrocytic microstructures and brain extracellular space after osmotic challenge. Glia 69:1605–1613. https://doi.org/10.1002/glia.23995
6. Arizono M, Inavalli VVGK, Panatier A et al (2020) Structural basis of astrocytic ca(2+) signals at tripartite synapses. Nat Commun 11:1906. https://doi.org/10.1038/s41467-020-15648-4

Chapter 12

NanoPlex: A Universal Strategy to Multiplex Immunostainings for Fluorescence Microscopy Using Secondary Nanobodies

Nikolaos Mougios, Parisa Dashtestani, and Felipe Opazo

Abstract

Fluorescence microscopy techniques rely strongly on revealing targets of interest using affinity binders, such as antibodies, through a process known as immunostaining or immunofluorescence. Immunostaining is typically achieved using a primary antibody against a specific target of interest, which is then detected by a fluorescently labeled secondary antibody that recognizes the primary antibody's species. This procedure, known as indirect immunofluorescence, constrains the detection of multiple targets by requiring primary antibodies from different species. Here, we describe NanoPlex, a methodology that allows multitarget detection in a species-independent fashion. NanoPlex employs secondary single-domain antibodies (sdAbs, or nanobodies) that enable the simultaneous use of multiple primary antibodies originating from the same species. In addition, these secondary sdAbs are engineered to allow for a mild and specific removal of the fluorescence signal after imaging, enabling a new round of immunostaining and imaging. NanoPlex includes secondary sdAbs with different functional groups, which support directed signal erasing upon application of either UV light (OptoPlex), proteolytic enzyme (EnzyPlex), or a reducing agent (ChemiPlex), fulfilling the demands of various experimental setups. This approach, which involves directed signal removal and re-staining of new targets on the same sample, permits the visualization of multiple targets under the same context in virtually any antibody-based application. Here, we present a detailed protocol for performing NanoPlex on cell cultures, including a practical example for 6-plex in ChemiPlex mode. In this practical example, three iterations are performed while employing dual-color confocal microscopy on each cycle. Notes have been included for step modifications required for EnzyPlex and OptoPlex.

Key words Nanobodies, sdAb, Antibodies, Immunofluorescence, Fluorescence microscopy, Secondary nanobodies, Multiplex

1 Introduction

Immunofluorescence (IF) is routinely used in various biomedical applications and typically relies on primary antibodies (1.Abs) to bind a protein of interest (POI). In indirect IF applications, 1.Abs are usually detected by secondary antibodies (2.Abs), conjugated to fluorophores, enzymes, or single-stranded DNA (ssDNA) for

Frédéric A. Meunier (ed.), *Super-Resolution Microscopy: Methods and Protocols*, Methods in Molecular Biology, vol. 3034,
https://doi.org/10.1007/978-1-0716-5268-8_12,

further signal detection. Typically, labeled 2.Abs are used as polyclonal mixtures, often resulting in multiple fluorescently labeled 2. Abs binding to a single 1.Ab. Therefore, indirect IF produces a stronger signal than staining with fluorescently labeled 1.Abs, but suffers from having an imprecise number of reporters (e.g., fluorophores) on each 2.Ab. This uncertainty, combined with the undefined number of 2.Abs attached to a given 1.Ab, renders reproducibility and quantification of the study target difficult. In addition, and significantly, in traditional indirect IF, the ability to detect multiple targets within a single sample (multiplexing) has been limited. This limitation arises from the need to use 1.Abs from different species (e.g., mouse, rabbit, goat) to reveal them selectively using species-specific fluorescently labeled 2.Abs (i.e., anti-mouse, anti-rabbit, anti-goat immunoglobulins). This limitation often results in using sub-optimal 1.Abs for a given target to match the different species for multiplex detection. Moreover, this species limitation, combined with the permitted spectral separation on the microscopy setups, has typically limited the maximal number of targets that can be imaged on the same sample to four (five to six targets have also been demonstrated using signal demixing post-processing analysis [1]).

Although indirect IF has contributed to important discoveries, increasing the multiplex ability and, thus, the possibility of detecting more than a handful of targets in the same sample is often desirable. Already in early 2000, steps were taken to make IF able to stain more than four targets in the same sample, where antibodies were stripped from the sample using chemical denaturation before performing a new IF cycle [2]. Alternatively, high-intensity illumination was used to bleach the signal [3], which, combined with antibody denaturation, was used in 2007 [4] to image 18 targets in the same sample. Further variations of these concepts were used by adding strong detergents [5, 6] or chaotropic salts [7, 8], using an extreme pH [5, 9], or microwaving the sample [10, 11]. Unfortunately, all these strategies resulted in a serious compromise of sample integrity, which can be tolerated if, for example, the aim is to determine the cell types in a tissue, but these procedures are by far not suitable if subcellular or molecular resolution is required.

Recent advancements have significantly improved the indirect IF technique by introducing secondary camelid single-domain antibodies (sdAbs, also known as nanobodies [12]). These secondary nanobodies (2.Nbs) offer several advantages over conventional 2. Abs [13, 14]. One key advantage is that 2.Nbs bind monovalently to 1.Abs, allowing them to be pre-mixed without causing agglutination (aggregation), which occurs if 1.Abs are pre-mixed with traditional divalent polyclonal 2.Abs. This apparent small advantage results in a significant gain by allowing species-independent IF, removing the need to find 1.Abs from different species to be used on the same sample [13]. Additionally, this allows a one-step IF (OneStep-IF), where incubation of 1.Ab and 2.Nb in a single step without extra washing and incubation times is possible, shortening protocols significantly. These advantages have been used to further leverage the multiplex capability of various imaging techniques [15, 16]. This chapter will focus on the recently designed and modified 2.Nbs that allow gentle and specific removal of the fluorescence signal after imaging developed by Mougios et al. [16], which permits species-independent OneStep-IF and iterative staining and imaging. This iterative approach, termed NanoPlex, is not only a universal methodology for any fluorescence microscopy technique, but its unique preservation of subcellular ultrastructure allows quality data acquisition with high multiplexing capabilities in various super-resolution techniques [16].

1.1 Fundaments of NanoPlex

NanoPlex is the general term for the methodology using 2.Nbs that have been modified to remove the fluorescence signal after imaging. Three modalities have been developed to do so. The first is OptoPlex, which uses 365 nm LED light to cleave the fluorescent moiety attached to the 2.Nb. The second is EnzyPlex, which uses an efficient and specific protease that cleaves the fluorescent moiety away. The last is ChemiPlex, which uses a small amount of reducing chemical to cleave the fluorescence signal on the secondary sdAb [16].

All NanoPlex-ready 2.Nbs have a 13 amino acid ALFA-tag [17] connected to the sdAb C-terminus via a cleavable domain. Therefore, to perform IF, it is necessary to pre-mix in a tube the 1.Ab with NanoPlex-ready 2.Nbs and the fluorescently labeled NbALFA (NbALFA*) (Fig. 1a). NanoPlex was designed this way for two main reasons. (1) Flexibility: This allows for easy choice of color by selecting NbALFAs with, for example, different fluorophores or single-stranded DNA. (2) Solubility: Cleaved, free fluorophores typically bind non-specifically to membranes, making them poorly washable. We knew fluorophores attached to NbALFA tended to be highly soluble, allowing for easy signal flushing after cleaving.

Therefore, the first common step for all NanoPlex modalities is to form a multimeric complex between 1.Ab, 2.Nb-cleavable-ALFA (i.e., Opto-2.Nb, Enzy-2.Nb or Chemi-2.Nb) and NbALFA* in a tube before being used to stain the sample (Fig. 1a). This pre-mixing of 1.Ab and 2.Nb allows the use of different 1.Abs from the same species in a OneStep-IF on the same sample if complexes are prepared in separate tubes (Fig. 1b) [13]. This unique feature of species-independent universal multiplex staining allows scientists to use their best 1.Abs for a given experiment, regardless of their host species, permitting an easy iterative multiplexing strategy on any fluorescence microscopy type or setup (Fig. 1c).

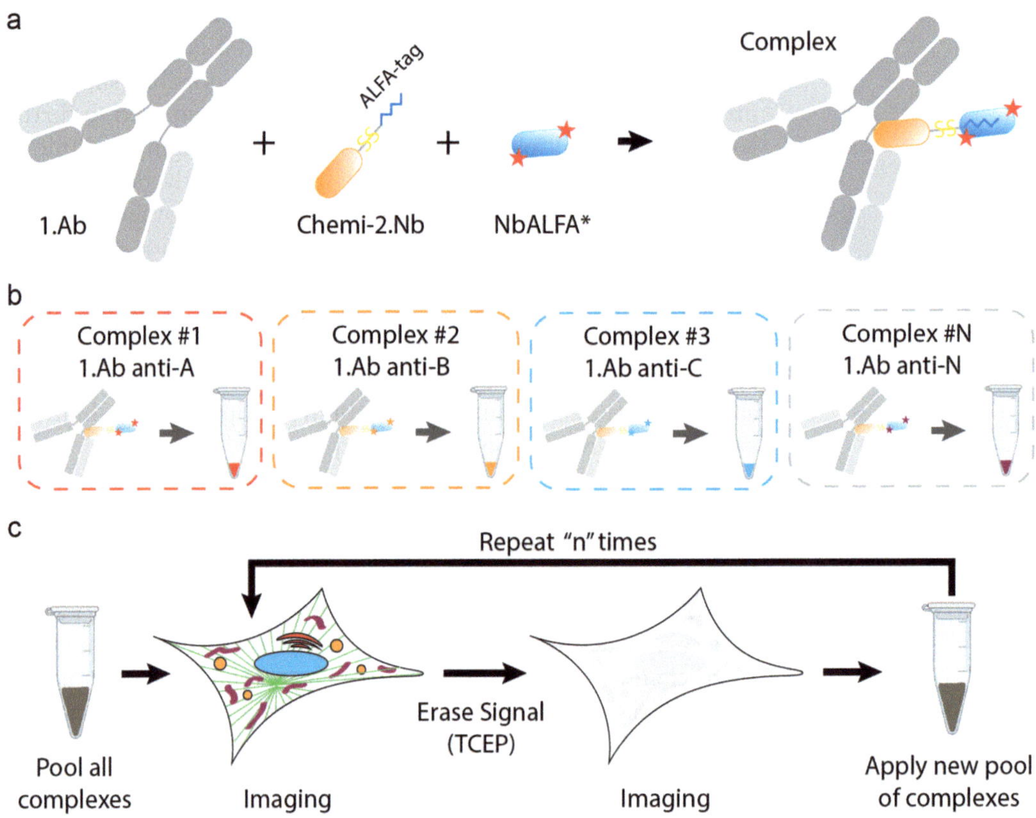

Fig. 1 Schemes depicting the fundamentals of ChemiPlex. (**a**) Complex formation using a primary antibody (1. Ab) combined with ChemiPlex cleavable 2.Nb (Chemi-2.Nb), which carries a disulfide bond (yellow S-S) connected at the C-terminal of the 2.Nb with an ALFA-tag (blue zig-zag line) and a fluorescently labeled sdAb anti-ALFA-tag (NbALFA*). (**b**) Complexes with 1.Abs with different specificities are formed in individual tubes, each carrying a 2.Nb and a NbALFA* with a different fluorophore. Pre-forming the complexes allows the use of 1.Abs from the same species, while the number of applicable complexes is limited only to the number of channels that a readout device can distinguish. (**c**) The combined complexes can be applied in a OneStep-IF, and the sample can be imaged. This is followed by an erasing step (here, for ChemiPlex, the reducing agent TCEP is used) and a OneStep-IF with a new pool of pre-formed complexes. This cycle can be repeated "n" times. The final number of targets imaged in the same sample is a function of the number of targets recognized in each imaging step and the number of cycles performed. For example, if three targets can be imaged confidently, nine different targets (9-plex) are imaged on the same sample after three cycles

In the following section, we will briefly overview NanoPlex's three modalities (OptoPlex, EnzyPlex, and ChemiPlex). This protocol focuses mainly on ChemiPlex, however, we have included notes for modification steps required for EnzyPlex and OptoPlex.

1.2 OptoPlex

OptoPlex uses a light-responsive linker between the 2.Nb and the ALFA-tag (Opto-2.Nb). This allows the specific removal of the ALFA-tag with the bound NbALFA* upon exposure to 365 nm light [16]. The light-responsive element was designed not to be cleaved when using commonly used excitation wavelengths, for example, 488 nm, 560 nm, or 640 nm laser lines. OptoPlex has the unique advantage that cleavage can be performed only in the illuminated region of the sample. This opens the possibility of using this approach on living cells, or in the case where only a selected area on the sample needs to be cleaved.

1.3 EnzyPlex

EnzyPlex takes advantage of a small ubiquitin-related modifier (SUMO) substrate derived from *Brachypodium distachyon* (bdSUMO) [18], between the 2.Nb and the ALFA-tag (Enzy-2. Nb). Cleavage is performed using its corresponding protease, bdSENP1 [18], which can cleave its substrate quickly and specifically. bdSENP1 can tolerate various buffers, including various temperatures, pHs, and salt concentrations. Cleavage will be highly efficient, reaching maximum efficiency after ~5 min after exposure [16] to moderate amounts of bdSENP1 in simple buffers like PBS, making EnzyPlex methodology an extremely mild signal removal since it won't affect anything else on the sample.

1.4 ChemiPlex

In ChemiPlex mode, the ALFA-tag is attached to the C-terminus of the 2.Nbs via a disulfide bond (Chemi-2.Nb). Therefore, a moderate concentration of a reducing agent suffices to cleave the ALFA-tag and the NbALFA* bound to it in less than 5 min at room temperature [16]. TCEP (tris(2-carboxyethyl)phosphine), was used as a reducing agent due to its small molecular size and high solubility in aqueous buffers, allowing a fast distribution of this agent, making this NanoPlex modality very fast and efficient in removing signals with readily available chemicals.

This Chapter provides a detailed protocol for performing 6-plex imaging using dual-color confocal microscopy in ChemiPlex mode. Notes have been included for step modifications required to perform OptoPlex and EnzyPlex, and aspects that require special care.

2 Materials

2.1 Cell Culture

NanoPlex has been established and validated on cultured cells (fibroblast and primary hippocampal neurons) [16]. It might also work on other sample types (e.g., tissue, plants, and whole organisms). However, each case will most likely require optimization.

1. COS-7 cells.
2. Culturing media: Complete Dulbecco's Modified Eagle Medium (DMEM) supplemented with 10% Fetal Bovine Serum (FBS), 4 mM L-glutamine, and 1% pen/strep.
3. Petri dishes.
4. 18 mm poly-L-lysine-coated coverslips in 12-well plates.

2.2 Sample Preparation Buffers

NanoPlex, especially the ChemiPlex approach, will typically be used on fixed samples. The samples should be fixed following a protocol that allows all the intended 1.Abs to recognize their targets. This might require optimizing the fixation procedure by using different chemicals (e.g., PFA, glutaraldehyde, glyoxal, methanol, etc.) and adjusting their concentrations and the timing and temperature of exposure. Critical remark: It is vital for successful multiplex imaging with NanoPlex that all available 1.Abs work in the selected fixation condition.

Prepare all buffers for each experiment using sterile Milli-Q water, unless otherwise stated.

1. 10x PBS: 1.27 M NaCl, 27 mM KCl, 100 mM Na_2HPO_4, 2 mM KH_2PO_4, pH 7.4.
2. Fixation buffer: 4% paraformaldehyde (PFA) in 1x PBS for fixation, pH 7.4.
3. Permeabilization buffer: 3% bovine serum albumin (BSA) and 0.2% Triton X-100 in 1x PBS (*see* **Note 1**).
4. Extraction Buffer (EB): 10 mM MES, 138 mM KCl, 3 mM $MgCl_2$, 2 mM EGTA, 320 mM sucrose, pH 6.8) supplemented with 0.1% Saponin.
5. Extraction Fixation Buffer (EFB): 4% PFA and 0.1% glutaraldehyde in EB.
6. Aldehyde quenching buffer: 0.1 M glycine in 1x PBS (*see* **Note 2**).

2.3 NanoPlex-Specific Buffers

1. Washing buffer: 0.1 M Glycine, 1x PBS, pH 7.5.
2. Imaging buffer: 1 mM PCA (Protocatechuic acid), 1.25 mM Trolox(6-Hydroxy-2,5,7,8-tetramethylchroman-2-carboxylic acid), 1x PCD (Protocatechuate 3,4-dioxygenase) (100x PCD: 0.7 mg/ml), 1x PBS, pH 7. To prepare stock solutions for the Imaging buffer: Trolox (100x) is made by adding 100 mg of Trolox to 430 μL of 100% methanol. After adding 3.2 mL of H_2O, 1 M NaOH is added (initially 350 μL, then, in 20 μL steps) until fully dissolved (480 μL total). PCA (40x) is made by mixing 154 mg of PCA in 10 mL of H_2O while bringing the pH to 2. The pH is adjusted to 9.0 by adding about 1 mL of NaOH in 100 μL increments while monitoring pH. PCD (100x) is made by adding 9.3 mg of PCD to 13.3 mL of the solution buffer (50% glycerol, 50 mM KCl, 1 mM EDTA, 100 mM Tris-HCl, pH = 8). All stock solutions are stored at −20 °C. To make 1x working solutions, they are combined in 100 μL of PBS.
3. TCEP buffer: 10 mM TCEP (tris(2-carboxyethyl)phosphine), 20 mM K_2CO_3, pH 6.8 (*see* **Note 3**).

4. SENP1 buffer: 5 μM SENP1, 0.1 M Glycine, pH 7 (*see* **Note 3**).
5. Thiol quenching buffer: 20 mM NEM (N-Ethylmaleimide), 0.1 M Glycine, 0.2 M NaCl, 1x PBS, pH 7.
6. Aldehyde quenching buffer: 0.1 M Glycine, 1x PBS, pH 7.5 (*see* **Note 2**).

2.4 Antibodies and Single-Domain Antibodies

Antibodies should be stored as suggested by the manufacturer, all sdAb (nanobody) aliquots should be stored in the dark at −20 °C for short periods (up to three months) or at −80 °C for extended periods, using 1xPBS, 50% glycerol. Aliquot them to needed volumes to avoid repetitive freezing and thawing cycles. Details of antibodies and sdAbs used in the practical example are listed in Tables 1, 2, and 3.

1. 1.Abs from rabbit or mouse against the proteins of interest (*see* **Note 4** and Table 1).
2. Chemi-2.Nbs (ChemiPlex anti-Mouse IgG1, ChemiPlex anti-Mouse IgG2 or ChemiPlex anti-Rabbit IgG; NanoTag Biotechnologies) (*see* **Note 5** and Table 2).
3. Fluorescently-labeled NbALFA (FluoTag®-X2 anti-ALFA; NanoTag Biotechnologies) (*see* Table 2).
4. Multiplexing Blocker Rabbit and/or Multiplexing Blocker Mouse (NanoTag Biotechnologies), selected according to the 1.Abs used simultaneously (*see* Table 2).
5. Unconjugated NbALFA (*see* Table 2).

2.5 Imaging Equipment

1. STED expert line microscope (Abberior Instruments, Göttingen, Germany) composed of an IX83 inverted microscope (Olympus, Hamburg, Germany) with a UPLSAPO 100x 1.4 NA oil immersion objective (Olympus).

Table 1
Primary antibodies used in the practical example

Target	Species, isotype	Company	Cat.#	RRID
Alpha-tubulin	Mouse, IgG1	Synaptic systems	302211	AB_887859
Lamp1	Rabbit, IgG	Cell signaling	9091	AB_2687579
PMP70	Rabbit, IgG	Abcam	ab85550	AB_10672335
Son	Rabbit, IgG	Atlas antibodies	HPA023535	AB_1857362
Tomm20	Rabbit, IgG	Abcam	Ab232589	AB_3065091
Clathrin heavy chain	Rabbit, IgG	Abcam	ab21679	AB_2083165

Table 2
Nanobodies used in the practical example

Target	Conjugation	Company	Cat. #	RRID
ChemiPlex anti-mouse IgG1	Cleavable ALFA-tag	NanoTag	N2007	
ChemiPlex anti-rabbit IgG	Cleavable ALFA-tag	NanoTag	N2407	
FluoTag®-X2 anti-ALFA	AbberiorStar635p	NanoTag	N1502	AB_3075979
FluoTag®-X2 anti-ALFA	Atto488	NanoTag	N1502	AB_3075982
Multiplexing blocker rabbit	N/A	NanoTag	K0202–50	AB_3075894
NbALFA	Unconjugated	Self-made	N/A	N/A

Table 3
Composition of pre-formed complexes used for the 6 targets ChemiPlex (Fig. 3)

Cycle #	1.Ab	[nM]	Chemi-2.Nbs	[nM]	FluoTag®-X2	[nM]
1	Anti-tubulin	7	Anti-mouse IgG1	21	Anti-ALFA AbbStar635p	28
1	Anti-Lamp1	15	Anti-rabbit IgG	45	Anti-ALFA Atto488	60
2	Anti-PMP70	10	Anti-rabbit IgG	30	Anti-ALFA Atto488	40
2	Anti-son	15	Anti-rabbit IgG	45	Anti-ALFA AbbStar635p	60
3	Anti-Tomm20	15	Anti-rabbit IgG	45	Anti-ALFA Atto488	60
3	Anti-Clathrin	10	Anti-rabbit IgG	30	Anti-ALFA AbbStar635p	40

2. Pre-installed autofocus system (Olympus, Hamburg, Germany) to compensate for Z-drift during 2D image acquisition in real-time.
3. Power and energy meter interface (PM100USB, Thorlabs, Inc.) and microscope slide power sensor (S170C, Thorlabs, Inc.) for measuring laser power at the sample stage.

2.6 Fluidics System

Here, we provide a practical example of using a peristaltic pump in conjunction with a homemade fluidic chamber to achieve multiple cycles on the microscope stage (*see* Fig. 2). However, any commercial or homemade solution that allows for the gentle exchange of liquids at the microscope stage should be effective. If no fluidics system is available, manual pipetting can be used as an alternative (*see* **Note 6**).

1. A standing platform or dedicated place near the microscope stage for placing the peristaltic pump.
2. 10 mL and 25 mL syringes (used to place the NanoPlex's buffers for repetitive immunolabeling cycles).

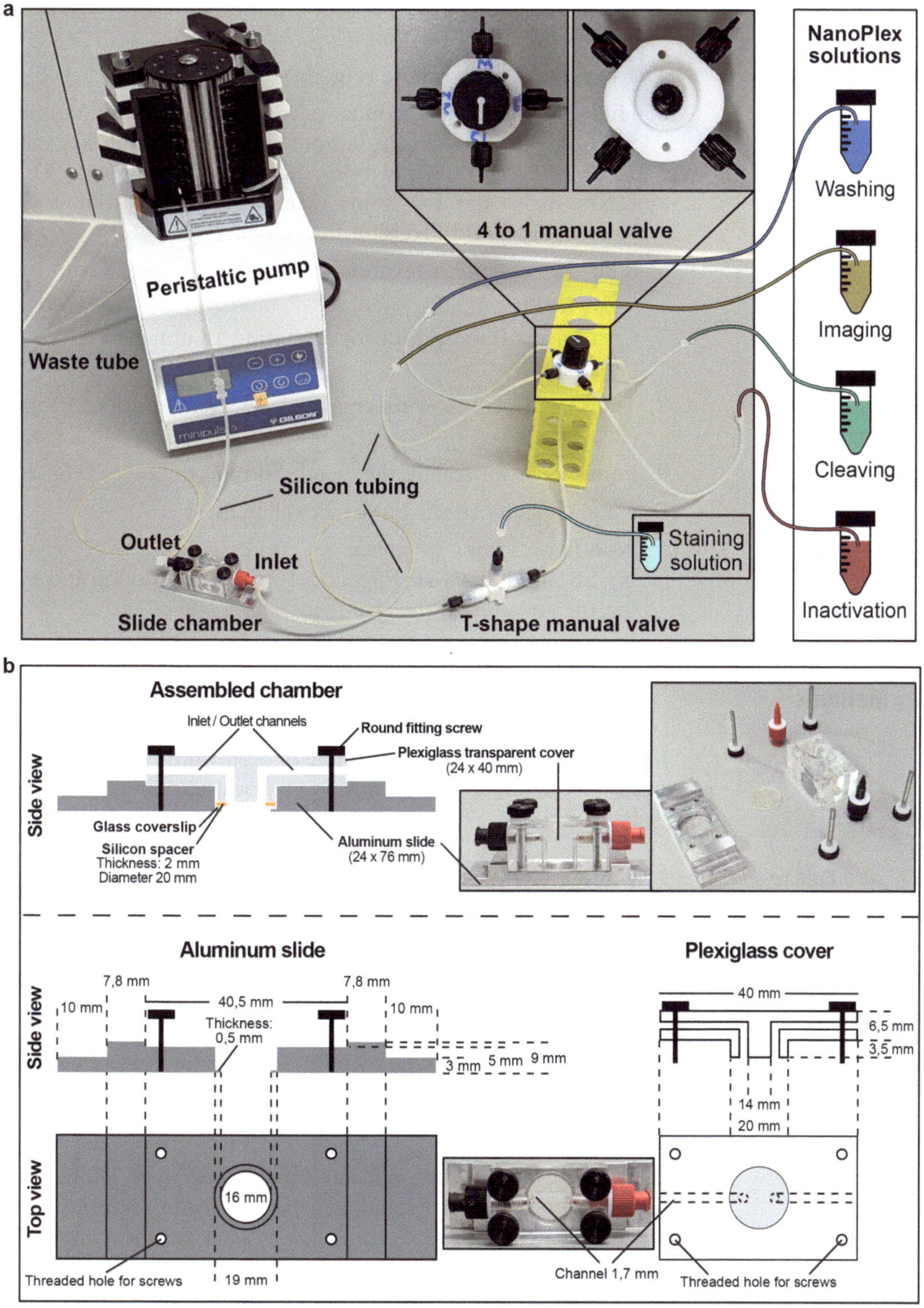

Fig. 2 Fluidic system and imaging chamber. (**a**) A picture of the elements needed to create a self-made fluidic system using a peristaltic pump (see Subheading 2.5 for more details). (**b**) Schemes of the flow chamber we built and used to apply NanoPlex on 18 mm glass coverslips. There are also several commercial solutions available

3. Peristaltic pump with peristaltic pump tubing (MINIPULS 3 Peristaltic pump, Gilson) (Fig. 2a).
4. 1 mm OD silicon tubing (Fig. 2a).
5. High-pressure 5-port manual valve for 0.5–4 mm OD tubing, 5 × 1/4″-28 UNF female, 0.8mm bore, 3-way single key valve, 1/4″-28 UNF, 0.8 mm bore, Barb acetal adapter, 1/4″-28 UNF male to 1.0 mm (Luer Valve Assortment kit, Nr. 14011, World Precision Instruments) (Fig. 2a).
6. 0.80*22 mm blunt cannula (B. Braun) for waste tube connection.
7. Custom-made metal slide for mounting 18 mm glass coverslips (Fig. 2b).
8. Transparent cover with engraved inlet and outlet channels (Fig. 2b).
9. Four screws for sealing the metal slide and the cover (Fig. 2b).
10. Channel-stamped silicon patches (Fig. 2b).
11. Non-fluorescent grease (Grease for pivots, Nr. 5810350.050–01, Eppendorf) for waterproofing and sealing the coverslip in the chamber.

3 Methods

Here, we provide a practical example of performing 6-plex imaging of cells in ChemiPlex mode under a confocal scanning microscope (*see* Fig. 3).

3.1 Cell Culture and Sample Preparation

1. Culture COS-7 cells in Petri dishes using culturing media.
2. Maintain the cells at 37 °C, 5% CO_2 in a humidified incubator.
3. Seed COS-7 cells on 18 mm poly-L-lysine-coated coverslips in 12-well plates.
4. Incubate for approximately 16 h at 37 °C, 5% CO_2 in a humidified incubator.
5. Examine the cell's health and density before proceeding with fixation.

3.2 Extraction-Fixation

This type of fixation removes part of cytoplasmic or soluble proteins, resulting in cleaner staining, especially for filamentous structures or organelles. Other fixations can be used; the most important thing is that all the 1.Abs intended to be used are compatible and working fine under the chosen fixation strategy.

1. Rapidly rinse the COS-7 cells with 37 °C pre-warmed culturing media.

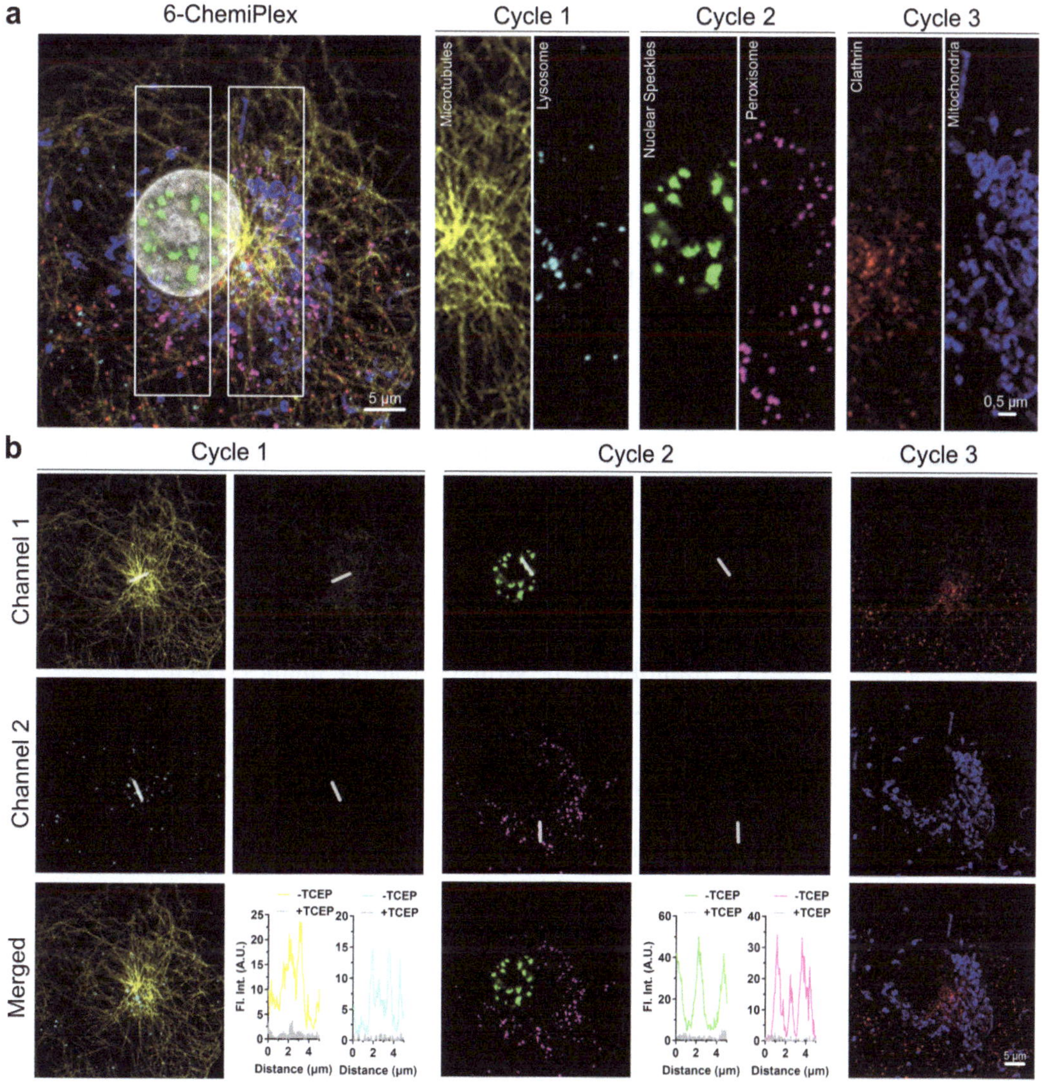

Fig. 3 Confocal 6-plex images from COS-7 fibroblasts undergoing three cycles of ChemiPlex and 2-channel confocal imaging per cycle. (**a**) The left image shows the overlay of 6 targets displayed individually in each panel at the right (microtubules, lysosomes, nuclear speckles, peroxisomes, clathrin, and mitochondria). (**b**) Images from 6 targets ChemiPlex and the corresponding images after erasing the signal using 10 mM TCEP buffer for 15 min. Images of microtubules, lysosomes, nuclear speckles, peroxisomes, clathrin, and mitochondria were acquired before and after TCEP treatment with equal microscopy settings and are displayed with equal gray value scaling. In channel 1, the complexes with AbberiorStar635p revealed microtubules, nuclear speckles, and clathrin. In channel 2, complexes with Atto488 revealed lysosomes, peroxisomes, and mitochondria. For each cycle, the merged image of Channels 1 and 2 is depicted at the bottom. The graphs indicate the line profile of the signal intensity before and after applying TCEP cleavage

2. Treat cells with 37 °C pre-warmed Extraction Buffer for no longer than 30 seconds (*see* **Note 7**).
3. Remove the Extraction buffer and fix the cells with a 37 °C pre-warmed Extraction Fixation Buffer for 15 min at room temperature.
4. Remove fixatives and rinse once with PBS.
5. Quench unreacted aldehyde groups using the Aldehyde Quenching buffer for 15 min (*see* **Note 2**).
6. Block and permeabilize cells using the Permeabilization buffer for 30 min at room temperature with mild shaking (*see* **Note 1**).

3.3 Assembling the Fluidics System

This protocol outlines the use of a peristaltic pump combined with a homemade fluidic chamber to achieve multiple cycles on the microscope stage (*see* Fig. 2). Any commercial or homemade solution that allows the gentle exchange of liquids at the microscope stage should work. If no fluidics chamber is available, manual pipetting can be used instead (see **Note 6**).

1. Immobilize the syringes near the fluidics system.
2. Connect the syringes with the four inlets of the high-pressure 5-port manual valve (Fig. 2a, upper right) using 1 mm OD tubing and barb acetal adapters. Rotate the valve to select the inlet of interest. The valve has four inlets (for the four syringes carrying the NanoPlex's buffers) and only one outlet.
3. Connect the high-pressure 5-port manual valve outlet with the 3-way T-shape single-key valve using 1 mm OD tubing and barb acetal adapters.
4. Connect the 3-way single key valve with the inlet channel of the chamber carrying the sample using 1 mm OD tubing and barb acetal adapters.
5. Connect 1 mm OD tubing to the chamber's outlet using barb acetal adapters and place the tubing within the peristaltic pump's rotational head (Fig. 2a, upper left). Fix it firmly using the respective screws of the rotational head.
6. Place the end of the outlet tubing in a waste vessel.
7. Move the peristaltic pump's rotational head clockwise to remove liquid from the sample. In the closed-environment fluidics system, the removed liquid will be replenished with the buffer of interest selected using the 5-port manual valve.

Each NanoPlex cycle has four main steps: OneStep-IF (Subheading 3.4.1), imaging (Subheading 3.4.3), signal erasing (Subheading 3.4.4), and inactivation (Subheading 3.4.5). These steps are repeated *n* times, depending on the total number of samples, and the number of samples that can be imaged together in each cycle. This practical example involves 3 ChemiPlex cycles (6-plex imaging where six 1.Abs are imaged in pairs) (*see* Table 3). The first cycle in this example involves the labeling of microtubules and lysosomes with pre-formed complexes of alpha-tubulin and Lamp1 1.Abs with their respective 2.Nbs (*see* Table 3 and Fig. 3).

3.4 NanoPlex: Iterative Immunolabeling, Imaging, Erasing Cycles

3.4.1 OneStep-IF

Best results are obtained if all 1.Abs have been tested and titrated individually before initiating a NanoPlex multiplex experiment (*see* **Note 8**). The single-target procedure (e.g., testing individual 1. Abs) is similar to the multi-target procedure described below. Notes have been provided where modifications to steps can be made to perform the single-target procedure.

1. Set the complex formation of all 1.Abs, cleavable 2.Nb and NbALFA* in a 1:3:4 molar ratio in a final volume of 20 μL, supplemented with PBS if necessary (*see* **Notes 9** and **10**). Each tube should carry a different 1.Ab and each complex should have different fluorophores on the NbALFA*, allowing a chromatic separation at the microscope. (e.g., Atto488, AzDye568 & Atto643).
2. Incubate at room temperature, protected from light for 30 min, allowing the complex to form (*see* **Note 11**). Alternatively, leave it at 4 °C for extended periods (e.g., overnight) (*see* **Note 12**).
3. If using 1.Abs of the same species on the OneStep-IF, add Rabbit or Mouse Multiplexing Blocker in tenfold molar excess relative to the 1.Ab to each tube, accordingly (only after the 30 min pre-mixing 1.Ab with 2.Nb and NbALFA*). Non-fluorescent NbALFA is added in tenfold molar excess relative to the 1.Ab (only after complex formation of 1.Ab, 2. Nb and NbALFA*). Incubate the pre-mixed complexes with the blockers for another 15 min (*see* **Note 13**).
4. Combine and mix gently all the different complexes.
5. Dilute the mix of complexes to the intended final concentration of 1.Abs using the Permeabilization buffer.
6. Add the mix of complexes to the sample and incubate for 60 min at room temperature (or the time titrated for the 1. Abs used) (*see* **Note 14**).
7. Wash the sample 3 times for 3–5 min using 1–2 mL of PBS each time.
8. Mount the sample in the imaging chamber (*see* **Note 15**).

3.4.2 Mounting the 18 mm Glass Coverslips for ChemiPlex

1. Carefully remove the bulk of the liquid from the coverslip by gently touching a paper towel to its edge.

2. Place the coverslip on the aluminum slide's circular inset, with the cells phasing upwards (Fig. 2b).
3. Carefully cover both sides of the channel-stamped silicon patch with a thin layer of non-fluorescent grease (*see* **Note 16**).
4. Place the channel-stamped silicon patch on the coverslip.
5. Place the transparent cover with the engraved inlet and outlet ports on top of the silicon patch (*see* **Note 17**).
6. Gently tighten the four screws of the cover to seal the chamber.
7. Connect the chamber to the two-way fluidics system to create a "closed" fluidic environment (*see* **Note 18**).
8. Add all NanoPlex buffers in different syringes on the standing platform. Use the 25 mL syringe for Washing buffer and the 10 mL syringes for imaging, cleaving, and quenching buffers.
9. Add or flow Washing buffer (*see* **Note 19**).
10. Replace the Washing buffer with the Imaging buffer. Use the peristaltic pump to fill the chamber with the Imaging buffer at a 0.5 mL/min flow rate (*see* **Note 20**).

3.4.3 Imaging

Confocal microscopy was used to generate the images in Fig. 3, for which laser lines used (i.e., 640 nm, 561 nm, 488 nm, 405 nm) were set to deliver powers of 1–2 μW, except for the 405 nm laser which was set to 0.2 μW, all measured at the sample level (slide power sensor S170C, Thorlabs, Inc.) (*see* **Note 21**), the pinhole was set to airy 1, the x-y pixel size was set to 60 nm, and the dwelling time per pixel was set to 10 μs; lines were scanned only once.

1. Stop the flow during image acquisition if using a flow chamber.
2. Acquire images using the imaging modality of interest (Confocal, epifluorescence, STED, etc.) (*see* **Note 22**). If 3D Volumetric imaging is needed, set confocal z-steps to ~500 nm, keeping the pinhole as airy 1 (*see* **Note 23**).

3.4.4 Signal Erasing

1. After imaging, wash the sample with Washing buffer for 2 min using a 1 mL/min flow speed.
2. Apply the TCEP buffer (under low flow) for 15 min. Decrease the flow speed to 0.5 mL/min and retain a constant flow of the buffer for the total duration of this step (*see* **Notes 24** and **25**).

3.4.5 Inactivation

1. Apply or flow for 5 min, Thiol quenching buffer, which reacts with reduced reactive thiols generated by TCEP during the cleaving steps (*see* **Note 26**).
2. Apply or flow the Washing buffer for 5 min.

3.4.6 Image After Signal Removal

1. Replace the Washing buffer with the Imaging buffer.
2. Acquire post-cleavage images using the same settings used in Subheading 3.4.3 to assess the quality and extent of signal removal (illumination intensity, detector gain, etc.) (*see* **Note 23**).

3.4.7 Perform the Next NanoPlex Cycle

Following the first cycle, a second cycle (OneStep-IF, Imaging, Signal Erasing, Inactivation) is performed for the next set of targets using the above protocol, with minor changes to some of the steps outlined in the Notes below. This process is repeated until all targets have been imaged. In this practical example, a second ChemiPlex cycle is performed to image nuclear speckles and peroxisomes using anti-PMP70 and anti-Son 1.Abs, respectively, followed by a third ChemiPlex cycle to image Clathrin and Mitochondria using anti-Tomm20 and anti-ClathrinHC 1.Abs, respectively (*see* Table 3 and Fig. 3).

1. Replace the Imaging buffer with the Washing buffer for 2 min.
2. Initiate the next cycle by applying the pre-formed complexes of the next set of targets to the sample, as explained in Subheading 3.4.1 (*see* **Note 27**).
3. Mount the sample as explained in Subheading 3.4.2.
4. Image the sample as explained in Subheading 3.4.3. (*see* **Note 28**).
5. Erase the signals as described in Subheading 3.4.4.
6. Inactivate the sample as described in Subheading 3.4.5.
7. Assess the removal efficiency as described in Subheading 3.4.6 before starting the next cycle (*see* **Notes 29** and **30**).

4 Notes

1. Filter the BSA-PBS solution using a syringe filter of 0.45 μm before adding Triton X-100. This helps to eliminate clumps of BSA, which adsorb antibodies, adding unspecific staining.
2. 0.1 M NH_4Cl can also be used for quenching, but it needs to be prepared fresh for optimal quenching of remaining aldehydes.
3. If performing ChemiPlex: prepare TCEP buffer. If performing EnzyPlex: prepare SENP1 buffer.
4. 1.Abs should ideally be monoclonal; if polyclonal, they must be antigen- or affinity-purified. Pre-mixing 1.Abs with 2.Nbs won't work efficiently if a 1.Ab used is whole serum or Protein A/G purified, which will include antibodies with other specificities.

5. If performing EnzyPlex: use Enzy-2.Nbs. If performing Opto-Plex: use Opti-2.Nbs.
6. Place the coverslip on an open imaging chamber that sustains liquid, allows access to exchanging liquids, and fits on your microscopy stage. To avoid the sample moving in the microscope stage between buffer exchange and OneStep-IF, it is preferable to set a fixed (using, for example, plasticine) needle or plastic tip connected with thin tubing to remove liquid using a syringe on the other end of the tubing. Liquids can be added simply by carefully pipetting them on the chamber or building a second fixed-tubing system where liquid can be added using a syringe.
7. Timing should be adjusted depending on the cell type. If artifacts are generated or cellular structures are affected (e.g., tubulin filaments are disrupted and non-continue), decrease the treatment duration and test this fixation before conducting an entire multiplexing experiment.
8. An iteration of NanoPlex cycles will work optimally if the primary antibodies are well validated and provide IF with bright target-specific fluorescent signals and minimal unspecific binding. We suggest testing all primary antibodies individually before using them in a multiplex approach. Ideally, testing and titration are performed using 2.Nbs, either directly labeled or with cleavable 2.Nbs plus NbALFA* (as it will be done in NanoPlex). If unfamiliar with OneStep-IF, perform the initial test using conventional indirect or sequential IF using 2.Nbs. This will ensure that the 1.Abs work and the signal from the 2. Nbs is specific and sufficient for the microscope setup utilized.
9. Depending on the application and sample, this ratio might need to be varied to maximize the signal and minimize the background. In general, avoid using a significant excess of 2. Nbs and NbALFA* to avoid background.
10. Pre-mixing the elements to form the complex needed for OneStep-IF should be performed in a small volume to increase the chances of molecules finding each other during a short incubation. The concentration of 1.Abs should be known.
11. If performing the single-target procedure (e.g., testing individual 1.Abs): following incubation (**Step 2**), skip the blocking step (**Step 3**) and complex combination step (**Step 4**), and go straight to the permeabilization step (**Step 5**).
12. Overnight formation of the labeling complex might become convenient for longer and already established protocols. However, when overnight complex formation is employed, avoid incubation on the sample for more than 16 h since the probes' binding efficiency might decrease. Lengthy multiplexing experiments (several cycles) have a higher chance of failing if

the 1.Abs used works poorly; for example, we advise against using 1.Abs that need overnight incubation. If poor-performing 1.Abs are used, extra effort will be necessary to adjust the protocol.

13. It is essential to add the blockers only after the complexes have been formed. Otherwise, the specific signal could be reduced or entirely blocked. These blockers ensure the blockage of potential crosstalk due to unsaturated complexes. For example, in case 2.Nb detaches from the 1.Abs; the epitope would not be available for the other 2.Nb of complexed with a color intended for a different target (i.e., unwanted crosstalk) since the significant excess of Multiplexing Blockers and non-fluorescent NbALFA will block this from happening.
14. Good working 1.Abs typically requires 30–60 min of incubation with the sample. Overnight incubation with multiple pre-mixed antibody complexes is not ideal and proper complex saturation and adding Multiplex blockers is vital. If overnight incubation cannot be avoided, the protocol should be optimized further.
15. If performing the single-target procedure (e.g., testing individual 1.Abs): samples can be mounted in another chamber, slide, or holder to which an aqueous buffer solution (e.g., PBS) can be added for imaging. Imaging during NanoPlex mainly takes place with non-embedded samples; thus, during testing of 1. Abs, imaging conditions should be as similar as possible to avoid over- or underestimating the fluorescent signal.
16. Non-fluorescent grease will provide a waterproof sealing when the chamber is finally assembled.
17. Align the inlet and outlet holes carefully on the transparent cover with the channel of the silicon patch.
18. Remove any potential bubbles from the fluidics system since they can affect the sample's integrity and image acquisition.
19. Rotate the valve to the assigned position and use the peristaltic pump at 1 mL/min speed to replace the buffers thoroughly. This should take 1–2 min.
20. The Imaging buffer contains a reactive oxygen species (ROS) scavenger to minimize the destruction of epitopes (for the following IF) or the potential crosslinking and retention of the cleaved NbALFA* caused by ROS generated upon excitation of the fluorophores.
21. It is important to use the minimum excitation power needed to minimize the photo-induced crosslinking of molecules upon excitation of fluorophores. Scavengers (Imaging buffer) should help in this respect, but if possible, preventing the formation of such adducts that won't be washed away after cleavage is highly recommended.

22. Avoid prolonged and high-intensity sample illuminations to minimize potential photochemistry and retention of cleaved fluorescent compartments. In the microscopy stage, record the coordinates of the imaged region to return for the following cycles and look for another location to image.
23. Nuclear DAPI or fiducials, like gold particles or fluorescent beads, can be used as a reference to refocus or register images from different cycles.
24. If using OptoPlex: Illuminate the sample with 365 nm light for 15 min (using ~20 μW at the sample, equivalent to ~60 mW/cm^2). We recommend illuminating while the Aldehyde quenching buffer is under constant flow on the sample to increase the removal efficiency of the cleaved fluorescent fragments. If using EnzyPlex: Apply SENP1 buffer for 15 min. Decrease the flow speed to 0.5 mL/min and retain a constant flow of the buffer for the total duration of this step.
25. If using manual pipetting, exchange the buffer at least ten times during the cleaving procedure (make the exchanges during the first 2 min fast, the others with some time interval, total of ~15 min). This ensures that the cleaved-off fluorescent parts are washed away.
26. If using OptoPlex: Add or flow the Washing buffer for 5 min. If using EnzyPlex (or ChemiPlex): Apply or flow for 5 min the Thiol quenching buffer, which inactivates potential traces of the SENP1 protease that remains in the sample.
27. Use only PBS and not blocking and Permeabilization buffer for the OneStep-IF taking place after the first cycle to minimize the extraction of membranes, reducing the quality of the ultrastructure.
28. New targets might need new imaging conditions (laser power, gain, etc.). Keep these new settings the same, assessing the signal removal efficiency in Subheading 3.4.6.
29. Use the same imaging parameters to assess the signal removal efficiency that were used to image after the cleaving and washing steps (*see* Subheading 3.4.3).
30. The signal removal efficiency is assessed only if there is a subsequent round of stainings. Therefore, in this practical example where 3 cycles are used, signal removal efficiency is only assessed twice.

Acknowledgments

We thank Jannik Hentze for his excellent technical support. F.O. acknowledges the support of the Deutsche Forschungsgemeinschaft (DFG) Collaborative Research Center 1286 on Quantitative Synaptology (CRC/SFB1286) project Z04.

References

1. Valm AM, Cohen S, Legant WR et al (2017) Applying systems-level spectral imaging and analysis to reveal the organelle interactome. Nature 546:162–167
2. Wählby C, Erlandsson F, Bengtsson E et al (2002) Sequential immunofluorescence staining and image analysis for detection of large numbers of antigens in individual cell nuclei. Cytometry 47:32–41. https://doi.org/10.1002/cyto.10026
3. Schubert W, Bonnekoh B, Pommer AJ et al (2006) Analyzing proteome topology and function by automated multidimensional fluorescence microscopy. Nat Biotechnol 24:1270–1278. https://doi.org/10.1038/nbt1250
4. Friedenberger M, Bode M, Krusche A et al (2007) Fluorescence detection of protein clusters in individual cells and tissue sections by using toponome imaging system: sample preparation and measuring procedures. Nat Protoc 2(9):2285–2294. https://doi.org/10.1038/nprot.2007.320
5. Pirici D, Mogoanta L, Kumar-Singh S et al (2009) Antibody elution method for multiple immunohistochemistry on primary antibodies raised in the same species and of the same subtype. JHistochem Cytochem 57(6):567–575. https://doi.org/10.1369/jhc.2009.953240
6. Klevanski M, Herrmannsdoerfer F, Sass S et al (2020) Automated highly multiplexed super-resolution imaging of protein nano-architecture in cells and tissues. Nat Commun 11(1):1552. https://doi.org/10.1038/s41467-020-15362-1
7. Gut G, Herrmann MD, Pelkmans L (2018) Multiplexed protein maps link subcellular organization to cellular states. Science 361(6401):eaar7042. https://doi.org/10.1126/science.aar7042
8. Cole JD, Sarabia Del Castillo J, Gut G et al (2022) Characterization of the neurogenic niche in the aging dentate gyrus using iterative immunofluorescence imaging. elife 11:e68000. https://doi.org/10.7554/eLife.68000
9. Stack EC, Wang C, Roman KA et al (2014) Multiplexed immunohistochemistry, imaging, and quantitation: a review, with an assessment of Tyramide signal amplification, multispectral imaging and multiplex analysis. Methods 70(10):46–58. https://doi.org/10.1016/j.ymeth.2014.08.016
10. Parra ER, Uraoka N, Jiang M et al (2017) Validation of multiplex immunofluorescence panels using multispectral microscopy for immune-profiling of formalin-fixed and paraffin-embedded human tumor tissues. Sci Rep 7:13380. https://doi.org/10.1038/s41598-017-13942-8
11. Gorris MAJ, Halilovic A, Rabold K et al (2018) Eight-color multiplex immunohistochemistry for simultaneous detection of multiple immune checkpoint molecules within the tumor microenvironment. J Immunol 200(1):347–354. https://doi.org/10.4049/jimmunol.1701262
12. Muyldermans S (2013) Nanobodies: natural single-domain antibodies. Annu Rev Biochem 82:775–797. https://doi.org/10.1146/annurev-biochem-063011-092449
13. Sograte-Idrissi S, Schlichthaerle T, Duque-Afonso CJ et al (2020) Circumvention of common labelling artefacts using secondary nanobodies. Nanoscale 12:10226–10239. https://doi.org/10.1039/D0NR00227E
14. Pleiner T, Bates M, Görlich D (2018) A toolbox of anti-mouse and anti-rabbit IgG secondary nanobodies. J Cell Biol 217(3):1143–1154. https://doi.org/10.1083/jcb.201709115
15. Unterauer EM, Boushehri SS, Jevdokimenko K et al (2024) Spatial proteomics in neurons at single-protein resolution. Cell 187:1785–1800. e16. https://doi.org/10.5281/zenodo.10212680
16. Mougios N, Cotroneo ER, Imse N et al (2024) NanoPlex: a universal strategy for fluorescence microscopy multiplexing using nanobodies with erasable signals. Nat Commun 15:8771. https://doi.org/10.1038/s41467-024-53030-w
17. Götzke H, Kilisch M, Martínez-Carranza M et al (2019) The ALFA-tag is a highly versatile tool for nanobody-based bioscience applications. Nat Commun 10:4403. https://doi.org/10.1038/s41467-019-12301-7
18. Frey S, Görlich D (2014) A new set of highly efficient, tag-cleaving proteases for purifying recombinant proteins. J Chromatogr A 1337:95–105. https://doi.org/10.1016/j.chroma.2014.02.029

Chapter 13

Spatiotemporal Analysis of Super-resolved Live Cell Molecular Trajectory Data Using Nanoscale Spatiotemporal Indexing Clustering

Alex J. McCann, Frédéric A. Meunier, and Tristan P. Wallis

Abstract

Most, if not all, cellular processes, exemplified by neuroexocytosis, rely on the precise interplay of specific molecules in both space and time. Super-resolution microscopy allows visualization of single-molecule trajectories far below the diffraction limit of visible light, which opens up possibilities for precise spatiotemporal analysis of these cellular processes. While a range of tools have been developed for assessing the *spatial* distribution of molecules in fixed cell data, the increasing sophistication of single particle tracking (spt) in live cells requires new approaches for *spatiotemporal* analysis, which can shed light on the nanoscale dynamics of molecular interaction and clustering. In this chapter, we will discuss a recently developed suite of novel tools—**NA**noscale **S**patio**T**emporal **I**ndexing **C**lustering (NASTIC), which uses the overlap of molecular trajectory bounding boxes to establish interaction in space and time, and will describe a workflow to allow the user to use NASTIC to derive spatiotemporal metrics from their own super-resolved live cell trajectory data.

Key words Spatiotemporal analysis, Clustering, Live cell

1 Introduction

Super-resolution microscopy, and in particular single particle tracking (spt) of molecules in live cells, is gaining increasing traction as a tool to investigate the intricate dynamics of molecules in living cells with unprecedented spatial and temporal resolution. These technologies are increasingly used to elegantly confirm the long-held hypothesis that proteins rarely act in isolation, but rather need to be brought together into functional assemblies with access to substrates and cofactors, at defined regions of the cell (membrane) and at defined times. This protein clustering behavior is dynamic and activity dependent and is exemplified at the neuronal synapse, where a highly spatially and temporally coordinated series of events are required for the neurotransmitter vesicle docking, priming,

Frédéric A. Meunier (ed.), *Super-Resolution Microscopy: Methods and Protocols*, Methods in Molecular Biology, vol. 3034,
https://doi.org/10.1007/978-1-0716-5268-8_13,

fusion and recycling at the heart of neuronal communication. In a sense, the Nobel Prize-winning discovery of SNARE (**SNAP REceptor**) mediated neurotransmitter vesicle fusion [1] kickstarted the modern era of spatiotemporal molecular neurobiology, with super-resolution microscopy playing a key role.

1.1 Why Perform Spatiotemporal Analysis?

The field now finds itself increasingly equipped to answer fundamental questions such as: (1) is the protein of interest uniformly distributed or clustered into discrete regions of higher density?; (2) how many molecules are recruited into these clusters and at what density and mobility?; (3) how long do the clusters last and how quickly do they form/dissociate?; (4) is clustering recurrent, such that "hotspots" of repeated clustering are observed at discrete regions of the cell membrane?; (5) is clustering dependent on clustering of another protein?; and overarchingly, (6) do these observed clustering metrics such as the number, size, density, lifetime and recurrence of clusters correlate with experimental or biological states? The key to these analyses is determining how the super-resolved molecules are distributed in both space and time to generate a functional outcome in the cell.

1.2 Different Algorithms Can Be Used for Spatial Analysis of Fixed Cell Data

For fixed cell imaging data with no temporal component, several well-established algorithms can be used to compute the distribution of molecules and establish whether they are clustering spatially. DBSCAN (density-based spatial clustering of applications with noise) determines whether each molecular detection has a user-defined number (**MinPts**) of other detections within a user-defined radius (ε) [2]. By iterating through all detections, clusters of higher-density molecular data can be quantified. Voronoï tesselation takes a fundamentally different image segmentation approach by drawing minimum size tiles around all detections, such that each tile contains one detection and none of the tile edges are closer to any other points [3, 4]. The higher the detection density, the smaller the tiles. By defining a tile area threshold below which a detection is considered to be in close proximity to other detections, regions of higher density clustered detections can be derived.

1.3 Spatiotemporal Analysis of Live Cell Data

Analysis of live cell trajectory data is more challenging. This data is inherently 3-dimensional, with each molecular trajectory described by spatial x, spatial y, and time [x, y, t] coordinates. If the time dimension is ignored, then purely spatial metrics can be derived using DBSCAN or Voronoï, along with other spatial algorithms such as Ripley's K function [5] and, increasingly, computer vision and AI-based approaches [6]. For spatiotemporal analysis of [x, y, t] trajectory data, it is possible to use one of the above-described techniques to define regions of clustering, and subsequently manually examine the temporal distribution of the trajectories within these regions, using techniques such as time correlated

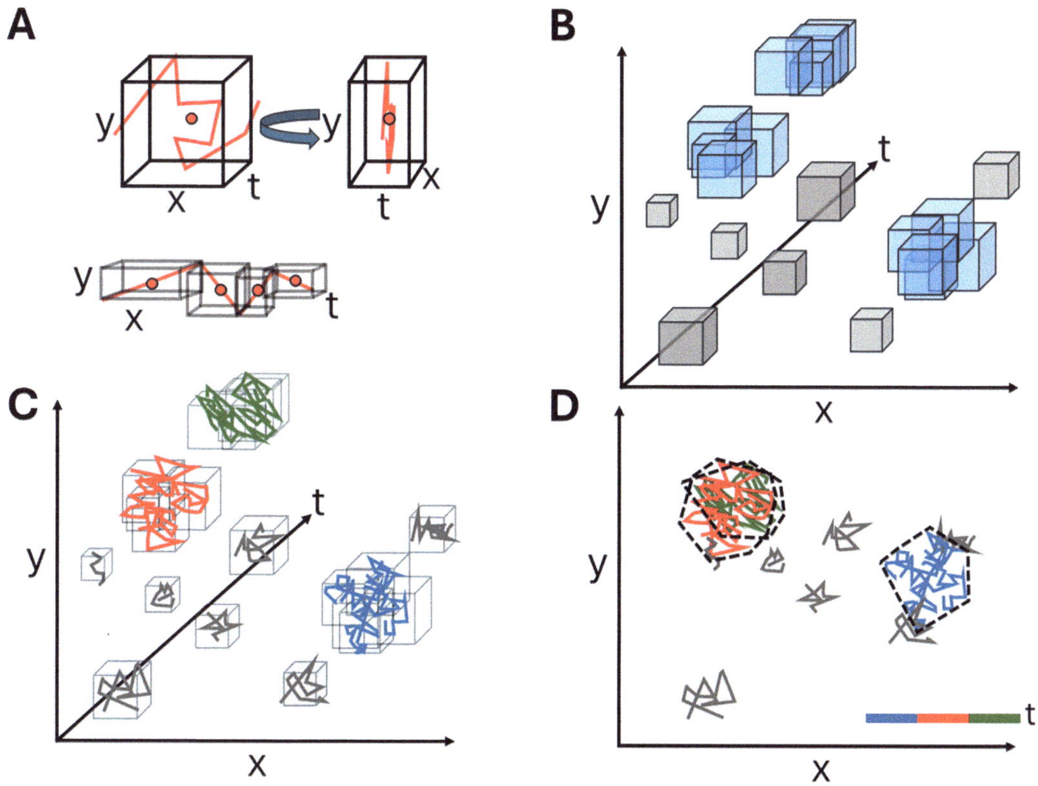

Fig. 1 Spatiotemporal indexing clustering. (**a**) For NASTIC, each molecular trajectory (red, with its centroid indicated by the red dot) is described by an idealized bounding box with a user-defined "thickness" in the time (t) dimension. Optionally, for segNASTIC, each segment of the trajectory is assigned a bounding box. (**b**) Bounding boxes are indexed into an R-tree spatial database, which allows determination of which boxes overlap (blue). (**c**) The trajectories associated with overlapping bounding boxes are considered as members of spatiotemporal clusters, colored according to the time of the acquisition. (**d**) Clusters are represented by a convex hull around the component spatiotemporally overlapping trajectories. Recurrent clusters appear in the same spatial extent but at different times

photoactivatable light microscopy (tcPALM) [7, 8]. The limitations of these post hoc approaches prompted our laboratory to develop alternative approaches that enable the simultaneous determination of molecular trajectory proximity to other trajectories in both space and time. The algorithm, Nanoscale Spatio Temporal Indexing Clustering (NASTIC) [9, 10], determines an idealized bounding box describing the spatial extent of all the molecular detections comprising the trajectory, and then assigns a user defined temporal "thickness" to each box (Fig. 1a). These 3D spatiotemporal bounding boxes are used to create an R-tree [11] index (Fig. 1b), which allows rapid determination of which boxes are overlapping spatially and temporally. Trajectories whose bounding boxes overlap are considered to be spatiotemporally clustered (Fig. 1c, d).

1.4 NASTIC

Although conceptually simple, the NASTIC approach is potentially complex and confusing to a user unfamiliar with its precepts, and NASTIC is thus implemented as a user-friendly Graphic User Interface (GUI) which can run on any computer architecture capable of supporting the Python programming language. NASTIC presents the user with a logical pipeline that streamlines: (1) the process of loading and visualizing super-resolution data; (2) selecting regions for analysis; (3) performing clustering with sensible default parameters; (4) visualizing a large range of spatiotemporal metrics derived from the clustering; and (5) saving this information for further meta-analysis using GUI tools provided as part of the NASTIC software suite. NASTIC has allowed our laboratory to derive unprecedented insights into the membrane dynamics of key proteins involved in neuroexocytosis and neuronal function, demonstrating how key metrics such as cluster density, size, molecular membership and lifetime change in response to activity, and for the first time providing evidence for recurrent clustering on the synaptic membrane [9, 10, 12–17]. In the following sections we will take the user through the process of:

- Downloading and installing the NASTIC suite (Python and the modules required for NASTIC operation).
- Processing super-resolution data through the NASTIC pipeline, including considerations for optimum analysis parameters.
- Performing meta-analysis on the NASTIC outputs from multiple samples across two experimental conditions to quantify how the clustering behavior of molecules differs between them.

Of note, existing trajectory files are required for this protocol. For information on the acquisition, localization and tracking of single-particle data (to obtain trajectory files), see Chapter 6.

2 Materials

2.1 Software

1. Python (v3.8 or greater).
2. Python modules: colorama, matplotlib, matplotlib-venn, numpy, pandas, Pillow, FreeSimpleGUI, rtree, scikit-learn, scipy, seaborn, statsmodels.
3. Super Res Data Wrangler (v1.0.12).
4. NASTIC (v1.0.12).
5. NASTIC Wrangler (v1.0.12).

2.2 Input Files (NASTIC)

1. Trajectory files in the TRXYT file format (for conditions 1 and 2) (*see* **Note 1**).
2. (Optional) Region of interest (ROI) files (one for each TRXYT file) (*see* **Note 2**).

2.3 Input Files (NASTIC Wrangler)

1. Folder containing NASTIC metrics.tsv files (for condition 1) (*see* **Note 3**).
2. Folder containing NASTIC metrics.tsv files (for condition 2) (*see* **Note 3**).

3 Methods

The following protocol comprises six main stages: (1) downloading and installing Python along with the required Python modules, (2) downloading the NASTIC software suite, (3) selecting the architecture-appropriate NASTIC script, (4) converting trajectory files into the trxyt file format using the Super Res Data Wrangler, (5) performing spatiotemporal cluster analysis in NASTIC and (6) conducting meta-analysis of cluster metrics in NASTIC Wrangler. For information on obtaining trajectory files (a prerequisite for **step 4**), see Chapter 6, which describes the acquisition, localization and tracking of single-particle data.

3.1 Download Python and Install Python Modules

1. Go to the Python downloads page https://www.python.org/downloads/. Click on the operating system that your computer uses (e.g., Windows, Linux/Unix or macOS) to go to the operating system-specific downloads page.
2. Click on the download link appropriate for your operating system architecture (e.g., 32bit or 64bit).
3. Once downloaded, run the Python installer. If applicable, tick the checkbox to save Python to PATH, which allows Python to be run by typing "python," rather than its full path, in a terminal. Alternatively, (Windows): go to Advanced System Settings > Environment Variables > PATH, then add the directory for Python to the PATH variable. (Mac): after the installation is complete, double-click on the "Install Certificates.command" file in the /Applications/Python folder.
4. To install the Python modules required for running NASTIC and NASTIC Wrangler, open a terminal window and type the following (followed by the return key): `python -m pip install FreeSimpleGUI colorama matplotlib matplotlib-venn numpy pandas Pillow rtree scikit-learn scipy seaborn statsmodels`.

3.2 Download NASTIC Software Suite

1. Go to the NASTIC GitHub https://github.com/tristanwallis/smlm_clustering. Navigate to the latest version of the NASTIC suite (v1.0.12 of writing) by clicking on the "NASTIC Latest" button on the right-*hand panel.

2. Download the latest version of the NASTIC suite by clicking on the NASTIC zip folder (e.g., NASTICv1_0_12.zip) under "Assets."
3. Extract the downloaded NASTIC folder to the desired directory (e.g., Desktop or Documents folder).

3.3 Select that NASTIC Software to Use

The NASTIC software suite package provides the user with the option of using different versions of the same software (e.g., *NASTIC* has a multi-threaded graphical user interface (GUI) version (nastic_gui.py), a single-threaded (compilable) GUI version (nastic_st_gui.py) and a single-threaded command line (CLI) version (nastic_cli.py).

1. Choose between the graphical user interface (GUI) and command line (CLI) versions of the *Super Res Data Wrangler* and *NASTIC* (*see* **Notes 4** and **5**).
2. **GUI:** if applicable, choose between the multi-threaded and single-threaded (ST) versions of *NASTIC* (*see* **Note 6**).

3.4 Super Res Data Wrangler

Super Res Data Wrangler helps bridge the gap between different spt analysis software by enabling the user to convert between various trajectory file formats.

3.4.1 Run Super Res Data Wrangler

1. Run *Super Res Data Wrangler* by either double-clicking on the script (e.g., double-click on super_res_data_wrangler_gui.py or super_res_data_wrangler_cli.py in the NASTIC folder) or by opening a terminal window, navigating to the location of the script, typing `python` followed by the name of the script (e.g., `python super_res_data_wrangler_gui.py` or `python super_res_data_wrangler_cli.py`) and pressing the return key.

3.4.2 "Select Filetypes" Tab

1. "Input file"/"file type to convert from": **GUI:** click on the arrow next to "Input filetype:" and select the format of trajectory file that will be converted. Make sure that the format of the trajectory file that you will be converting follows the formatting described on the selected input filetype tab in the Super Res Data Wrangler. **CLI:** Type the letter corresponding to the format of the trajectory file that you want to convert (letter surrounded by square brackets), followed by the return key (*see* **Note 7**)
2. "Output file"/"file type to convert to": **GUI:** click on the arrow next to "Output filetype:" and select the format of the trajectory file to convert to. **CLI:** Type the letter corresponding to the trajectory file format that you want to obtain (letter surrounded by square brackets), followed by the return key (*see* **Note 8**).

3. **GUI:** Click on the "CONFIRM FILETYPES" button to confirm the input and output trajectory file formats, and swap to the next tab.

3.4.3 "Load Files" Tab

1. **GUI:** (Convert a single file): Click on the "Browse file" radio, followed by the "Browse file" button to open a Browse file window. Using the Browse file window, select the trajectory file to convert. (Convert multiple files): Click on the "Browse folder (directory)" radio, followed by the "Browse folder" button to open a Browse Folder window. Using the Browse folder window, select the top-level folder containing the trajectory files that you wish to convert.
2. **GUI:** (Convert multiple files): Click on the "Find files" button to populate the table below with trajectory files found in the selected directory.
3. **GUI:** (Convert multiple files): Tick/untick the checkboxes next to the files that you wish to include/exclude. **CLI:** Type in the numbers corresponding to the files that you wish to convert (with each file number separated by a comma), before pressing the return key to confirm.
4. **GUI:** Click on the "LOAD DATA" button to confirm input file selection and swap to the next tab.

3.4.4 "Set Parameters" Tab

1. "X,Y units": **GUI:** If applicable, the "X,Y units" parameter will be enabled. Click on the arrow next to "X,Y units:" and select the appropriate units from the list (px = pixels, um = microns, nm = nanometers).
2. "Pixel size (um/px)": **GUI:** If applicable, the "Pixel size (um/px)" parameter will appear. Click on the text box and type in the pixel size of the trajectory file. **CLI:** If applicable, a prompt will appear asking the user to enter the pixel size. Type in the pixel size, followed by the return key (*see* **Note 9**).
3. "Time units": **GUI:** If applicable, the "Time units" parameter will become enabled. Select the units of the column containing time data (Frames = frame number, s = seconds).
4. "Acquisition frequency (Hz)": **GUI:** If applicable, the "Acquisition frequency (Hz)" parameter will appear. Click on the textbox and type in the acquisition frequency. **CLI**: If applicable, a prompt will appear asking the user to enter the acquisition frequency. Type in the acquisition frequency, followed by the return key (*see* **Note 10**).
5. "Number of headers": **GUI:** If applicable, the "Number of headers" parameter will be enabled. Select the number of rows at the top of the trajectory file that contain non-data information (e.g., are blank or contain headers).

6. "Delimiter": **GUI:** If applicable, select the character that separates each row of the trajectory file into separate columns (comma, tab, 1 space, 2 spaces, 3 spaces, 4 spaces, semicolon).
7. "Trajectory col": **GUI:** If applicable, select the column number that contains the trajectory number information.
8. "X-position col": **GUI:** If applicable, select the column number that contains the x-coordinate information.
9. Y-position col.: **GUI:** If applicable, select the column number that contains the y-coordinate information.
10. "Time col"/"Frame col.:" **GUI:** If applicable, select the column that contains the time information (in seconds) or the frame information (in frame numbers).
11. "Starting frame": **GUI:** If applicable, select whether the numbering of the first frame starts with 0 or 1.

3.5 NASTIC

Nanoscale SpatioTemporal Indexing Clustering (NASTIC) is used to perform cluster analysis on trajectory files in the trxyt file format and output various cluster metrics. *NASTIC* comes with a comprehensive user manual that is located in the same folder as *NASTIC* itself. Additionally, in the GUI version of *NASTIC*, hovering over each parameter and button will display a tooltip with helpful information. Of note, *NASTIC* is updated regularly with new and improved features. Make sure to visit GitHub regularly to check for the latest version of the NASTIC suite: https://github.com/tristanwallis/smlm_clusteringClustering.

3.5.1 Run NASTIC

1. **GUI:** Run *NASTIC* by either double-clicking on the script (e.g., double-click on nastic_gui.py or nastic_st_gui.py) or by opening a terminal window, navigating to the location of the script, typing `python` followed by the name of the script (e.g., `python nastic_gui.py` or `python nastic_st_gui.py`) and pressing the return key. A graphical user interface will appear shortly after. **CLI:** Run *NASTIC* by opening a terminal window, navigating to the location of the script, typing `python` followed by the name of the script and the input trajectory filename, and pressing the return key (e.g., python nastic_cli.py filename.trxyt). The script will generate a metrics.tsv file using hardcoded clustering parameters for use in NASTIC Wrangler (Subheading 3.6). The remaining steps in this section (Subheadings 3.5.2, 3.5.3, 3.5.4, 3.5.5, and 3.5.6) are for the GUI version only.

3.5.2 "File" Tab—Load a Trxyt File

1. Click on the "Browse" button to open a file browser window. Use the file browser window to navigate to and select the trxyt file that you would like to perform cluster analysis on.

2. Set the "Minimum trajectory length" and "Maximum trajectory length" parameters to 8 and 1000, respectively (*see* **Note 11**).
3. Press the "PLOT RAW DETECTIONS" button to plot the detections in the trxyt file and change to the next tab. The raw detections will be displayed in a new window titled "Main display window—DO NOT CLOSE!". If this window is closed at any point, the analysis will have to be restarted with the replotting of the raw detections.

3.5.3 "ROI" Tab—Select a Region of Interest (ROI)

1. Select a region of interest (ROI, only use detections within this area), using one of the following options. Option 1: Load a previously saved ROI by clicking on the "Load" button to open a Browse file window and selecting the ROI file (*see* **Note 2**). Make sure the pixel size is correct before pressing the "Replot ROI" button to replot the ROI with the new pixel size (*see* **Note 9**). Option 2: Select the entire acquisition by pressing the "All" button. Option 3: Draw a rectangle ROI around a part of the acquisition by clicking on the magnifying glass symbol at the bottom of the "Main display window" and then drawing the rectangle directly on the "Main display window." Press the "Add" button to confirm ROI selection. Option 4: Hand draw a ROI by making sure that the magnifying glass symbol at the bottom of the "Main display window" is unticked, and then free-hand drawing the ROI directly on the "Main display window." Press the "Add" button to confirm ROI selection.
2. (Optional): Save the ROI(s) by clicking on either the "Save" button (if present multiple ROIs will be saved in the same ROI file) or the "Save Separately" button (each ROI will be saved as a separate ROI file).
3. Make sure that the "Cluster immediately" button is unticked before pressing the "SELECT DATA IN ROIS" button to confirm the ROI(s) and swap to the next tab.

3.5.4 "Clustering" Tab—Define the Clustering Parameters

1. Set the "Acquisition time (s)" parameter to the length of time taken to acquire the entire acquisition (*see* **Note 12**).
2. Set the "Frame time (s)" parameter to the length of time taken to acquire each individual frame (*see* **Note 9**).
3. Set the "Time threshold (s)" parameter to the desired value. This is the window of time in which to look for overlapping trajectories (*see* **Note 13**).
4. Set the "Radius factor" parameter to the radius around each trajectory centroid to look for centroids from other trajectories (overlap) (*see* **Note 14**).

5. Set the "Cluster threshold" parameter to the minimum number of trajectories that must have overlapping bounding boxes (within both time threshold and the radius defined by radius factor) to be considered clustered (*see* **Note 15**).
6. Set the "Minimum cluster radius" and "Maximum cluster radius" parameters to the smallest radius and largest radius (respectively) that a cluster is allowed to be (*see* **Note 16**).
7. (Optional): Tick the "MSD screen" checkbox to exclude trajectories with a mean square displacement (MSD) greater than the average MSD of all trajectories. Trajectories that are excluded by this screen are still present but are considered to be unclustered. (*see* **Note 17**).
8. (Optional): Tick the "Diffusion coefficient screen (log10)" checkbox to filter out trajectories with log10 diffusion coefficient values that are lower than the "Min" value and higher than the "Max" value. Trajectories with diffusion coefficient values outside of these parameters are removed completely. Set the minimum and maximum log10 diffusion coefficient values to the desired values (*see* **Note 18**).
9. Press the "CLUSTER SELECTED DATA" button to cluster the detections in the defined ROI using the selected clustering parameters, and swap to the next tab. If the "Plot immediately" checkbox is ticked, the 2D plot of clustered trajectories will replace that of the raw detections in the "Main display window—DO NOT CLOSE!" plot. If this window is closed at any point, the analysis will have to be restarted with the replotting of the raw detections (Subheading 3.4.2, **step 3**).

3.5.5 "Display" Tab—Adjust the Aesthetics of the 2D Plot

Various aspects of the 2D plot shown in the "Main display window—DO NOT CLOSE" window can be altered by adjusting the parameters in each of the subtabs (Trajectory, Centroid, Cluster, Hotspot, Colorbar, Scalebar) and pressing the "PLOT CLUSTERED DATA" button to display the changes. If the "Main display window—DO NOT CLOSE" window is closed at any point, analysis will have to be restarted at Subheading 3.4.2, **step 3.**

1. (Optional): Choose which features to display/hide on the 2D plot by ticking/unticking the "Traj.", "Centr.", "Clust.", "Hotsp.", "Colorbar" and "Scalebar" checkboxes. (*see* **Note 19**).
2. (Optional): Adjust the aesthetics of clustered and unclustered trajectories by using the options in the "Trajectory" subtab (*see* **Notes 20–22**).
3. (Optional): Adjust the aesthetic of trajectory centroids using the "Centroid" subtab (*see* **Note 23**).

4. (Optional): Adjust the aesthetics of clusters using the "Cluster" subtab (*see* **Note 24**).
5. (Optional): Adjust the aesthetics of hotspots using the "Hotspot" subtab (*see* **Note 25**).
6. (Optional): Adjust the aesthetics of the colorbar using the "Colorbar" subtab (*see* **Note 26**).
7. (Optional): Adjust the aesthetics of the scalebar using the "Scalebar" subtab (*see* **Note 27**).
8. (Optional): Adjust how each plot is saved using the "Export" subtab (*see* **Note 28**).
9. Click on the "PLOT CLUSTERED DATA" button to view any changes made in the "Display" tab on the 2D plot, and click on the "SAVE PLOT" button to save the current version of the 2D plot.

3.5.6 "Metrics" Tab—Plot Cluster Metrics

Do not close any plot windows that you want to keep (only plot windows that are currently open will be saved upon pressing the "SAVE ANALYSES" button).

1. "MSD": Plot the Mean Square Displacement (MSD) plot by clicking on the "MSD" button (*see* **Notes 22** and **29**).
2. "Hotspot": Plot hotspot metrics by clicking on the "Hotspot" button (*see* **Note 30**).
3. "PCA": Plot a 3D principal component analysis (PCA) plot by clicking on the "PCA" button (*see* **Note 31**).
4. "3D": plot an interactive 3D version of the 2D plot shown in the "Main display plot—DO NOT CLOSE!" window by clicking on the "3D" button (*see* **Note 32**).
5. "KDE": plot a 2D Kernel Density Estimation (KDE) plot by clicking on the "KDE" button (*see* **Note 33**).
6. "Diffusion coefficient": plot a 1D plot, 2D plot and frequency distribution histogram by clicking on the "Diffusion coefficient" button (*see* **Notes 18** and **34**).
7. "Density over time": plot the changes in density (trajectories/μm^2) over the entire acquisition by clicking on the "Density over time" button (*see* **Note 35**).
8. "Vector autoregression": Plot a Venn diagram of "Clustered" and "Confined" trajectories and another Venn diagram of "Unclustered" and "Unconfined" trajectories (as determined by VAR) by clicking on the "Vector autoregression" button (*see* **Notes 21** and **36**).
9. Press the "SAVE ANALYSES" button to save all open plots, cluster metrics and ROIs used in a date-stamped folder (*see* **Note 37**).

3.6 NASTIC Wrangler

NASTIC Wrangler is used to compare the clustering metrics of two different conditions (metrics.tsv files output by NASTIC). Of note, the processed_metrics.tsv file generated by *NASTIC Wrangler* will be saved in the same location as the NASTIC Wrangler script, rather than in the folders containing the original metrics.tsv files. Therefore, it is useful to copy and paste the NASTIC Wrangler script in the same directory as the two top-level folders (directory that contains both the Condition 1 folder and the Condition 2 folder).

3.6.1 Organize Files into Folders

1. Create a new folder/directory. Copy-paste the nastic_wrangler_gui.py script from the downloaded NASTIC folder to this new folder.
2. In this new directory, create 2 appropriately named folders—one for Condition 1 and one for Condition 2.
3. Copy-paste all the NASTIC output folders that contain metrics.tsv files for Condition 1 into the new Condition 1 folder.
4. Copy-paste all the NASTIC output folders that contain metrics.tsv files for Condition 2 into the new Condition 2 folder.

3.6.2 Run NASTIC Wrangler

1. Run NASTIC Wrangler by either double-clicking on the script (e.g., double-click on nastic_wrangler_gui.py in the new folder/directory) or by opening a terminal window, navigating to the location of the script, typing `python` followed by the name of the script (e.g., `python nastic_wrangler_gui.py`) and pressing the return key.

3.6.3 "Load Files" Tab

1. Condition 1 (left-hand side): Press the "Browse" button to open a Browse Folder window. Navigate to the location of the folder containing the NASTIC metrics.tsv files for Condition 1.
2. Condition 1 (left-hand side): Press the "Find files" button to populate the table below with a list of NASTIC metrics.tsv files found in the folder that was selected for Condition 1.
3. Condition 1 (left-hand side): Tick/untick the checkboxes for the files in the table to include/exclude them from analysis.
4. Condition 2 (right-hand side): Press the "Browse" button to open a Browse Folder window. Navigate to the location of the folder containing the NASTIC metrics.tsv files for Condition 2.
5. Condition 2 (right-hand side): Press the "Find files" button to populate the table below with a list of NASTIC metrics.tsv files found in the folder that was selected for Condition 2.

6. Condition 2 (right-hand side): Tick/untick the checkboxes for the files in the table to include/exclude them from analysis.
7. Press "LOAD DATA" button to confirm the file selection for Conditions 1 and 2, and swap to the next tab.

3.6.4 "Plot Data" Tab

1. Replace "COND1" with a short-hand name for Condition 1 that will appear on the plots (*see* **Note 38**).
2. (Optional): click on the "Condition 1" button next to "Select color" to change the color for Condition 1 on the plots.
3. Replace "COND2" with a short-hand name for Condition 2 that will appear on the plots (*see* **Note 38**).
4. (Optional): click on the "Condition 2" button next to "Select color" to change the color for Condition 2 on the plots.
5. (Optional): Tick the "Outlier Removal" checkbox to perform median filtering outlier reduction on aggregate data.
6. (Optional): Tick the "Aggregate Data" checkbox to generate plots that combine metrics from all clusters observed across all files for each condition (N = total number of clusters) (*see* **Note 39**).
7. (Optional): Tick the "Average Data" checkbox to generate plots that derive an average for each metric across the number of samples for each condition (N = number of samples) (*see* **Note 40**).
8. (Optional): Tick the "2D PCA analysis" checkbox to generate principal component analysis (PCA) plots where each datapoint represents a subset of normalized average metrics (*see* **Note 41**).
9. (Optional): Tick the "Frequency distributions" checkbox to generate frequency distribution histograms of each cluster metric, in addition to average trajectory length, trajectory step length and $\log_{10}$ diffusion coefficient values for each condition (*see* **Note 42**).
10. Press the "PLOT DATA" button to perform meta-analysis using the above settings. The processed_metrics.tsv file and each of the selected plots will be saved in the same directory as the NASTIC Wrangler script.

4 Notes

1. The TRXYT file format does not contain any headers (no header or blank rows before start of data rows), and contains 4 space-separated columns. Column 1: trajectory number, Column 2: x-coordinate (μm), Column 3: y-coordinate (μm), Column 4: time (s). Various trajectory filetypes can be

converted into the trxyt file format using the Super Res Data Wrangler (part of the NASTIC software package), which is covered in Subheading 3.4. Acquisition, localization and tracking of trajectory files is covered in Chapter 6.

2. ROI filetypes accepted by *NASTIC*: (1) NASTIC (roi_coordinates*.tsv); (2) PalmTracer (*.rgn); (3) FIJI/TrackMate (*XY_Coordinates*.csv). * = any character.

3. *NASTIC* performs spatiotemporal clustering analysis on trxyt trajectory files, and outputs the results of the analysis in metrics.tsv files. *NASTIC Wrangler* performs meta-analysis on metrics.tsv files from two conditions (e.g., Condition 1 = untreated cells and Condition 2 = treated cells). These metrics.tsv files should be organized into two separate folders (e.g., Condition 1 and Condition 2).

4. CLI: perform rapid batch conversion of trajectory files using predefined file formatting parameters (PalmTracer.txt, PalmTracer.trc, SharpViSu.ascii, NASTIC.trxyt and TrackMate.csv). Assumes TrackMate.csv files are in pixels (XY units) and frames (Time units). GUI: user-friendly interface that enables batch conversion of trajectory files using predefined file formatting parameters (PalmTracer.txt, PalmTracer.trc, SharpViSu.ascii and NASTIC.trxyt) or user-defined file formatting parameters (TrackMate.csv files and other trajectory filetypes not listed above).

5. CLI: perform rapid spatiotemporal cluster analysis on a trajectory file (.trxyt format) using hardcoded clustering parameters. Outputs a metrics.tsv for meta-analysis in *NASTIC Wrangler*. GUI: user-friendly interface that performs spatiotemporal cluster analysis on a trajectory file (.trxyt format) using user-defined clustering parameters. Enables the user to select a region of interest (ROI) file for defining the selection area on which to perform cluster analysis. Outputs a metrics.tsv file for meta-analysis in *NASTIC Wrangler*, the ROI used for clustering, 2D and (optional) 3D clustering plots with aesthetics that can be adjusted by the user, and (optional) an array of other cluster metrics plots.

6. Select the multi-threaded version of the GUI for use on physical hardware (e.g., nastic_gui.py). Alternatively, select the single-threaded (compilable) version for use on virtual hardware (e.g., nastic_st_gui.py).

7. There are several trajectory types with preset parameters (e.g., preset xy-coordinate and time units) available to choose from: (1) PalmTracer.txt file (*.txt); (2) PalmTracer.trc file (*.trc); (3) NASTIC.trxyt file (*.trxyt); (4) SharpViSu.ascii file (before drift correction) (*.ascii, requires *.id file with same name); (5) SharpViSu.ascii file (after drift correction) (*.ascii, requires

.id file with same name); (6) TrackMate.csv file (.csv). CLI: a = uncorrected SharpVisu.ascii file; d = drift corrected SharpViSu.ascii file; t = PalmTracer.txt file; r = PalmTracer.trc file; x = NASTIC.trxyt file; c = TrackMate.csv file (XY units = microns; Time units = Frames). Conversion of TrackMate.csv files with XY units in pixels or Time units in seconds, and trajectory files in file formats other than the ones listed above, can be performed using the GUI (other(manual) option).

8. CLI: a = SharpVisu.ascii file (before drift correction); d = SharpViSu.ascii file (after drift correction); t = PalmTracer.txt file; r = PalmTracer.trc file; x = NASTIC.trxyt file. An.id file will be generated for each.ascii file that is generated—this.id file will be needed for converting back from the.ascii file format to other file formats.

9. The pixel size (microns per pixel) parameter is used as a conversion factor for converting between pixels and microns. It is also used for calculating the threshold for the $\log_{10}$ diffusion coefficient frequency distribution histogram that separates the molecules into two populations: the "immobile" fraction ($\log_{10}$ diffusion coefficient values lower than the threshold) and the "mobile" fraction ($\log_{10}$ diffusion coefficient values higher than the threshold).

10. The acquisition frequency (Hz) and frame time (s) parameters refer to the frequency and length of time taken to acquire each individual frame respectively. Acquisition frequency (Hz) = 1/frame time (s); Frame time (s) = 1/Acquisition frequency (Hz). E.g., if frame time = 0.02 s (20 ms): acquisition frequency = 1/0.02 = 50 Hz.

11. We recommend screening trajectories by using a minimum trajectory length of 8 and a maximum trajectory length of 1000. Very short trajectories often contain too few data points to reliably characterize particle mobility, while excessively long trajectories may be a result of fluorescent artifacts or background signals and are more prone to drift, which can distort diffusion coefficient and mean square displacement (MSD) calculations.

12. Acquisition time = frame time (amount of time taken to acquire an individual frame) multiplied by the total number of frames in the acquisition. E.g., if 16,000 frames were acquired with each one taken at 20 ms: acquisition time = 0.02 s * 16000 = 320 s.

13. The default time threshold is 20 s, which means for each trajectory centroid, NASTIC searches for overlapping trajectories occurring within 10 s before and 10 s after the trajectory centroid timepoint. Increasing this value will increase the

likelihood of trajectories overlapping in time, whereas decreasing it reduces this likelihood. It may be useful to decrease the time threshold in the case of high-density data in order to prevent spurious overlap.

14. The default radius factor is 1.2. This value is applied as a multiplier to the bounding box radius of each trajectory to determine the radius used when searching for spatially overlapping trajectories. Decreasing the radius factor may be useful when analyzing high-density data, as it limits the chance of spurious overlap.
15. The default cluster threshold is 3, meaning that at least three trajectories must have bounding boxes that overlap in both space and time to be considered part of a cluster. Increasing this value can be useful when analyzing high-density data, as it imposes a more stringent criterion for clustering.
16. After clusters have been identified, they are further screened based on cluster radius. The default minimum and maximum cluster radius values are 0.05 μm and 0.2 μm, respectively. Clusters with a radius outside of this range (i.e., smaller than 0.05 μm or larger than 0.2 μm) are excluded from subsequent analyses. Increasing the maximum cluster radius may improve the retention of larger clusters; however, it can also lead to the retention of spurious clusters formed by overlapping trajectories with high mobilities, which cover larger areas due to their higher speeds.
17. Mean square displacement (MSD) quantifies a molecule's mobility over time. Molecules with higher mobilities typically travel larger distances, resulting in larger MSD values, whereas molecules with lower mobilities travel shorter distances, yielding smaller MSD values.
18. We recommend screening trajectories based on their $\log_{10}$ diffusion coefficients using a minimum value of -5 and a maximum value of 1. This is because trajectories with diffusion coefficient values lower than this might be due to artefacts, background or trapped molecules whose behavior does not reflect that of the broader population. Alternatively, trajectories with diffusion coefficient values that are higher than this range might be a result of mistracking.
19. "Traj." = show trajectories; "Centr." = show trajectory centroids, "Clust." = show clusters, "Hotsp." = show cluster hotspots, "Colorbar" = show colorbar and "Scalebar" = show scalebar. Trajectories and trajectory centroids can be toggled on/off for the 2D, 3D and KDE plots. Clusters and hotspots can be toggled on/off for the 2D and 3D plots. Colorbars and scalebars can be toggled on/off for the 2D, 3D, KDE and Diffusion coefficient plots.

20. “Trajectory” subtab: “Width” = trajectory line width/thickness, “Opacity” = trajectory line opacity/transparency. Trajectory color options (only one option/checkbox can be selected at a time): “Cluster” = color clustered trajectories by the color of the cluster they belong to (unclustered trajectories colored by the “Color Choose” button); “Time” = Color all trajectories by time of appearance in the acquisition; “Random” = arbitrarily color trajectories using 8 bright colors; “Random(pastel)” = arbitrarily color trajectories using 14 pastel colors; “VAR” = color trajectories based on whether they are considered “confined” or “unconfined” based on vector autoregression (VAR) analysis; “MSD” = color trajectories based on whether they are higher or lower than the average mean square displacement (MSD). Change the colors of the VAR and MSD trajectory color options by clicking on the “<” and “>” buttons and selecting a new color. The “Traj.” checkbox must be ticked to display trajectories on the 2D, 3D and KDE plots. KDE plot trajectory color is unaffected by the color options selected in the “Display” tab.
21. NASTIC applies vector autoregression (VAR) analysis to classify each trajectory as either “confined” or “unconfined.” This classification is based on how strongly each trajectory step depends on the previous step, as well as on the overall range of motion of the trajectory. Comparing VAR classifications with identified clusters can help validate clustering accuracy, as trajectories within clusters typically exhibit confined motion, whereas unclustered trajectories tend to diffuse more freely.
22. NASTIC calculates the mean square displacement (MSD) of each trajectory to quantify its mobility over time. As with VAR analysis, comparing MSD plots with cluster maps can help assess clustering accuracy. Trajectories with lower MSD values generally correspond to molecules in confined or clustered regions, while higher MSD values indicate more freely diffusing unclustered molecules.
23. “Centroid” subtab: “Size” = centroid size, “Opacity” = centroid opacity/transparency. Click on the “Color Choose” button to open a colorpicker window and select a new centroid color. The “Centr.” checkbox must be ticked to display the centroids on the 2D, 3D and KDE plots. KDE plot centroid color is unaffected by the color options selected in the “Display” tab.
24. “Cluster” subtab: “Opacity” = cluster opacity/transparency, “Line width” = cluster outline line width/thickness; “Line type” = line style of the cluster outline. Tick/untick the “Filled” checkbox to toggle between displaying clusters as a filled shape or as a colored outline. Click on the “Color

Choose" button to open a colorpicker window and select a new cluster color. This option is overridden if the "Time" checkbox is ticked (color trajectories by time of appearance in the acquisition). The "Clust." checkbox must be ticked to display clusters on the 2D and 3D plots.

25. "Hotspot" subtab: "Radius" = distance threshold for grouping neighboring clusters into the same hotspot, "Opacity" = hotspot outline opacity/transparency, "Line width" = hotspot outline width/thickness, "Line type" = style of hotspot outline. Click on the "Color Choose" button to open a colorpicker window and select a new hotspot outline color. The "Hotsp." checkbox must be ticked to display hotspots on the 2D plot. The default value for Radius is 1, meaning that two cluster centroids must be closer that the average cluster radius to be considered part of the same hotspot. If Radius > 1, cluster centroids can be further apart than the average cluster radius and still be grouped into the same hotspot. If Radius < 1, the cluster centroids must be closer together than the average cluster radius to qualify as part of the same hotspot.
26. "Colorbar" subtab: Tick the "Border" checkbox to include a colored border (white or black) around the colorbar on 2D, KDE and Diffusion coefficient plots. "Color" = choose between white and black colorbar borders. "Line width" = width of the colored border around the colorbar. Tick the "Values" checkbox to display minimum and maximum values on the 2D, MSD and KDE plots. "Text size" = size of colorbar value text. "Text font" = font style of colorbar value text. "Colorbar" checkbox must be ticked to display colorbars on the 2D, KDE and Diffusion coefficient plots. Colorbars will only appear on the 2D plot if trajectories are colored by "Cluster" or "Time." Colorbar values will change depending on the type of plot.
27. "Scalebar" subtab: "Length" = size of scalebar, "Units" = scalebar units. Tick the "Value" checkbox to display numerical values on the scalebar. "Text size" = size of the value text, "Text font" = font style of value text. Tick the "Decimal" checkbox to write the "Value" text as a decimal (rather than integer). "Color" = color of scalebar (white or black), "Thickness" = height of scalebar. The "Scalebar" checkbox must be ticked to display a scalebar on the 2D, KDE and diffusion coefficient plots. KDE and diffusion coefficient plot scalebar color is unaffected by the color options selected in the "Display" tab.
28. "Export" subtab: "Format" = file format of saved plots. Tick "Transparent background" to save plots with a transparent background. "DPI" = resolution at which to save plots (dots

per inch). Options apply to 2D cluster plot and all open metrics plots. Press the "Choose" button next to "Directory" to select a folder to save the 2D main display plot.

29. "MSD" metrics plot: The default plot shows the MSDs of clustered (yellow) and unclustered (blue) trajectories. To generate an MSD plot with the clustered and unclustered trajectories combined (all trajectories, black), click on the "Plot All MSD" checkbox before pressing the "MSD" button. To hide/unhide the figure legend on the MSD plots (for figure making purposes), make sure the "Plot legend" checkbox is ticked/unticked respectively before pressing the "MSD" button.
30. The "Hotspot metrics" plot is divided into 4 subplots. Overlap probability (top left): indicates the likelihood that a cluster overlaps with a neighboring cluster to form a hotspot. Hotspot membership (top right): shows the average number of clusters that make up each hotspot. Hotspot intercluster time (bottom left): represents the average time interval (in seconds) between the appearance of successive clusters within the same hotspot region (i.e., periods when no clusters are present at that location). Cluster number (bottom right): displays changes in the density of clusters (number of clusters per square micron) over the course of the acquisition. In the Overlap Probability, Hotspot Membership and Hotspot Intercluster Time plots, the leftmost dotted vertical line on the Distance (nm) axis denotes the cluster center, while the rightmost dotted vertical line marks the average cluster radius.
31. The Principal Component Analysis (PCA) metric performs dimensionality reduction on a subset of cluster metrics to identify potential subpopulations of clusters with distinct characteristics.
32. "3D" metrics plot: "Tmin" and "Tmax" = minimum and maximum values to show on the Time axis. Tick/untick the "Axes" checkbox to toggle between a gridded plot with marked axes on a grey background, and a gridless plot with a background determined by the "Canvas" parameter in the "Display" tab.
33. The "Kernel Density Estimation" (KDE) plot provides a top-down view of a 2D histogram of detection density, which is useful for identifying regions of high detection density and comparing them with cluster regions on the 2D cluster map. This plot can be very slow—initially try plotting with a small region of interest (e.g., 2 × 2μm) and a lower resolution (e.g., 0.5 or lower). "Res" = resolution of the KDE plot. Tick the "Alt." checkbox to display the KDE plot with alternative colors.

34. "Diffusion coefficient" metrics plot: generates 3 plots: (1) Diffusion coefficient time plot: compares the appearance of clusters over time with their $\log_{10}$ diffusion coefficient values, (2) 2D diffusion coefficient plot: colors trajectories by their $\log_{10}$ diffusion coefficient values, (3) frequency distribution histogram: shows the distribution of $\log_{10}$ diffusion coefficients with the mobile-to-immobile threshold shown. Make sure that the pixel size of the "Pixel(um)" parameter is correct before pressing the "Diffusion coefficient" button.
35. The "Density over time" metrics plot displays the change in the trajectory density over the duration of the acquisition using 1 s time bins.
36. "VAR" metrics plot: Tick the "Plot legend" checkbox before pressing the "Vector autoregression" button to display a legend on the plot.
37. The cluster metrics and parameters selected for analysis will be saved in a metrics.tsv file, which NASTIC Wrangler uses to perform meta-analysis across multiple files for two different conditions.
38. Keep the short-hand names to as few characters as possible to prevent overlapping of names on the plots. The short-hand names for Condition 1 and Condition 2 must be different.
39. Aggregate data metrics include: Area (average cluster area), Avmsd (average cluster mean square displacement; MSD), Avtime (average cluster time centroid), Density (average density of trajectories within clusters), Lifetime (average cluster lifetime), Membership (average number of trajectories per cluster), Radius (average cluster radius) and Rate (average rate of detections within clusters).
40. Average data metrics include: Average cluster area, Average cluster lifetime, Average cluster membership, Average cluster radius, Average cluster rate, Average cluster time centroid, Average clusters per hotspot, Average density of clustered trajectories, Average MSD of clustered trajectories, Average trajectory length, Average trajectory steplength, Cluster density, Clustered trajectories, Clustered trajectories instantaneous diffusion coefficient, Hotspots, Percentage clustered trajectories, Percentage clusters in hotspots, Selected trajectories, Selection area, Total clusters, Total clusters in hotspots, Unclustered trajectories and Unclustered trajectories instantaneous diffusion coefficient.
41. Similarly to *NASTIC*, *NASTIC Wrangler* performs Principal Component Analysis (PCA) on a subset of cluster metrics to identify subpopulations; however, this is done at the level of each sample, rather than at the level of individual cluster (i.e., each point in the PCA plot represents a cell or sample).

42. Histograms are saved in two separate folders; "histogram_plots" and "histogram_plots_2. "histogram_plots" folder: five frequency distribution histogram plots are generated for each cluster metric used to create aggregate data plots; (1) Condition 1 histogram, (2) Condition 1 smoothed line (Savgol) histogram, (3) Condition 2 histogram, (4) Condition 2 smoothed line (Savgol) histogram and (5) combined Condition 1 and 2 smoothed line (Savgol) histogram. "histogram_plots_2" folder: eighteen smoothed line (Savgol) frequency distribution histogram plots are generated, including individual and combined histograms for each condition as follows: $\log_{10}$ diffusion coefficient histograms (all = all trajectories, no diffusion coefficient filtering; capped = trajectories filtered using user-defined minimum and maximum $\log_{10}$ diffusion coefficient values), trajectory length histograms (all = all trajectories, no filtering; capped = trajectories filtered using user-defined minimum and maximum trajectory lengths; outliers = trajectories filtered within 2 standard deviations) and trajectory steplength histograms (trajectories filtered using data-defined minimum and maximum steplengths).

Acknowledgments

We thank the IT department at the Queensland Brain Institute (QBI), Rumelo Amor and all past and present members of the Advanced Microscopy and Microanalysis Facility at QBI for their outstanding microscopy support. The work was supported by an Australian Research Council (ARC) Discovery Project grant (DP190100674); an ARC Linkage Infrastructure Equipment and Facilities grant (LE130100078); a National Health and Medical Research Council (NHMRC) Senior Research Fellowship (GNT1155794); a National Institute On Aging of the National Institutes of Health (NIH) Award R21AG080435; and funding from the University of Queensland Strategic Initiatives Fund (DVCR22052A).

Conflict of Interest The authors of this chapter declare no conflict of interest.

References

1. Sollner T, Bennett MK, Whiteheart SW, Scheller RH, Rothman JE (1993) A protein assembly-disassembly pathway in vitro that may correspond to sequential steps of synaptic vesicle docking, activation, and fusion. Cell 75: 409–418
2. Ester M, Kriegel H-P, Sander J et al (1996) A density-based algorithm for discovering clusters in large spatial databases with noise. In: Proceedings of the second international conference on knowledge discovery and data mining (KDD'96). AAAI Press, pp 226–231

3. Levet F, Hosy E, Kechkar A et al (2015) SR-Tesseler: a method to segment and quantify localization-based super-resolution microscopy data. Nat Methods 12(11):1065–1071
4. Andronov L, Orlov I, Lutz Y et al (2016) ClusterViSu, a method for clustering of protein complexes by Voronoi tessellation in super-resolution microscopy. Sci Rep 6:24084
5. Ripley BD (1977) Modeling spatial patterns. J R Stat Soc B Stat Methodol 39:172–212
6. Khater IM, Nabi IR, Hamarneh G (2020) A review of super-resolution single-molecule localization microscopy cluster analysis and quantification methods. Patterns (NY). 1(3): 100038
7. Gormal RS, Padmanabhan P, Kasula R et al (2020) Modular transient nanoclustering of activated beta2-adrenergic receptors revealed by single-molecule tracking of conformation-specific nanobodies. Proc Natl Acad Sci USA 117(48):30476–30487
8. Cisse II, Izeddin I, Causse SZ et al (2013) Real-time dynamics of RNA polymerase II clustering in live human cells. Science 341(6146):664–667
9. Wallis TP, Jiang A, Young K et al (2023) Super-resolved trajectory-derived nanoclustering analysis using spatiotemporal indexing. Nat Commun 14:3353
10. Gormal RS, Wallis TP, McCann AJ et al (2025) Nanoscale spatiotemporal cluster analysis of expressed and endogenous proteins. Nat Protoc. https://doi.org/10.1038/s41596-025-01209-w
11. Gutmann A (1984) R-trees: a dynamic index structure for spatial searching. In: Proceedings of the 1984 ACM SIGMOD international conference on management of data – SIGMOD'84, p 47
12. Shohayeb B, Sempert K, Wallis TP, Meunier FA, Durisic N, O'Brien EA, Flores C, Cooper HM (2024) BDNF-dependent nano-organization of Neogenin and the WAVE regulatory complex promotes actin remodeling in dendritic spines. iScience 27(9):110621
13. Small C, Harper C, Jiang A, Kontaxi C, Pronot M, Yak N, Malapaka A, Davenport EC, Wallis TP, Gormal RS, Joensuu M, Martínez-Mármol R, Cousin MA, Meunier FA (2024) SV2A controls the surface nanoclustering and endocytic recruitment of Syt1 during synaptic vesicle recycling. J Neurochem 168(9):3188–3208
14. Jiang A, Kudo K, Gormal RS, Ellis S, Guo S, Wallis TP, Longfield SF, Robinson PJ, Johnson ME, Joensuu M, Meunier FA (2024) Dynamin1 long- and short-tail isoforms exploit distinct recruitment and spatial patterns to form endocytic nanoclusters. Nat Commun 15(1): 4060
15. Longfield SF, Gormal RS, Feller M, Parutto P, Reingruber J, Wallis TP, Joensuu M, Augustine GJ, Martínez-Mármol R, Holcman D, Meunier FA (2024) Synapsin 2a tetramerization selectively controls the presynaptic nanoscale organization of reserve synaptic vesicles. Nat Commun 15(1):2217
16. Longfield SF, Mollazade M, Wallis TP, Gormal RS, Joensuu M, Wark JR, van Waardenberg AJ, Small C, Graham ME, Meunier FA, Martínez-Mármol R (2023) Tau forms synaptic nano-biomolecular condensates controlling the dynamic clustering of recycling synaptic vesicles. Nat Commun 14(1):7277
17. Joensuu M, Syed P, Saber SH, Lanoue V, Wallis TP, Rae J, Blum A, Gormal RS, Small C, Sanders S, Jiang A, Mahrhold S, Krez N, Cousin MA, Cooper-White R, Cooper-White JJ, Collins BM, Parton RG, Balistreri G, Rummel A, Meunier FA (2023) Presynaptic targeting of botulinum neurotoxin type a requires a tripartite PSG- Syt1-SV2 plasma membrane nanocluster for synaptic vesicle entry. EMBO J 42(13):e112095

Chapter 14

Case Study: Illuminating the Nanoscale World of Microbiology

Parnayan Syed, Mark A. Schembri, Alex J. McCann, and Frédéric A. Meunier

Abstract

Super-resolution microscopy has emerged as a groundbreaking tool in microbiology, enabling the visualization of structures and molecules far below the diffraction limit of light. We provide an overview of the various techniques employed in super-resolution microscopy, including deterministic (scanning and structured illumination) and stochastic (single-molecule localization microscopy and fluctuation-based computation) methods, summarizing their respective advantages and limitations. Applications in microbiology are presented, including investigations of cellular processes, physiology, cell wall biology, and elucidation of bacterial toxin pathogenesis mechanisms at nanoscale resolution. New and emerging uses of super-resolution microscopy are also explored, focusing on addressing critical challenges in the field: pathogen-host interactions, antibiotic resistance, mode of action of bacterial toxins, and nanoscale detail of bacterial secretion systems and their translocated effectors. With the expanding accessibility of resources and increasing availability of open-source software to democratize data analysis, super-resolution microscopy is poised to revolutionize our understanding of the nanoscale world in microbiology, paving the way for new discoveries.

Key words Microbiology, Super-resolution microscopy, Botulinum neurotoxins, SARS-CoV-2, MINFLUX, 3D-SIM, PALM, STORM, DNA-PAINT, Single-particle tracking

1 Introduction

"Progress in science depends on new techniques, new discoveries, and new ideas, probably in that order." Sydney Brenner's famous insights resonate when highlighting the progress in biology currently stemming from the use of super-resolution microscopy techniques. Optical fluorescence microscopy has played a pivotal role in unveiling the complex world of microbes and their cellular structures. Visualization of these structures is inherent to the nature of biological studies, which strive to determine the localization of key molecules involved in physiological and pathological processes. Ever since the discovery of fluorescence in 1845 [1], its potential

Frédéric A. Meunier (ed.), *Super-Resolution Microscopy: Methods and Protocols*, Methods in Molecular Biology, vol. 3034, https://doi.org/10.1007/978-1-0716-5268-8_14,

in microscopy has been evident, resulting in the development of a myriad of fluorescence microscopes [2, 3]. The field of fluorescence microscopy has undergone rapid expansion and diversification since the development of the first microscopes in the early twentieth century. Whilst an undoubtedly powerful and popular imaging technique, fluorescence microscopy, in common with its transmission and reflection light microscopy forebears, has a limit to which it can discern small details below the micrometer (μm) scale. Ernst Abbe determined the limit that defines the minimum distance at which two adjacent fluorescent molecules can be resolved (Fig. 1a) [4]. This is because when light emitted from point sources (e.g., fluorescently tagged molecules) is focused through a microscope objective, it undergoes diffraction, producing a pattern of concentric bright and dark rings in the image plane, known as the Airy pattern. When two fluorophores are closer together than the center-to-center distance of their Airy disks, they cannot be resolved as individual fluorophores. For visible light, this limit, referred to as Abbe's limit or diffraction limitation, is roughly 250 nm. Unfortunately, this is larger than the size of macromolecules and many subcellular structures of interest in biology. For example, DNA is 2 nm in diameter, porins are roughly 2 nm [5], antibodies are about 10 nm and botulinum neurotoxin type-A1 (BoNT/A1) produced by the bacterial pathogen *Clostridium botulinum* is 10–13 nm [6]. This necessitates techniques that can resolve these fluorescent signals below the diffraction limit of light. While several diffraction-limited techniques have been used to improve axial resolution, such as confocal microscopy (for review, *see* Ref. [7]), spatial resolution has remained limited by the diffraction of light. Electron microscopy (EM) has historically been used since its advent in 1930 as an alternative to fluorescence microscopy to resolve nanometer-scale cellular structures. However, EM (excluding cryo-EM) presents its own unique challenges: it can only be used on fixed samples which may involve additional harsh pre- and post-embedding treatments; it is not amenable to the identification of dynamic protein interactions; and it is prone to electrical and magnetic interference due to its high sensitivity to environmental fluctuations (for a review of the technique, *see* Refs. [8, 9]). These limitations, in particular the incompatibility of EM with live samples, presented a unique challenge that could only be resolved using fluorescence microscopy. Indeed, this set the stage for the advent of super-resolution microscopy as a key discovery to fill the technological gap between electron and fluorescence microscopy.

Super-resolution microscopy has overcome the diffraction barrier of light, allowing unprecedented access to subcellular processes at spatial resolutions down to a few nanometers. At its simplest, super-resolution microscopy relies on sparse labelling and selective activation of fluorophores such that two adjacent molecules are

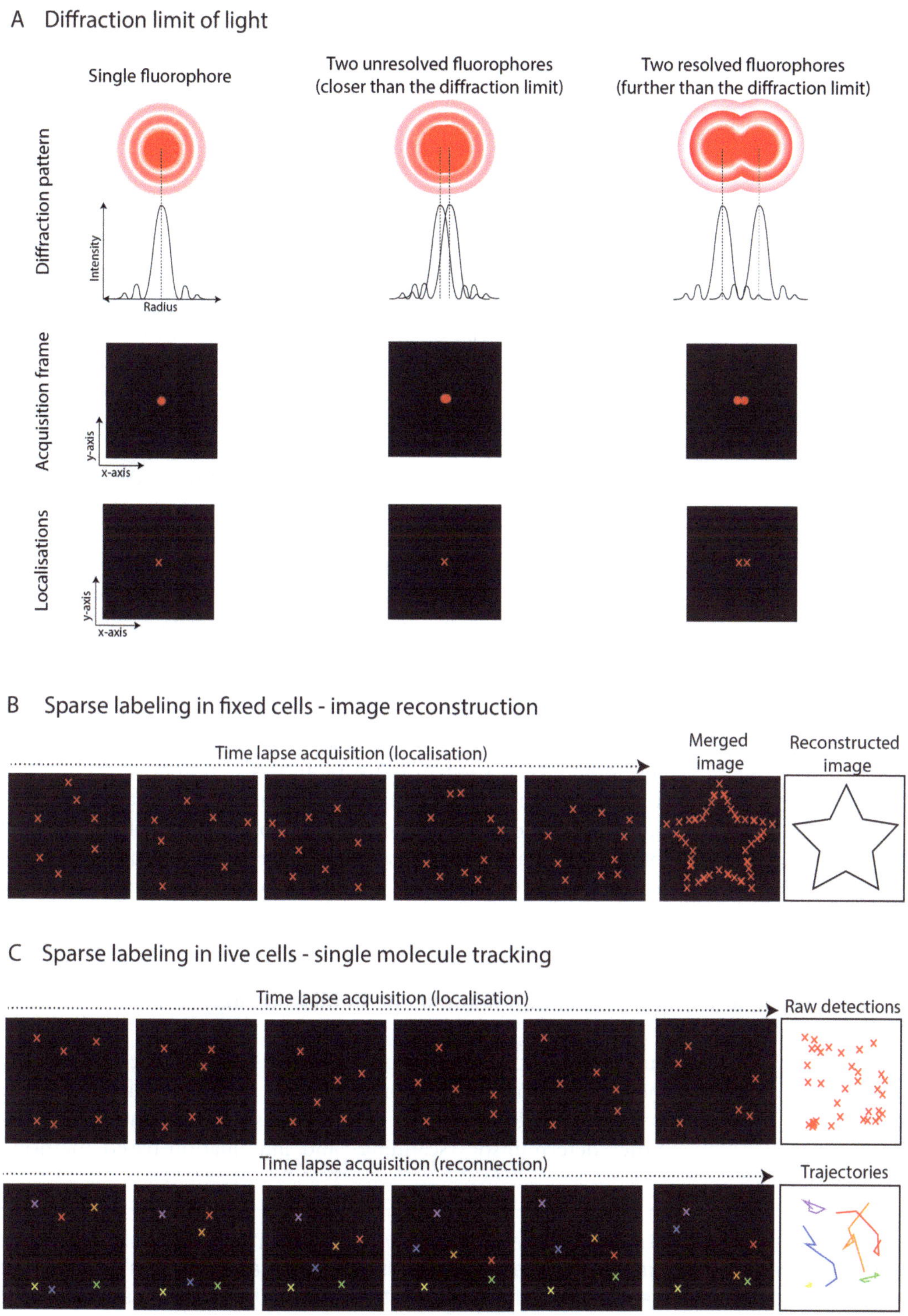

Fig. 1 Super-resolution microscopy concepts. (**a**) Imaging of individual molecules and complex biological structures is often limited by Abbe's resolution limit, which defines the best possible spatial resolution due to the diffraction of light. Fluorophores that are closer than this limit can only be resolved if they are activated and imaged sequentially (i.e., at separate time points). For both image reconstruction and single-molecule

never "on" at the same time. This process is repeated in cycles as a time-lapse: molecules in each frame are localized, and then either reconstructed to generate nanometer-resolved images (Fig. 1b), or linked to produce single-molecule trajectories (Fig. 1c). These techniques take advantage of the point spread function (PSF) of the emitted fluorophore, fitting a Gaussian to it in order to localize the emitter with sub-pixel precision ranging from 1 to 10 nm, depending on the type of super-resolution microscopy used [10]. They can be divided into deterministic techniques, which include scanning methods that use structured illumination patterns to resolve images below the diffraction limit, and stochastic techniques, which include single-molecule localization microscopy (SMLM) methods that employ stochastic activation of fluorophores, as well as fluctuation-based computational techniques. Each type of super-resolution microscopy technique is based on distinct principles with advantages and disadvantages [10–25], which are listed in Table 1 (deterministic techniques) and Table 2 (stochastic techniques).

All super-resolution microscopy techniques have the potential to advance research in microbiology and answer critical open questions in the field. Accordingly, in this perspective, we present the pros and cons of commonly used super-resolution techniques and discuss several recently published works to highlight current applications. We also explore a case study of super-resolution microscopy used to reveal the molecular mechanism of action of botulinum neurotoxin. Finally, we end with a perspective on recent and future developments in super-resolution and raise potential applications that could be explored with super-resolution techniques to further our understanding of the nanoscale world of microbes and their interactions with host cells. Visualizing these interactions at the nanoscale could open up new therapeutic avenues for clinical applications.

2 The Potential of Super-Resolution Microscopy in Microbiology

Super-resolution microscopy techniques can be used to localize molecules with nanoscale precision and to track their dynamics in live cells (*see* Fig. 2). These techniques are broadly classified as either deterministic (scanning- and illumination-based methods,

Fig. 1 (continued) tracking, the molecule of interest is sparsely labelled with a fluorescent probe, and a series of images (time-lapse) is acquired. (**b**) Image reconstruction: fluorescently labelled molecules are localized in each frame in the time-lapse acquisition, and these localizations are then merged into a single image to reconstruct the biological structure being imaged. (**c**) Single-molecule tracking: fluorescently-labelled molecules are localized in each frame in the time-lapse acquisition, and these localizations are then linked between frames in order to generate trajectories

Table 1
Pros and cons of various deterministic super-resolution techniques

Deterministic (scanning and structured illumination) techniques				
Basic principle	**Pros**	**Cons**	**Usage**	**Refs**
STED				
Uses two diffraction patterns, for excitation and depletion via a doughnut-shaped depletion laser that reduces the PSF of fluorescence below the diffraction limit	Resolution: 30–70 nm (XY), 50 nm (Z) Good axial resolution Suitable for resolving intercellular structures	System complexity Laser alignment issues Phototoxicity from the quenching beam Not suitable for high-speed applications (35 ms / image)	High-resolution imaging of structures in live or fixed cells	[20, 26, 39]
SIM				
Uses an illumination grid with an interference pattern that rotates to spatially resolve fluorophores. The images are postprocessed to generate super-resolved images	Resolution: 90–140 nm (XY) 300 nm (Z) Compatible with commonly used fluorophores Faster acquisition speeds with latest developments (TIRF-SIM: 8 Hz, 2D: 10 Hz, 3D SIM: 4 Hz)	Limited super-resolution capability Not suitable for thick samples	Moderate resolution imaging of structures in live cells at faster acquisition speeds using standard fluorescent labels	[21, 41, 43, 44, 53]
SoRa / pixel re-assignment				
Uses a detector array instead of a point detector with a scanning laser. The PSF of the excited fluorophore is reassigned between illumination and detection spots and reconstructed. The detector array collects most of the emitted photons, resulting in superior axial resolution and low SNR	Resolution: 120 nm (XY), 400 nm (Z) Compatible with any sample and fluorophore High speed compared to other confocal systems Suitable for high-throughput imaging	Phototoxicity in live samples due to higher laser power Prone to higher lateral background Can requires post-acquisition processing to achieve super-resolution	Moderate resolution imaging of structures in live cells using standard labelling strategies	[16]

(continued)

Table 1 (continued)

Deterministic (scanning and structured illumination) techniques				
Basic principle	**Pros**	**Cons**	**Usage**	**Refs**
RESOLFT				
Uses reversible switchable fluorophores (rsFPs) to saturate the OFF state using a light pattern with an intensity minima (ring-shaped), where only rsFPs in the direct vicinity of the minima are ON and can be resolved	Resolution: 30–40 nm (XY), 70–90 nm (Z) Suitable for observing structures in deep tissues as it suppresses autofluorescence noise Requires low light doses, making it suitable for live cell imaging Fast acquisitions	Not compatible with organic dyes Lower resolution than some other nanoscopy techniques	Long time-lapse imaging in live cells with reduced phototoxicity	[17, 100]

see Table 1) or stochastic (SMLM and fluctuation-based computational methods, *see* Table 2). Various super-resolution systems can be suited for different purposes, depending on the sample and resolution requirements (for an in-depth comparison of super-resolution techniques, *see* review [10]). We focus on broad technical concepts to contextualize their application in microbiology. The unique characteristics of each system make them suitable for different experimental approaches (*see* Tables 1 and 2 for details).

Deterministic imaging techniques such as Stimulated Emission Depletion (STED) [20], Structured Illumination Microscopy (SIM) [21], Super-resolution via optical Re-assignment (SoRa) [16] and Reversible Saturable/Switchable Optical Fluorescence Transition (RESOLFT) [17] rely on the known spatial distribution of the excitation light to selectively activate and localize fluorophores. STED relies on a linear excitation beam that scans across a sample, along with a depletion beam that quenches all signal except that in the center of the beam. This donut-shaped STED beam allows nanoscale determination of the location of the fluorophores, offering up to 30 nm resolution in live cells, and 15 nm in fixed cells [26]. Similarly, SIM uses a pattern of fine parallel lines to illuminate the sample and takes advantage of the Moiré effect to resolve the fine details of the sample.

Stochastic imaging techniques rely on either the iterative activation of spatially isolated target fluorophores (as in SMLM) or on fluctuations in fluorescence intensity (as in fluctuation-based methods) over an acquisition period to reconstruct a super-resolved map of the molecules of interest. SMLM techniques include (direct)

Table 2
Pros and cons of various stochastic super-resolution techniques

Stochastic (single-molecule localization microscopy) techniques				
Basic principle	**Pros**	**Cons**	**Usage**	**Refs**
(d)STORM				
Uses a buffer to induce stochastic photoswitching of fluorophores bound to a molecule of interest, enabling sequential imaging of nearby emitters as they blink asynchronously	Resolution: ~20 nm (XY) Uses highly photostable organic dyes Does not require genetically engineered fluorescent tags Multiplexing and 3D imaging possible	Sensitive to labelling density Oxygen-scavenging system required for certain dyes, which are not compatible with live cell imaging Better suited for fixed cells Slow acquisitions	Very high resolution imaging of fixed cells	[22, 27, 37, 47, 101, 102]
TALM/sptPALM				
Relies on the activation of a photoswitchable fluorophore that emits sufficient photons for reliable localization, before bleaching (or going into the dark state)	Resolution: 20 nm (XY), 50–100 nm (Z) 100% efficiency of labelling 3D imaging compatible	Requires genetically engineered fluorescent proteins Molecules can only be imaged once Phototoxicity from multiple quenching cycles Imaging target needs to be close to the coverslip for TIRF mode imaging	Tracking the dynamics of single proteins in live cells, enables high spatiotemporal resolution of molecular trajectories	[23, 35, 36, 38, 51, 52]
(u)PAINT				
Uses the transient association of fluorophore probes with a target molecule. The probe's concentration is adjusted so at any time only a few fluorophores are imaged, allowing for the construction of super-resolved images	Resolution: 5–10 nm (XY) High-speed acquisitions Capable of tracking moving molecules in live cells Easy to implement and no photoswitching or special dyes are required	Requires well-designed probes or labelled nanobodies Relies on undersampling the target, therefore unable to quantify the total number of molecules	Very high resolution multiplexed imaging of cellular structures in fixed cells	[12, 24, 31, 36, 75, 101, 103]

(continued)

Table 2 (continued)

Stochastic (single-molecule localization microscopy) techniques				
Basic principle	**Pros**	**Cons**	**Usage**	**Refs**
MINFLUX				
A new class of super-resolution microscopy that works by activating emitters one by one to obtain single molecule separation and an excitation pattern with a central intensity-zero. It minimizes fluorescence flux by aligning the intensity zero with the molecule's location and calculating the number of detected photons at each minimum	Resolution: 1–3 nm (XYZ) Compatible with various common fluorophores Live cell imaging possible High-speed acquisitions Low photon numbers required, making long temporal acquisitions possible	Very high resolution only possible in fixed samples Sensitive to labelling density Require in-depth post-processing analysis	Ultra-high resolution imaging of fixed cells and tracking of single protein dynamics in live cells with improved spatial and temporal resolution	[15, 28, 64, 104]
SOFI				
Computational reconstruction of time-dependent fluctuations from fluorophores, using higher-order cumulants to resolve the correlated signal and exclude any background noise	Resolution: 100–150 nm (XYZ) Compatible with other super-resolution acquisition methods Does not necessitate photoswitching/ activation Capable of high resolution in short acquisition times and low light exposure Can be performed on any imaging platform Well-adapted to quantum-dot use Fast acquisition compatible	Requires emitters that exhibit optically distinguishable states Susceptible to brightness bias (less bright emitters might be masked or omitted in the reconstruction) Not suitable for use with organic dyes prone to photobleaching	Moderate resolution long-time-lapse imaging in live cells in low light conditions	[25, 105]

STochastic Optical Reconstruction Microscopy ((d)STORM) [22, 27], Photo-activated Localization Microscopy (PALM) [23], (universal) Points Accumulation for Imaging in Nanoscale Topography ((u)PAINT) [12, 24] and Minimal photon Fluxes (MINFLUX) [28]. An example of a fluctuation-based computation method is Super-resolution Optical Fluctuation Imaging (SOFI) [25]. Due to their nanoscale localization precision, PALM and dSTORM can be used on fixed samples to determine the precise positions of key proteins and their interaction partners.

In addition to the localization of biological structures and molecules, SMLM techniques can also be used to track fluorescently tagged molecules in live samples. This enables real-time tracking of the dynamics of biological molecules and their interactions in living cells, revealing novel insights into the nanoscale organization of molecules involved in cellular processes. Figure 2a shows an example of using single-particle tracking (spt) to capture single-molecule detections in live *E. coli*, which are then localized and constructed into trajectories [29]. Examples of super-resolution imaging techniques that incorporate spt include sptPALM [30] and uPAINT [12]. These methods allow for the tracking of molecules within cells or molecules that are interacting with cells, and the visualization of ensemble nanoscale behavior to reveal membrane interactions, molecular uptake, and intracellular dynamics. Importantly, fluorescently labelled single molecules, such as bacterial neurotoxins, can be tracked as they bind to the plasma membrane of a host cell and subsequently transit within the cell, thus providing insights into the potency and dose-dependent mechanism of action of the toxin [31, 32]. Figure 2b shows an example of using spt to track the transport of the heavy chain of the Botulinum neurotoxin type A (BoNT/A-HC) in hippocampal neurons [32]. Localization precision of 20–40 nm can be achieved with these single-molecule tracking techniques [31, 33, 34]. The recently developed MINFLUX [28] concept has pushed the resolution even further to an unprecedented 1–3 nm in 3 dimensions. MINFLUX combines the principles of STED with other SMLM techniques by using a donut-shaped excitation beam with a central intensity minimum to localize single fluorophores. By positioning the beam at several defined locations around the emitter and measuring the resulting fluorescence intensities, the position of the fluorophore can be determined with nanometer precision. This technique is particularly relevant to microbiology, given the small scale of bacterial cells and their target molecules.

Below, we present selected examples to highlight how super-resolution microscopy techniques have provided unparalleled new detail to understand complex processes in microbiology—including DNA and RNA synthesis and metabolism, mechanisms underlying antibiotic resistance, host-pathogen interactions, and bacterial secretion systems.

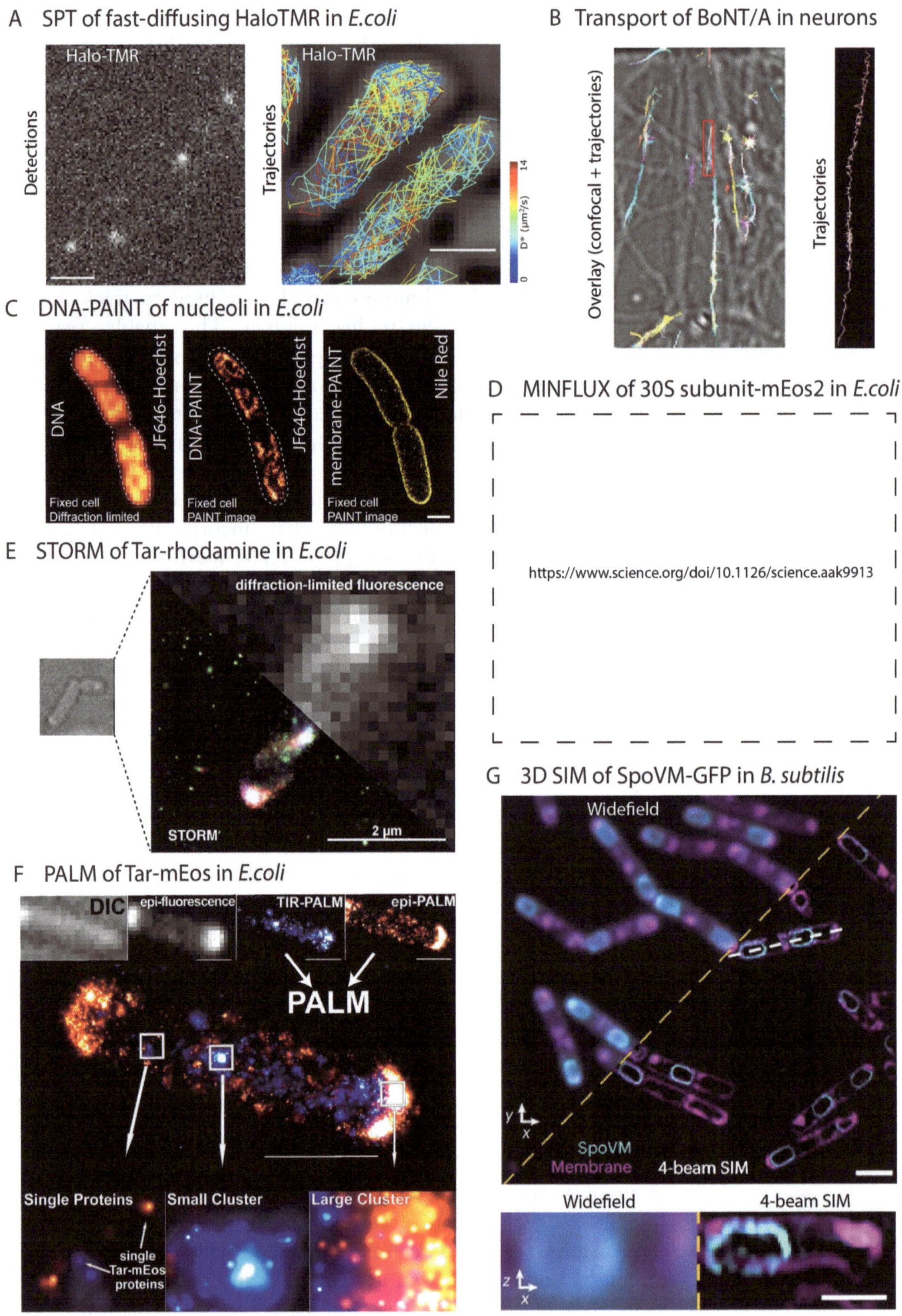

Fig. 2 Example images of super-resolution microscopy techniques applicable to bacterial toxin research. (**a**) Single particle tracking (SPT) of Halo-TMR in live *E. coli* cells, with raw detections shown on the left (scalebar: 2 μm), and trajectories colored by their diffusion coefficients shown on the right (scalebar: 1 μm). Adapted-from: Banaz et al. [29]. Used under a Creative Commons Attribution 3.0 licence (https:/creativecommons.org/

3 Applications in Microbiology

3.1 Cellular Processes of DNA and RNA Metabolism

Super-resolution microscopy techniques have been used to visualize essential processes involved in the synthesis and modification of nucleic acids (DNA and RNA) within bacterial cells. Using PALM combined with single-molecule tracking, individual RNA polymerase molecules can be mapped as DNA-bound or mobile subpopulations within the nucleoid [35]. Localization of single RNA polymerase molecules at transcription sites on DNA using 3D SIM has also enabled mapping of the spatial reorganization of the nucleoid during bacterial growth, revealing that strongly transcribed genes dissociate from the bulk of nucleoid DNA

Fig. 2 (continued) licenses/by/3.0/). Changes were made. SPT can be used to track molecules in live cells, enabling the determination of the mobilities. (**b**) SPT of the heavy chain of internalized Botulinum Neurotoxin Type A (BoNT/A-HC) tagged with Atto647N, as it travels in the axons of hippocampal neurons. The trajectory highlighted by the red box outline is shown magnified on the right. Adapted from: Wang et al. [32]. Used under a Creative Commons Attribution 4.0 licence (https:/creativecommons.org/licenses/by/4.0/). Changes were made. (**c**) Comparison of diffraction-limited images with PAINT images, taken of the same *E. coli* cell before and after fixation, using highly inclined and laminated optical sheet (HILO) microscopy. From left to right: diffraction-limited image of JF_{646}-Hoechst (nucleoid marker) following fixation, DNA-PAINT image of JF_{646}-Hoechst (nucleoid marker) following fixation and membrane-PAINT image of Nile Red following fixation. Scale bar: 1 μm. Multiplexed PAINT allows for the detection of ultrastructures within cells. Adapted from: Spahn et al. [36]. Used under a Creative Commons Attribution 4.0 license (https:/creativecommons.org/licenses/by/4.0/). Changes were made. (**d**) Figure can be found at: https:/www.science.org/doi/10.1126/science.aak9913. Description of figure: overlay of the widefield image taken of an mEos2-tagged 30S ribosomal subunit protein in *E. coli* with the trajectories obtained from Minimal emission fluxes (MINFLUX) (scalebar: 500 nm) [28]. MINFLUX enables SPT to be performed at unprecedented spatial and temporal resolutions. (**e**) Comparison of a diffraction-limited image (top) with a Stochastic Optical Reconstruction Microscopy (STORM) image (bottom) taken of Tar protein labelled with a cell-permeant photoswitchable rhodamine spirolactam dye in an *E. coli* cell [37]. Scale bar: 2 μm. STORM can resolve a high number of localizations within a diffraction-limited spot, enabling precise molecular counting of the target protein. Adapted from: https:/pubs.acs.org/doi/10.1021/acscentsci.9b00617. Changes were made. (**f**) Top row panels, from left to right: Comparisons of diffraction-limited differential interference contrast (DIC) and epi-fluorescence images with total internal reflection-photoactivated localization microscopy (TIR-PALM) and epi-PALM images (scalebar: 1 μm) [38]. Middle panel: Overlay of TIR-PALM and epi-PALM images (scalebar: 1 μm) [38]. White box outlines are shown magnified in the respective panels in the bottom row. Bottom row panels, from left to right: Examples of (left) individual Tar-mEos molecules, (middle) small clusters of Tar-mEos molecules and (right), and large clusters of Tar-mEos molecules. PALM can be extended to visualize both single molecules and clusters within cells, which have important biological implications. Adapted from: Greenfield et al. [38]. Used under a Creative Commons Attribution 4.0 License (https:/creativecommons.org/licenses/by/4.0/). Changes were made. (**g**) Top panel: Comparison of widefield and 3D structured illumination microscopy (SIM) images of SpoVM-GFP (sporulation marker) and CellBrite Fix 555 (membrane marker) in live *B. subtilis* (scalebar: 2 μm) [44]. The white dashed outline is shown magnified in the bottom panels. Bottom panels: widefield image (left) and 3D SIM image (right) (scalebar: 1 μm). SIM is well-suited for live-cell imaging due to its low level of phototoxicity and fast acquisition rates, enabling the visualization of biological structures and assemblies with enhanced spatial detail. Adapted from: Li et al. [44]. Used under a Creative Commons Attribution 4.0 License (https:/creativecommons.org/licenses/by/4.0/). Changes were made

[35]. Multiplexing techniques, combined with super-resolution microscopy, can discern spatial correlations between different cellular processes. Using a palette of fluorophores with well-separated spectra, Spahn et al. developed multiplexed single-molecule imaging as a method to analyze co-localization patterns in the cell membrane or nucleoid (*see* Fig. 2c) [36]. Using this labelling technique for PALM and PAINT imaging, they also showed that changes in the nanoscale distribution and interaction of RNA polymerase occur in the nucleoid of *E. coli* in response to various antibiotic treatments [36]. MINFLUX has also been used to track the dynamics of individual 30S ribosomal subunits at ultra-high spatial and temporal resolution, with example trajectories shown in Fig. 2d [28]. Such approaches offer the potential to discover new antibiotic targets across the landscape of the bacterial cell.

3.2 Chemoreceptor Function and Clustering

The visualization of protein organization into clusters is important for understanding the spatial regulation of cellular signaling processes, such as those involved in chemotaxis, and can be achieved using super-resolution microscopy techniques such as PALM and STORM. In particular, chemotactic receptors such as Tar have been shown to form distinct nanoscale assemblies that play a crucial role in signal detection and amplification. Examples of Tar clustering are shown in Fig. 2e, which uses STORM [37], and Fig. 2f, which uses PALM [38].

3.3 Cell Wall Biology and Antibiotic Resistance

The bacterial cell wall is a target for many antibiotics, and thus understanding the spatial organization and localization of peptidoglycan and key transpeptidase enzymes involved in peptidoglycan synthesis is vital for the development of new therapeutics. Super-resolution microscopy techniques have enabled the tracking of components of the cell wall synthesis machinery at the single-molecule level. For example, high-resolution STED nanoscopy has been used to track peptidoglycan synthesis using fluorescently labelled D-amino acids [39], and combining the optical resolution capabilities of STORM and SIM with atomic force microscopy has been used to map the synthesis of peptidoglycan during the growth of *Bacillus subtilis* [40]. Fluorescently labelled antibiotics that bind to peptidoglycan synthesis enzymes can also be used to investigate the dynamics of cell wall synthesis. For example, vancomycin probes fluorescently imaged using SIM/3D-SIM were localized to the dividing septum in a range of Gram-positive bacteria [41]. In another example, teixobactin, a new class of antibiotic that selectively targets lipid II in the bacterial membrane of Gram-positive bacteria [42], was fluorescently tagged to image its interaction with *B. subtilis*. SIM imaging revealed that teixobactin interaction with essential precursors in the cell wall synthesis pathway results in the formation of localized aggregated nanoclusters. Combined with Fluorescence Resonance Energy Transfer (FRET), this work highlights the potential of using super-resolution microscopy to unveil the mechanism of action of antibiotics [43]. An example of using 3D-SIM to investigate cell wall biology in *B. subtilis* is shown in Fig. 2g [44]. Additional work using single particle tracking techniques can shed light on the dynamics of such supramolecular assemblies and how they correlate with cell wall lysis in real-time, critical discovery steps required for the development of new antibiotics that target pathogens such as methicillin-resistant *Staphylococcus aureus* (MRSA) and vancomycin-resistant *Enterococcus* (VRE) that cause life-threatening human infections.

3.4 Pathogen-Host Interactions

The elucidation of mechanisms by which viruses enter host cells is of fundamental significance in microbial pathogenesis and has direct translational outcomes in medicine. This became ever more relevant during the COVID-19 pandemic, where a rapid understanding of the mechanism of infection of the SARS-CoV-2 virus provided a discovery platform for the development of new antivirals [45, 46]. Super-resolution microscopy employing single-molecule tracking techniques was used to map the pathogenic trajectory of single SARS-CoV-2 virus particles by visualizing viral receptor targeting and uptake into host cells [47]. By using lipid intercalating dyes to distinctly label the membrane of host cells and the viral envelope, a specific mutation (D614G) was identified in the viral spike protein that increased interaction of the virus with its receptor ACE2. Using STORM imaging on fixed cells at single-molecule resolution, viral variants were shown to exhibit decreased reliance on ACE2 for uptake into host cells [47], possibly accounting for their increased transmissibility [45, 46]. A more recent study by Pearson et al. [48] employed Fluorescence Lifetime Imaging Microscopy-FRET (FLIM-FRET) to localize single molecules of SARS-CoV-2 Envelope (E) proteins to the Golgi and endoplasmic reticulum, identifying the guanosine triphosphatase (GTPase) adenosine diphosphate ribosylation factor-related protein 1 (ARFRP1) as a key protein required for the recruitment of adaptor protein-1 (AP-1). Demonstration that AP-1 promotes traffic of E proteins to lysosomes using FLIM-FRET enabled the identification of key mechanisms of SARS-CoV-2 replication and release in complementation with biochemical methods.

3.5 Bacterial Secretion Systems (Injectosomes) and Effector Signaling

Super-resolution microscopy techniques have delivered major advances in our understanding of the structure and function of bacterial secretion systems. Secretion systems can range in complexity from single proteins (e.g., autotransporters) [49] to large needle-like organelle structures containing multiple different proteins (e.g., type 3 secretion systems; T3SS). Here, the power of super-resolution microscopy techniques can be showcased through discoveries including the visualization of individual components of the T3SS from *Salmonella* Typhimurium using 2D and 3D SMLM

[50], the application of STED and SIM to image the pore complex proteins YopB and YopD as integral components of the T3SS of *Yersinia enterocolitica* during host cell infection [50], and the use of single particle tracking techniques like Tracking and Localization Microscopy (TALM) [51] which combine localization and sptPALM, to track the translocation of *Salmonella* T3SS effectors and their interaction with host endomembranes [52]. Additional examples can be evidenced through studies of bacterial type VI secretion systems (T6SS), which are multicomponent needle-like organelles that inject toxins across the membrane of proximal target eukaryotic and prokaryotic cells. Using 3D SIM, Lin et al. revealed new insight into the spatial and temporal assembly of the T6SS injectosome produced by *Acinetobacter* and *Burkholderia* species during contact with prey bacteria [53].

4 Super-Resolution Microscopy to Study Bacterial Toxins

Bacterial toxins play a critical role in pathology outcomes caused by infection with a range of pathogens. Secreted bacterial toxins bind to specific eukaryotic cell receptors to elicit their effect, which can affect cell signaling, cause membrane disruption and pore formation, or disrupt retrograde trafficking following their internalization. Nanoscale-level super-resolution techniques can be used to define the mode of action of toxins, offering the potential to develop new selective and efficacious therapeutic interventions. These techniques can also enhance our understanding of toxin function and reveal biological insight to define host susceptibility. A better understanding of bacterial toxins through high-resolution imaging could also allow us to harness their application for therapeutic purposes, such as cancer immunotherapies, spasticity and pain treatments [54–56].

Super-resolution microscopy has been used with great success to unravel the biology of bacterial toxins. An example involves *C. botulinum* BoNT/A, a potent neurotoxin that targets synaptic vesicles by binding to a tripartite receptor complex on the nerve terminal surface [31]. Single-molecule tracking in live neurons was used to unravel how BoNT/A binds to the preassembled tripartite complex composed of polysialogangliosides1 (PSG), synaptotagmin-1 (Syt1), and with synaptic vesicle protein 2 (SV2) in order to promote its internalization in synaptic vesicles and ultimately block synaptic transmission (Fig. 3) [31]. This study demonstrates that Syt1, which does not directly bind the toxin, is nonetheless a key component of the nanoscale surface cluster allowing "coincidence detection" for targeted internalization in synaptic vesicles. Simultaneous single-molecule tracking of the toxin and its receptors, SV2A, in live hippocampal neurons revealed that BoNT/A not only binds to these functional nanoplatforms but also

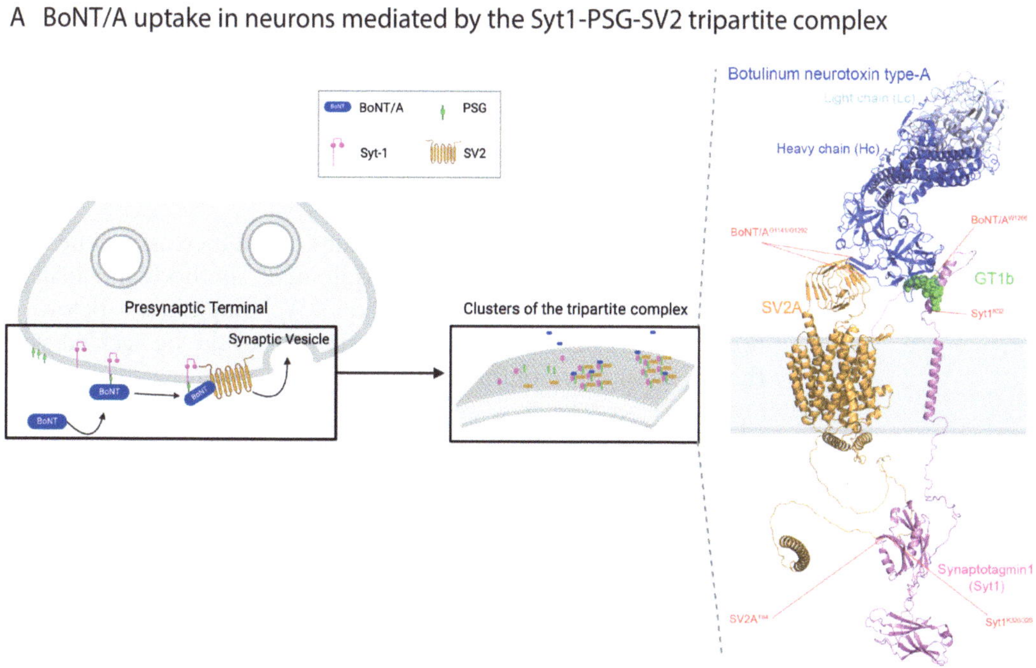

Fig. 3 Schematic and structural representation of BoNT/A uptake in neuronal presynapses via the Syt1–PSG–SV2 tripartite complex. (**a**) Left panels: BoNT/A (blue) binds to and hijacks the Syt1-PSG-SV2 tripartite complex (pink/green/yellow) on the plasma membrane to promote its own internalization into the presynaptic terminal of neurons. The Syt1-PSG-SV2 tripartite complex forms clusters on the plasma membrane that are exploited by BoNT/A. Right panel: Structure of BoNT/A (blue), highlighting heavy (dark blue) and light (light blue) chains as well as binding partners in the Syt1-PSG-SV2 tripartite complex (pink, green and yellow, respectively). Mutations that disrupt binding are shown in red. Initial structural models: full-length BoNT/A crystal structure was obtained from PDB (2NYY) [106]. Full-length SV2A and Syt1 models were obtained from the AlphaFold2 database (https:/alphafold.ebi.ac.uk) [107, 108]. Structural alignment for model building: full-length SV2A and BoNT/A structures were aligned with the SV2C-BoNT/A-HC co-crystal structure obtained from PDB (PDB 4JRA) [109]. Full-length Syt1 was aligned with the GT1b-BoNT/A-HC co-crystal structure [110], and with the Syt1 C2B domain/SV2A peptide phosphorylated at Thr84 co-crystal structure [111]. Model refinement: manual adjustments were made in PyMOL (https:/pymol.org/2/), and optimization of bond lengths and angles was made in Coot (https:/www2.mrc-lmb.cam.ac.uk/personal/pemsley/coot/). Structural images: made in PyMOL. Reproduced with permission from Joensuu et al [31]. Used under a Creative Commons Attribution 4.0 License (https:/creativecommons.org/licenses/by/4.0/). Changes were made

manipulates them by dramatically increasing their number on the plasma membrane to promote its targeted internalization into synaptic vesicles at the presynapse [31]. Further, BoNT/A triggered the immobilization of the tripartite complex, likely conducive to recruiting endocytic effectors, ultimately promoting synaptic vesicle internalization [31]. The inner workings of the tripartite complex could only be deciphered using single-molecule technologies and are likely to represent a general mechanism by which botulinum neurotoxins hijack motor nerve terminals. The techniques that facilitated this conceptual advance—revealing the dynamic nanoscale landscape and molecular events by which a single neurotoxin can intoxicate neurons—can now be applied to investigate the nanoscale mode of action of other toxins [57].

5 Future Perspectives

We anticipate super-resolution techniques will increasingly be used in molecular microbiology as the equipment, methods and analyses become more widely accessible. We highlight below a non-extensive list of open questions that could be tackled with the use of super-resolution microscopy.

1. Bacterial pathogens use adhesins to bind to host cell receptors in a lock-and-key manner and facilitate colonization, a process that defines tissue tropism. Uropathogenic *Escherichia coli* (UPEC), which are the primary cause of urinary tract infections (UTIs), use type 1 fimbriae (via the tip-located FimH adhesin) and a second fimbriae termed Ucl fimbriae (via a tip-located UclD adhesin) as key mechanisms of adherence to the colonic epithelium to form intestinal reservoirs that seed recurrent UTI [58]. Using the purified lectin binding domain of each adhesin in combination with confocal microscopy, specific binding to different receptors was demonstrated: FimH to N-linked oligosaccharides and UclD to O-linked oligosaccharides at different spatial locations within the colonic mucosa [58]. These findings now offer a prime opportunity to employ super-resolution microscopy to tag the adhesins and track bacterial binding, internalization and host intracellular interactions using sptPALM and/or uPAINT [12, 57]. In neurobiology, postsynaptic receptors such as the α-amino-3-hydroxy-5-methyl-4-isoxazolepropionic acid (AMPA) receptors in mammalian neurons, responsible for excitatory glutamatergic responses, have been shown to diffuse in and out of synapses, thereby regulating the size of the postsynaptic response [59–62]. Similar plastic responses could occur following binding of bacterial adhesins to target receptors, leading to remodeling of host cell surfaces and advancing infection cascade pathways, all of which can now be tracked by single-molecule super-resolution microscopy.
2. Structural studies are crucial to enhancing our understanding of microbial biology and host interactions. With the advances in 3D super-resolution techniques [63–65], coupled with deep learning analysis techniques [66–68], it is now possible to use super-resolution microscopy in combination with Cryo-EM [69] and X-ray crystallization to resolve the orientation and arrangement of proteins of interest within subcellular structures in live cells.

3. The study of biofilms, including interactions between composite organisms, their extracellular components, and the diffusion of antibiotics, offers new opportunities for super-resolution microscopy in complementation with other methods. For example, single-plane illumination microscopy fluorescence correlation spectroscopy (SPIM-FCS) has been used to obtain 3D diffusion maps with micrometer spatial and millisecond temporal resolution to track the movement of antimicrobials through *Pseudomonas aeruginosa* colony biofilms [70]. Combining such approaches with super-resolution (e.g., STED-FCS) may achieve sub-diffraction resolution in complex biofilm structures [71]. In another study, a hybrid of fluorescence in situ hybridization (FISH) with super-resolution imaging enabled the multiplexing of various bacterial taxa and the visualization of bacterial ribosome reorganization in the human oral microbiome [72]. Compact bacterial aggregates and biofilms have been difficult to resolve using single-molecule localization techniques due to steric hindrance for the imaging probes [73]. Expansion microscopy (ExM), which has been successfully used to enable super-resolution of densely packed structures like organotypic brain slices [74], has recently been repurposed to resolve detailed structures within the architecture of *Proteus mirabillis* biofilms [75]; see Chapter 2 for information on how to perform ExM. The development of these methods opens new avenues for further research to characterize inter- and extracellular features of biofilms and their interaction with antimicrobials [76].
4. The development of the MINFLUX technique has opened new avenues for studying nanoscale localization and interaction of microbes and toxins on lipid membranes and other cellular structures in 3D with unprecedented detail (1–3 nm in 3D) [15, 64]. When combined with multiplexing super-resolution techniques [36, 77], MINFLUX can reveal unique interactions that could provide a foundation for the development of novel treatments. Although MINFLUX currently has limited accessibility due to its technical complexity and expensive instrumentation, it is likely to become a powerful method for investigating a wide range of microbiological processes in the future.
5. Super-resolution microscopy techniques offer exciting new opportunities to understand the biology of toxins such as BoNT/A [31]. Another example is hemolysin, a potent pore-forming toxin produced by many UPEC strains that cause UTI, sepsis and neonatal meningitis [78]. A recent study identified a key role for the membrane-bound purinergic receptor P2XR during infection by hemolysin-producing UPEC [79], paving the way for further studies to track the mechanism of action of hemolysin in nanoscale resolution and detail.

These applications provide examples of how super-resolution microscopy can further advance the field of microbiology at the nanoscale level. As the quest for superior temporal and spatial resolution spurs the development of more precise microscopy techniques, we anticipate that microbiologists will be at the forefront of their application to dissect complex interactions between microbes and their interactions with eukaryotic cells. The tools and solutions offered by super-resolution microscopy offer enormous potential in microbiology and are primed to catalyze a new wave of breakthroughs in the field.

6 Analysis Techniques

Super-resolution microscopy is a data-intensive technique that requires sophisticated analysis to extract valuable biological insights from the acquisitions. Although some methods, like trajectory reconstruction of single fluorophores using PALMTracer or TrackMate, are widely available and used [80, 81], recent developments in the analysis of single-molecule tracking and clustering data can now advance the utility of techniques even further; see [82] for a review of different single particle tracking methods, and [83] for a comparison of the performance of various single particle tracking algorithms.

1. *Spatial clustering*: Density-based spatial clustering of applications with noise (DBSCAN) is a density- and proximity-based clustering algorithm that clusters points (such as trajectory detections) in space using the neighborhood radius (Epsilon) and minimum number of points (MinPts) parameters [84]. If the number of points within Epsilon around a central point meets or exceeds MinPts, the central point is considered a core point, and neighboring points within this radius are considered part of the same cluster as the core point. Voronoï tessellation is a segmentation-based algorithm used in clustering analyses, where points are assigned to tiles based on spatial proximity. Space is divided into tiles, each of which contains a point at its center, and includes all of the space that is closer to that point than to any other neighboring point. Points can be considered clustered if their corresponding tile areas fall below a defined size threshold (i.e., clustered points are closer together, resulting in smaller tile areas) [85, 86]; see [87, 88] for a comparison of various spatial clustering algorithms, and [89] for their performance in different scenarios.
2. *Spatiotemporal clustering*: Nanoscale spatiotemporal indexing clustering (NASTIC), a clustering software [90], can resolve trajectory clusters both in space and time, giving valuable information on how molecules interact in biological systems.

NASTIC generates a spatiotemporal bounding box around each trajectory and then uses a bounding box radius factor (r) and time window (t) to search for neighboring trajectory bounding boxes. Trajectories are considered to be clustered if the number of overlapping trajectory bounding boxes meets or exceeds the Cluster threshold (minimum number of trajectories required to define a cluster). See Chapter 13 for information on how to perform spatiotemporal cluster analysis using NASTIC.

3. *Deep learning and neural networks*: These methods have enabled the resolution of single molecules from image acquisitions of a few frames, which enables the use of low phototoxicity to generate high-resolution images [91–93]. The techniques have the potential to further enhance the axial and temporal resolution of existing microscopy techniques described above. They are also particularly useful for imaging high-density labelling, which often needs to be undersampled in single-fluorophore detection techniques. Deep learning algorithms can extrapolate this information to reconstruct the localization maps of molecules, which is particularly useful for structures like biofilms, which are difficult to image at high resolution using conventional microscopy techniques.
4. *Rationalized deep learning*: This represents another technique that can enhance the details that can be extracted from super-resolution microscopy tenfold. The method can be applied not just to localization data but can also enhance the tracking of motion in live cells [68]. This technique has great applicability in tracking toxin-host interactions and extracting structural information in a dynamic manner that is not possible with techniques such as Cryo-EM.

7 Considerations for Reproducibility and Potential Challenges

The following must be taken into consideration to achieve high standards of data reporting and reproducibility of work done using super-resolution microscopy:

1. Due to variability in microscope setup and sample preparation, it is good practice to optimize imaging conditions using standard fluorescent beads to determine the correct point spread function (PSF) and to test various temporal resolutions for live-cell imaging. The size and manufacturer of the beads, as well as laser power and Nyquist sampling rates of the motion of the molecule of interest, should be reported clearly in the Methods [94, 95].

2. When reporting single-molecule localization data, the localization precision should be included, as the pixel-based resolution is a less meaningful parameter for super-resolved images [96, 97]. It is also good practice to show widefield and super-resolved images of the same field of view to establish the extent of localization precision (*see* Fig. 2 for examples).
3. Labelling density is critical for techniques such as PALM, STORM and PAINT. These densities should be established by determining a probe-titration curve [98, 99], and these rationalizations should be reported when describing the experiment.

Researchers must also consider the following potential challenges when adopting super-resolution microscopy into their research repertoire:

1. Most techniques require the setup of expensive equipment. However, in this review, we have also highlighted low-cost setups or techniques that are compatible with multiple imaging infrastructures, such as Super-resolution Optical Fluctuation Imaging (SOFI) [25] and Super-resolution via optical Re-assignment (SoRa) [16].
2. The techniques also require careful training and optimization of imaging conditions to obtain reliable results. With their increasing popularity, these resources are now more widely and equitably available for research groups to use. There are also many open-source software tools available online that allow for accurate analysis of super-resolution microscopy data [80, 90].
3. Super-resolution imaging techniques generate large amounts of data (ranging from 1 to several hundred gigabytes per experiment), often with many replicates, requiring effective data handling, storage, and analysis practices. Time-lapse and z-stack acquisitions, in particular, significantly increase the volume of data that is generated. Raw data is often initially captured locally on a communal microscope. This data should be transferred as soon as possible to a high-capacity storage device with regular backups, both to free up space on the microscope and to prevent data loss before transfer. Data files should be named using a well-structured and consistent naming convention that includes details such as the date, sample identity, and imaging parameters, and organized within clearly defined folder structures. Metadata (e.g., laser power, exposure time, fluorophore used) should also be retained with the imaging data. Super-resolution data often requires extensive analysis post-acquisition. Copies of the raw data should be stored in a separate analysis folder to preserve the original data in its unaltered format. Data analysis should follow a consistent,

well-documented workflow. All imaging processing steps should be recorded, including the date of analysis, the type of analysis performed, and any parameters used.

Acknowledgments

We thank Dr. Rachel Gormal for providing the illustrations of the BoNT/A structure in Fig. 3, and Dr. Tristan Wallis for assistance with editorial revisions. Schematic images in Fig. 3 were generated using BioRender. A.J.M. generated Fig. 1 in Adobe Illustrator. This work was supported by an Australian Research Council Discovery Project grant (170100125), an Australian Research Council Linkage Infrastructure, Equipment, and Facilities grant (LE130100078), a National Health and Medical Research Council (NHMRC) grant (1139316) and funding from the University of Queensland Strategic Initiatives Fund (DVCR22052A) awarded to F.A.M. P.S. is supported by a postdoctoral research fellowship from Allergan/AbbVie. F.A.M. is an NHMRC Senior Research Fellow (1155794). M. A. S. is supported by an ARC Discovery Project grant (DP230101930).

References

1. Frederick J, Herschel W (1845) IV. Αμόρφωτα, no. I. on a case of superficial colour presented by a homogeneous liquid internally colourless. Philos Trans R Soc Lond 135:143–145. https://doi.org/10.1098/rstl.1845.0004
2. Masters BR (2010) The development of fluorescence microscopy. In: eLS. https://doi.org/10.1002/9780470015902.a0022093
3. The fluorescence-microscope.-Web of Science Core Collection. https://www.webofscience.com/wos/woscc/full-record/WOS:000201029400229?SID=EUW1ED0EB151HkWOrF3McnjlTpZ4q
4. Abbe E (1873) Beiträge zur Theorie des Mikroskops und der mikroskopischen Wahrnehmung. Arch Mikrosk Anat 9:413–418
5. Nikaido H (1992) Porins and specific channels of bacterial outer membranes. Mol Microbiol 6(4):435–442
6. Winner BM, Bodt SML, McNutt PM (2020) Special delivery: potential mechanisms of botulinum neurotoxin uptake and trafficking within motor nerve terminals. Int J Mol Sci 21(22):1–14
7. Renz M (2013) Fluorescence microscopy-a historical and technical perspective. Cytometry A 83(9):767–779
8. Gordon RE (2014) Electron microscopy: a brief history and review of current clinical application. Methods Mol Biol 1180:119–135
9. Haguenau F, Hawkes PW, Hutchison JL et al (2003) Key events in the history of electron microscopy. Microsc Microanal 9(2):96–138. https://doi.org/10.1017/S1431927603030113
10. Valli J, Garcia-Burgos A, Rooney LM et al (2021) Seeing beyond the limit: a guide to choosing the right super-resolution microscopy technique. J Biol Chem 297(1):100791
11. Gookin SE, Taylor MR, Schwartz SL et al (2022) Complementary use of super-resolution imaging modalities to study the nanoscale architecture of inhibitory synapses. Front Synaptic Neurosci 14:1–10
12. Giannone G, Hosy E, Sibarita J-B et al (2013) High-content super-resolution imaging of live cell by uPAINT. Methods Mol Biol 950:95–110
13. Simoncelli S, Griffié J, Williamson DJ et al (2020) Multi-color molecular visualization of signaling proteins reveals how C-terminal Src kinase nanoclusters regulate T cell receptor activation. Cell Rep 33(22):108523

14. Coltharp C, Xiao J (2012) Superresolution microscopy for microbiology. Cell Microbiol 14(12):1808–1818
15. Gwosch KC, Pape JK, Balzarotti F et al (2020) MINFLUX nanoscopy delivers 3D multicolor nanometer resolution in cells. Nat Methods 17(2):217–224
16. Azuma T, Kei T (2015) Super-resolution spinning-disk confocal microscopy using optical photon reassignment. Opt Express 23(11):15003–15011
17. Schnorrenberg S, Ghareeb H, Frahm L et al (2020) Live-cell RESOLFT nanoscopy of transgenic Arabidopsis thaliana. Plant Direct 4(9):e00261
18. Salditt T, Egner A, Luke DR (2020) Nanoscale photonic imaging. Top Appl Phys 134
19. Glogger M, Spahn C, Enderlein J et al (2021) Multi-color, bleaching-resistant super-resolution optical fluctuation imaging with oligonucleotide-based exchangeable fluorophores. Angew Chem Int Ed Engl 60(12): 6310–6313
20. Hell SW, Wichmann J (1994) Breaking the diffraction resolution limit by stimulated emission: stimulated emission-depletion fluorescence microscopy. Opt Lett 19(11): 780–782
21. Gustafsson MG (2000) Surpassing the lateral resolution limit by a factor of two using structured illumination microscopy. J Microsc 198(Pt 2):82–87
22. Rust MJ, Bates M, Zhuang X (2006) Sub-diffraction-limit imaging by stochastic optical reconstruction microscopy (STORM). Nat Methods 3(10):793–795
23. Betzig E, Patterson GH, Sougrat R et al (2006) Imaging intracellular fluorescent proteins at nanometer resolution. Science 313(5793):1642–1645
24. Sharonov A, Hochstrasser RM (2006) Wide-field subdiffraction imaging by accumulated binding of diffusing probes. Proc Nat Acad Sci U S A 103(50):18911–18916
25. Dertinger T, Colyer R, Iyer G et al (2009) Fast, background-free, 3D super-resolution optical fluctuation imaging (SOFI). Proc Natl Acad Sci USA 106(52):22287–22292
26. Wu Z, Xu X, Xi P (2021) Stimulated emission depletion microscopy for biological imaging in four dimensions: a review. Microsc Res Tech 84(9):1947–1958
27. Heilemann M, van de Linde S, Schüttpelz M et al (2008) Subdiffraction-resolution fluorescence imaging with conventional fluorescent probes. Angew Chem Int Ed Engl 47(33): 6172–6176
28. Balzarotti F, Eilers Y, Gwosh KC et al (2017) Nanometer resolution imaging and tracking of fluorescent molecules with minimal photon fluxes. Science 355(6325):606–612
29. Banaz N, Mäkelä J, Uphoff S (2019) Choosing the right label for single-molecule tracking in live bacteria: side-by-side comparison of photoactivatable fluorescent protein and halo tag dyes. J Phys D Appl Phys 52(6):064002. https://doi.org/10.1088/1361-6463/aaf255
30. Manley S, Gillette JM, Patterson GH et al (2008) High-density mapping of single-molecule trajectories with photoactivated localization microscopy. Nat Methods 5(2): 155–157
31. Joensuu M, Syed P, Saber SH et al (2023) Presynaptic targeting of botulinum neurotoxin type a requires a tripartite PSG-Syt1-SV2 plasma membrane nanocluster for synaptic vesicle entry. EMBO J 42(13):e112095. https://doi.org/10.15252/embj.2022112095
32. Wang T, Martin S, Papadopulos A et al (2015) Control of autophagosome axonal retrograde flux by presynaptic activity unveiled using botulinum neurotoxin type a. J Neurosci 25(15):6179–6194
33. Liu S, Hoess P, Ries J (2002) Super-resolution microscopy for structural cell biology. Annu Rev Biophys 51:301–326. https://doi.org/10.1146/annurev-biophys-102521
34. Carsten A, Wolters M, Aepfelbacher M (2024) Super-resolution fluorescence microscopy for investigating bacterial cell biology. Mol Microbiol 121(4):646–658
35. Stracy M, Lesterlin C, Garza de Leon F et al (2015) Live-cell superresolution microscopy reveals the organization of RNA polymerase in the bacterial nucleoid. Proc Natl Acad Sci USA 112(32):E4390–E4399
36. Spahn CK, Glaesmann M, Grimm JB et al (2018) A toolbox for multiplexed super-resolution imaging of the E. coli nucleoid and membrane using novel PAINT labels. Sci Rep 8(1):1–12
37. Ho SH, Tirrell DA (2019) Enzymatic labeling of bacterial proteins for super-resolution imaging in live cells. ACS Cent Sci 5(12): 1911–1919
38. Greenfield D, McEnvoy AL, Shroff H et al (2009) Self-organization of the Escherichia coli chemotaxis network imaged with super-resolution light microscopy. PLoS Biol 7(6): e10000137
39. Söderström B, Ruda A, Widmalm G et al (2020) An oregongreen488-labelled

d-amino acid for visualizing peptidoglycan by super-resolution sted nanoscopy. Microbiology (Reading) 166(12):1129–1135
40. Tank RKG, Lund VA, Kumar S et al (2021) Correlative super-resolution optical and atomic force microscopy reveals relationships between bacterial cell wall architecture and synthesis in Bacillus subtilis. ACS Nano 15(10):16011–16018
41. Zhang B, Phetsang W, Stone MRL et al (2023) Synthesis of vancomycin fluorescent probes that retain antimicrobial activity, identify gram-positive bacteria, and detect gram-negative outer membrane damage. Commun Biol 6(1):1–16
42. Shukla R, Lavore F, Maity S et al (2022) Teixobactin kills bacteria by a two-pronged attack on the cell envelope. Nature 608(7922):390–396
43. Morris MA, Vallmitjana A, Grein F et al (2022) Visualizing the mode of action and supramolecular assembly of teixobactin analogues in Bacillus subtilis. Chem Sci 13(26): 7747–7754
44. Li X, Wu Y, Su Y et al (2023) Three-dimensional structured illumination microscopy with enhanced axial resolution. Nat Biotechnol 41(9):1307–1319. https://doi.org/10.1038/s41587-022-01651-1
45. Cantuti-Castelvetri L, Ojha R, Pedro LD et al (2020) Neuropilin-1 facilitates SARS-CoV-2 cell entry and infectivity. Science 370(6518):856–860
46. Martínez-Mármol R, Giordano-Santini R, Kaulich E et al (2023) SARS-CoV-2 infection and viral fusogens cause neuronal and glial fusion that compromises neuronal activity. Sci Adv 9(23):eadg2248
47. Christie SM, Tada T, Yin Y et al (2022) Single-virus tracking reveals variant SARS-CoV-2 spike proteins induce ACE2-independent membrane interactions. Sci Adv 8(49):1–11
48. Pearson GJ, Mears HV, Broncel M et al (2024) ER-export and ARFRP1/AP-1–dependent delivery of SARS-CoV-2 envelope to lysosomes controls late stages of viral replication. Sci Adv 10(14):eadl5012
49. Heras B, Totsika M, Peters KM et al (2014) The antigen 43 structure reveals a molecular Velcro-like mechanism of autotransporter-mediated bacterial clumping. Proc Natl Acad Sci USA 111(1):457–462
50. Zhang Y, Lara-Tejero M, Bewersdorf J et al (2017) Visualization and characterization of individual type III protein secretion machines in live bacteria. Proc Natl Acad Sci USA 114(23):6098–6103
51. Appelhans T, Beinlich FRM, Richter CP et al (2018) Multi-color localization microscopy of single membrane proteins in organelles of live mammalian cells. J Vis Exp 136:57690
52. Göser V, Sander N, Schulte M et al (2023) Single molecule analyses reveal dynamics of salmonella translocated effector proteins in host cell endomembranes. Nat Commun 14(1):1240
53. Lin L, Capozzoli R, Ferrand A et al (2022) Subcellular localization of type VI secretion system assembly in response to cell–cell contact. EMBO J 41(13):e108595
54. Moore AA, Nelson M, Wickware C et al (2023) OnabotulinumtoxinA effects on trigeminal nociceptors. Cephalalgia 43(2): 3331024221141683
55. Dodick DW, Mauskop A, Elkind AH et al (2005) Botulinum toxin type a for the prophylaxis of chronic daily headache: subgroup analysis of patients not receiving other prophylactic medications: a randomized double-blind, Placebo-Controlled Study. Headache 45(4):315–324
56. James NG, Malik S, Sanstrum BJ et al (2021) Characterization of clostridium botulinum neurotoxin serotype a (BoNT/A) and fibroblast growth factor receptor interactions using novel receptor dimerization assay. Sci Rep 11: 1–12
57. Manley S, Gillette JM, Lippincott-Schwartz J (2010) Single-particle tracking photoactivated localization microscopy for mapping single-molecule dynamics. Methods Enzymol 475:109–120
58. Spaulding CN, Klein RD, Ruer S et al (2017) Selective depletion of uropathogenic E. coli from the gut by a FimH antagonist. Nature 546(7659):528–532
59. Ehlers MD, Heine M, Groc L et al (2007) Diffusional trapping of GluR1 AMPA receptors by input-specific synaptic activity. Neuron 54(3):447–460
60. Klaassen RV, Stroeder J, Coussen F et al (2016) Shisa6 traps AMPA receptors at postsynaptic sites and prevents their desensitization during synaptic activity. Nat Commun 7: 1–12
61. Groc L, Heine M, Cognet L et al (2004) Differential activity-dependent regulation of the lateral mobilities of AMPA and NMDA receptors. Nat Neurosci 7(7):695–696
62. Makino H, Malinow R (2009) AMPA receptor incorporation into synapses during LTP: the role of lateral movement and exocytosis. Neuron 64(3):381–390
63. Albrecht NE, Jiang D, Akhanov V et al (2022) Rapid 3D-STORM imaging of diverse

molecular targets in tissue. Cell Rep Methods 2(7):100253
64. Schmidt R, Weihs T, Wurm CA et al (2021) MINFLUX nanometer-scale 3D imaging and microsecond-range tracking on a common fluorescence microscope. Nat Commun 12(1):1–12
65. Schueder F, Lara-Gutiérrez J, Beliveau BJ et al (2017) Multiplexed 3D super-resolution imaging of whole cells using spinning disk confocal microscopy and DNA-PAINT. Nat Commun 8(1):2090
66. Granik N, Weiss LE, Nehme E et al (2019) Single-particle diffusion characterization by deep learning. Biophys J 117(2):185–192
67. Moen E, Bannon D, Kudo T et al (2019) Deep learning for cellular image analysis. Nat Methods 16(12):1233–1246
68. Qiao C, Li D, Liu Y et al (2023) Rationalized deep learning super-resolution microscopy for sustained live imaging of rapid subcellular processes. Nat Biotechnol 41(3):367–377. https://doi.org/10.1038/s41587-022-01471-3
69. Szymborska A, de Marco A, Daigle N et al (2013) Nuclear pore scaffold structure analyzed by super-resolution microscopy and particle averaging. Science 341(6146):655–658
70. Sankaran J, Tan NJHJ, But KP et al (2019) Single microcolony diffusion analysis in Pseudomonas aeruginosa biofilms. NPJ Biofilms Microbiomes 5(1):35
71. Sankaran J, Wohland T (2023) Current capabilities and future perspectives of FCS: super-resolution microscopy, machine learning, and in vivo applications. Commun Biol 6(1):699
72. Shi H, Shi Q, Grodner B et al (2020) Highly multiplexed spatial mapping of microbial communities. Nature 588(7839):676–681
73. Yan J, Sharo AG, Stone HA et al (2016) Vibrio cholerae biofilm growth program and architecture revealed by single-cell live imaging. Proc Natl Acad Sci USA 113(36):E5337–E5343
74. Zwettler FU, Reinhard S, Gambarotto D et al (2020) Molecular resolution imaging by post-labeling expansion single-molecule localization microscopy (ex-SMLM). Nat Commun 11(1):3388
75. Castagnini D, Palma K, Jara-Wilde J et al (2024) Proteus mirabilis biofilm expansion microscopy yields over 4-fold magnification for super-resolution of biofilm structure and subcellular DNA organization. J Microbiol Methods 220:106927
76. Melia CE, Bolla JR, Katharios-Lanwermeyer S et al (2021) Architecture of cell-cell junctions in situ reveals a mechanism for bacterial biofilm inhibition. Proc Natl Acad Sci USA 118(31):e2109940118
77. Civitci F, Shangguan J, Zheng T et al (2020) Fast and multiplexed superresolution imaging with DNA-PAINT-ERS. Nat Commun 11:4339
78. Nhu NTK, Phan M-D, Forde BM et al (2019) Complex multilevel control of hemolysin production by uropathogenic Escherichia coli. MBio 10(5):e02248–19
79. Chambers CA, Dadelahi AS, Moley CR et al (2022) Nucleotide receptors mediate protection against neonatal sepsis and meningitis caused by alpha-hemolysin expressing Escherichia coli K1. FASEB J 36(3):e22197
80. Tinevez J-Y, Perry N, Schindelin J et al (2017) TrackMate: an open and extensible platform for single-particle tracking. Methods 115:80–90
81. Kechkar A, Nair D, Heilemann M et al (2013) Real-time analysis and visualization for single-molecule based super-resolution microscopy. PLoS One 8(4):e62918
82. Simon F, Weiss LE, van Teeffelen S (2024) A guide to single-particle tracking. Nat Rev Methods Primers 4:66
83. Chenouard N, Smal I, de Chaumont F et al (2014) Objective comparison of particle tracking methods. Nat Methods 11:281–289
84. Ester M, Kriegel H-P, Sander J et al (1996) A density-based algorithm for discovering clusters in large spatial databases with noise. In: Proceedings of the second international conference on knowledge discovery and data mining (KDD'96). AAAI Press, pp 226–231
85. Levet F, Hosy E, Kechkar A et al (2015) SR-Tesseler: a method to segment and quantify localization-based super-resolution microscopy data. Nat Methods 12(11):1065–1071
86. Andronov L, Orlov I, Lutz Y et al (2016) ClusterViSu, a method for clustering of protein complexes by Voronoi tessellation in super-resolution microscopy. Sci Rep 6:24084
87. Khater IM, Nabi IR, Hamarneh G (2020) A review of super-resolution single-molecule localization microscopy cluster analysis and quantification methods. Patterns (NY) 1(3):100038
88. Hyun Y, Kim D (2023) Recent development of computational cluster analysis methods for single-molecule localization microscopy images. Comput Struct Biotechnol J 21:879–888

89. Nieves DJ, Pike JA, Levet F et al (2023) A framework for evaluating the performance of SMLM cluster analysis algorithms. Nat Methods 20:259–267
90. Wallis TP, Jiang A, Young K et al (2023) Super-resolved trajectory-derived nanoclustering analysis using spatiotemporal indexing. Nat Commun 14(1):3353
91. Fu S, Shi W, Luo T et al (2023) Field-dependent deep learning enables high-throughput whole-cell 3D super-resolution imaging. Nat Methods 20(3):459–468
92. Speiser A, Müller L-R, Hoess P et al (2021) Deep learning enables fast and dense single-molecule localization with high accuracy. Nat Methods 18(9):1082–1090
93. Kim T, Moon S, Xu K (2019) Information-rich localization microscopy through machine learning. Nat Commun 10:1996
94. Diekmann R, Kahnwald M, Schoenit A et al (2020) Optimizing imaging speed and excitation intensity for single-molecule localization microscopy. Nat Methods 17(9):909–912
95. Good IJ (1960) Effective sampling rates for signal detection; or can the Gaussian model be salvaged? Inf Control 3:116–140
96. Montero Llopis P, Senft RA, Ross-Elliot TJ et al (2021) Best practices and tools for reporting reproducible fluorescence microscopy methods. Nat Methods 18(12): 1463–1476
97. Munda M, Ries J (2020) How good are my data? Reference standards in superresolution microscopy. Mol Biol Cell 31(19): 2093–2096
98. Lin Y, Long JJ, Huang F et al (2015) Quantifying and optimizing single-molecule switching nanoscopy at high speeds. PLoS One 10(5):e0128135
99. Gagnon L (2023) A beginner's guide to dye selection and labeling strategies for single-molecule localization microscopy. Micros Today 31(1):14–23
100. Grotjohann T, Testa I, Leutenegger M et al (2011) Diffraction-unlimited all-optical imaging and writing with a photochromic GFP. Nature 478(7368):204
101. Spahn C, Middlemiss S, Gómez-de-Mariscal E et al (2025) The nucleoid of rapidly growing Escherichia coli localizes close to the inner membrane and is organized by transcription, translation, and cell geometry. Nat Commun 16(1):3732
102. Trouve J, Zapun A, Arthaud C et al (2021) Nanoscale dynamics of peptidoglycan assembly during the cell cycle of Streptococcus pneumoniae. Curr Biol 31(13):2844–2856
103. Cramer K, Reinhardt SCM, Auer A et al (2024) Comparing divisome organization between vegetative and sporulating *Bacillus subtilis* at the nanoscale using DNA-PAINT. Sci Adv 10(2):eadk5847
104. Carsten A, Rudolph M, Weihs T et al (2022) MINFLUX imaging of a bacterial molecular machine at nanometer resolution. Methods Appl Fluoresc 11(1)
105. Rozario AM, Zwettler F, Duwé S et al (2020) 'Live and large': super-resolution optical fluctuation imaging (SOFI) and expansion microscopy (ExM) of microtubule remodelling by rabies virus P protein. Aust J Chem 78(3):686–692
106. Garcia-Rodriguez C, Levy R, Arndt JW et al (2007) Molecular evolution of antibody cross-reactivity for two subtypes of type a botulinum neurotoxin. Nat Biotechnol 25(1):107–116
107. Jumper J, Evans R, Pritzel A et al (2021) Highly accurate protein structure prediction with AlphaFold. Nature 596(7873):583–589
108. Tunyasuvunakool K, Adler J, Wu Z et al (2021) Highly accurate protein structure prediction for the human proteome. Nature 596: 590–596
109. Benoit RM, Frey D, Hilbert M et al (2014) Structural basis for recognition of synaptic vesicle protein 2C by botulinum neurotoxin a. Nature 505(7481):108–111
110. Stenmark P, Dupuy J, Imamura A et al (2008) Crystal structure of botulinum neurotoxin type a in complex with the cell surface co-receptor GT1b-insight into the toxin-neuron interaction. PLoS Pathog 4(8): e1000129
111. Zhang N, Gordon SL, Fritsch MJ et al (2015) Phosphorylation of synaptic vesicle protein 2A at Thr84 by casein kinase 1 family kinases controls the specific retrieval of synaptotagmin-1. J Neurosci 35(6): 2492–2507

Index

Frédéric A. Meunier (ed.), *Super-Resolution Microscopy: Methods and Protocols*, Methods in Molecular Biology, vol. 3034, https://doi.org/10.1007/978-1-0716-5268-8, © The University of Queensland 2026

O

P

S

T

V

MIX
Papier aus verantwortungsvollen Quellen
Paper from responsible sources
FSC® C105338

If you have any concerns about our products,
you can contact us on
ProductSafety@springernature.com

In case Publisher is established outside the EU,
the EU authorized representative is:
Springer Nature Customer Service Center GmbH
Europaplatz 3, 69115 Heidelberg, Germany

Printed by Libri Plureos GmbH
in Hamburg, Germany